CRITICAL THINKING AND WRITING FOR SCIENCE AND TECHNOLOGY

CRITICAL THINKING AND WRITING FOR SCIENCE AND TECHNOLOGY

THOMAS H. MILES
West Virginia University

HBJ

Harcourt Brace Jovanovich, Publishers
San Diego New York Chicago Austin Washington, D.C.
London Sydney Tokyo Toronto

To Martha, with love

Cover: © Mike Dobel/Masterfile

ISBN: 0-15-516156-3
Printed in the United States of America

Copyrights and acknowledgments appear at the bottom of the first page of each selection. Illustration credits appear on page 341.

Preface

Critical Thinking and Writing for Science and Technology is designed for students in the sciences and engineering. This technical writing textbook combines a description of the writing process, an anthology of readings from science and technology, and a review of the important resources and methods for creating technical reports and scientific papers. It gives students an understanding of and practice in the writing process, and it introduces them to the extraordinarily rich field of scientific and technical literature.

The anthology is organized according to the major modes of thinking in science and technology: communicating traditional scientific and technical knowledge, exploring new areas and making discoveries, and reflecting about the significance of scientific and technological advances. The traditional rhetorical methods—description, explanation, analysis, and persuasion—are discussed within the framework of these cognitive modes. That is, the rhetorical methods are presented as means for furthering critical thinking, not just as ends in themselves.

THE ORGANIZATION OF THIS BOOK

Critical Thinking and Writing for Science and Technology has three parts and three appendices. Part One gives an overview of writing in science and technology. Chapter 1 discusses the importance of simplicity and clarity in scientific and technical writing, the importance of audience, and the problem of relativity in determining what is fact; it ends by describing the various disciplines of science and technology and their potential fusion. Chapter 2 investigates how professional scientists and engineers write for general audiences. And Chapter 3 presents an overview of the writing process and applies the process to a professional, case-study topic (the collapse of the Hyatt-Regency Walkway in Kansas City).

Part Two provides an anthology of scientific and technical writing arranged by modes of thinking. The writing assignments that follow each set of readings are broad and flexible so professors can adapt them easily to their own course requirements. The purpose of this anthology is twofold: to show students the history of scientific and technical writing, by providing examples from the seventeenth century to the present, and to broaden students' awareness of the literature of science and technology, from traditional works to works of discovery, innovation, exploration, and reflection. This is a literature to which students are not often introduced in their undergraduate curricula. All of the readings reflect the cross-disciplinary nature of this text and serve to emphasize that professionals should not limit themselves to single disciplines.

Part Three reviews other activities involved in the writing process: doing library research efficiently, documenting sources accurately, revising one's writing for readability and clarity of style, using computers in the writing process, and using graphics effectively.

Appendix A is a bibliography of recent literature in science and technology. It complements the anthology and offers a broad base of reference for additional student research. Appendix B covers memos, letters, and oral presentations, complete with formats and recommendations for effective style. And Appendix C reviews the basics of grammar and punctuation, with emphasis on scientific and technical conventions.

An Instructor's Manual accompanies the text.

INVITATION

I encourage readers to send suggestions to me about improving this book, care of the College Department, Harcourt Brace Jovanovich, Inc., 1250 Sixth Avenue, San Diego, CA 92101. And I invite professors to send exemplary student writing for possible inclusion in subsequent editions.

A WORD OF THANKS

Many helped bring this book to life. Among those I particularly want to thank are Dr. Rudolph Almasy, chair of the West Virginia University English Department, for supporting my work and for arranging a timely sabbatical; Dr. David Farkas, of the University of Washington, who introduced me to the world of technical writing; Lucy Miskin, of Copley Publishing, and Michael Miskin, of Tapestry Press, who showed me the ropes of individualized publishing; and the several colleagues who have taught technical writing with me and the students who have endured it.

I wish to express my gratitude to the reviewers who made many helpful suggestions at the early stages of the manuscript: Jerome Bump of the University of Texas at Austin; Daniel Jones of the University of Central Florida; and Don Ross of the University of Minnesota.

I also give my thanks to the extraordinary professionals at Harcourt Brace Jovanovich: Marlane A. Miriello, acquisitions editor, who gave this book the shape and tone it now has; Sarah Helyar Smith, manuscript editor, who adroitly ushered me through a rigorous editing; Cynthia Sheridan, production editor, who saw that the book reached the printer with all parts intact; Gina Sample, designer, who created the handsome cover and interior design of the book; Avery Hallowell, art editor, who searched far and wide for the unusual pieces of art I requested; and Sarah Randall, production manager, who patiently worked with an awkward schedule.

Most of all, I want to thank my wife, Martha, who patiently read too many versions of this book and suggested many creative changes, and who, not incidentally, offered more love than I often deserved.

THOMAS H. MILES

Contents

Part One

An Introduction to Writing in Science and Technology

1

Writing in Your Professional Career

In your odyssey as a student, you have undoubtedly written more than you can remember—short, impromptu pieces in grade school; book reports and research summaries in junior high and high school, and maybe some fiction and journalism if you were lucky; and in college, longer research reports, lab reports, and perhaps some personal writing as well. Now is the time to focus all this practice in many different areas into one concentrated point: writing about what you know best—your major field of study. You need to be able to write effectively about your academic area to succeed not only in preparing junior- and senior-level research projects but also, after you graduate, in organizing, writing, and producing real, on-the-job reports.

THE TESTIMONY OF PROFESSIONALS

As a major in science or engineering, you might believe that the knowledge you are acquiring and your growing mastery of how to apply mathematical calculations and computer routines to that knowledge are all you will need to be successful in your career. After all, most of your instruction concentrates on these factors. Professionals, however, will tell you that you must also be able to express that knowledge and mathematical facility, both orally and in writing, if you plan to progress in your career.

Recent surveys of scientists and engineers reveal that writing occupies about one-quarter of a professional's time. Richard M. Davis (1978) surveyed 245 successful engineers, who reported that 24.4 percent of their time was spent writing. Moreover, 95.5 percent said the ability to write effectively was either very important or of critical importance in their present position. Almost the same number (94 percent) said they had spent more and more time on writing as their professional responsibilities increased, and claimed that the ability to write was critical in assessing candidates for advancement. Davis concluded that the vast majority of the respondents indicated that all scientific and engineering students should be either encouraged or required to take a technical writing course (212). In another research study, Lester Faigley and Thomas P. Miller (1982) confirmed Davis's conclusion: on the average, professionals in their study spent 23.1 percent of their total work time writing; that is more than one day per week.

Reinforcing these studies is a survey of graduates who had taken technical writing courses. Paul V. Anderson (1985) found that 38 percent spent more than 20 percent of their professional time in writing and that 15 percent spent more than 40 percent of their time writing.

In addition, many scientists and engineers work in a business environment, where writing is no less important. Wanda S. Sharplin, Theda P. Birdsong, and Arthur Sharplin (1986) report from a survey of business managers that 86.7 percent of the respondents believed that writing skills (good organization, grammatical correctness, and persuasive writing) were important or very important in their success and in the success of their employees.

The Broader Base for Writing

This research makes the case that writing is a professional skill. But how *does* one become a better writer, especially given the rigorous demands of a curriculum in engineering or the sciences?

In an article in the *Harvard Business Review,* L. C. Keyes laments the dull and pompous tone of much business and technical writing and says that many managers have never been trained to write "clearly, rigorously, and concisely" (106). Keyes believes that managers can learn to write effectively only if they read books written with "clearness, force, and freshness" (110). That is, instead of concentrating solely on grammar, managers should read and imitate the style of what they have read.

The books he recommends are outside the field of traditional business studies. They include *The Elements of Style* by William Strunk, Jr., and E. B. White; *English Prose Style* by Herbert Read; *A New Way to Better English* by Rudolf Flesch; *Walden* by Henry David Thoreau; *The Education of Henry Adams, An Autobiography; Treasure Island* by Robert Louis Stevenson; *The Forsyte Saga* by John Galsworthy; *The Bridge of San Luis Rey* by Thornton Wilder; *Grapes of Wrath* by John Steinbeck; *The Old Man and The Sea* by Ernest Hemingway; *The Summing Up* by W. Somerset Maugham; *The Second Tree from the Corner* by E. B. White; and *The Years with Ross* by James Thurber. Keyes believes that we learn to write powerful, lively prose by reading such quality works. Managers need to broaden their perspectives on language, he says, and then bring that new knowledge to bear on their own writing.

Learning how to avoid the dullness and pomposity that Keyes describes, however, is no easy task. One of the best methods is to keep your prose style simple and reflective of your personality.

On Remaining Human When Writing Technical Prose

F. Peter Woodford eloquently argues for a scientific and engineering writing style that retains its "humanness," by which he means that it is lively and unpretentious. Writing in the journal *Science,* Woodford describes the low quality of writing found in most professional journals in the sciences and engineering.

Sounder Thinking Through Clearer Writing

F. Peter Woodford

All are agreed that the articles in our journals—even the journals with the highest standards—are, by and large, poorly written. Some of the worst are produced by the kind of author who consciously pretends to a "scientific scholarly" style. He takes what should be lively, inspiring, and beautiful and, in an attempt to make it seem dignified, chokes it to death with stately abstract nouns; next, in the name of scientific impartiality, he fits it with a complete set of passive constructions to drain away any remaining life's blood or excitement; then he embalms the remains in molasses of polysyllable, wraps the corpse in an impenetrable veil of vogue words, and buries the stiff old mummy with much pomp and circumstance in the most distinguished journal that will take it. Considered either as a piece of scholarly work or as a vehicle of communication, the product is appalling. (743)

Woodford identifies the faults he finds in this morgue of scientific discourse:

The blemishes may include ungrammatical constructions, confused thought, ambiguity, unjustifiable interpretation, subspecialty jargon, concealed hedging, inadequate description of statistical treatment, or imperfect controls. (743)

He then dramatizes his point by giving a before and after example:

When science students enter graduate school they often write with admirable directness and clarity of purpose, like this:

> *In order to determine the molecular size and shape of A and B, I measured their sedimentation and diffusion constants. The results are given in Table 1. They show that A is a roughly spherical molecule of molecular weight 36,000. The molecular weight of B remains uncertain since the sample seems to be impure. This is being further investigated.*

Two years later, these same students' writing is verbose, pompous, full of fashionable circumlocutions as well as dangling constructions, and painfully polysyllabic, like this:

> *In order to evaluate the possible significance of certain molecular parameters at the subcellular level, and to shed light on the conceivable role of structural configuration in spatial relationships of intracellular macromolecules, an integrated approach [see 1] to the problem of cell diffusivity has been devised and developed. The results, which are in a preliminary stage, are discussed here in some detail because of their possible implication in mechanisms of diffusivity in a wider sphere.*

The student can no longer write: he pontificates.

Woodford implies that one's writing style telegraphs one's character.

One way of "remaining human" when writing about science and technology is to use the first person ("I" and "we") at crucial junctures in your document. In the example above, Woodford demonstrates the advantage of this usage. Notice, in the selections in Part Two, how the first-person pronoun is used by such distinguished writers as Wendell Berry, Michael Faraday, Samuel Florman, Benjamin Franklin, Jane Goodall, Robert Hooke, Ralph A. Raimi, Richard Selzer, Hans Selye, J. D. Watson and F. H. C. Crick, and others. The use of "I" and "we" is not unscientific: when authors

---

From Woodford, F. P. 12 May 1967. Sounder thinking through clearer writing. *Science* 156, no. 3776:743–45.

use them judiciously, these first-person pronouns keep the reader tuned to the human element.

Writing to Think More Clearly

From the title of Woodford's article ("Sounder Thinking Through Clearer Writing") you can infer his belief that writing is the best way to sort out and order your thoughts. To write clearly means that you have succeeded in thinking clearly:

> My experience is . . . that the discipline of marshalling words into formal sentences, writing them down, and examining the written statement is bound to clarify thought. Once ideas have been written down, they can be analyzed critically and dispassionately; they can be examined at another time, in another mood, by another expert. Thoughts can therefore be developed, and if they are not precise at the first written formulation, they can be made so at a second attempt. . . .
>
> The power of writing as an aid in thinking is not often appreciated. Everyone knows that someone who writes successfully gets his thoughts completely in order before he publishes. But it is seldom pointed out that the very act of writing can help to clarify thinking. Put down woolly thoughts on paper, and their woolliness is immediately exposed. If students come to realize this, they will write willingly and frequently at all stages of their work, instead of relegating "writing up" to the very end and regarding it as a dreadful chore that has very little to do with their "real" work. (744)

Woodford, an affiliate of the Rockefeller University and executive secretary of *The Journal of Lipid Research* when he wrote this article, concludes by making a plea for the teaching of writing in the curricula of science and engineering:

> The process of educating scientists is becoming increasingly complex. The student has to learn more and more facts, study exceedingly complex theories that are out of date before he can master them, and become adept at using more and more machines. We seldom make him, or even let him, write—which is the only way for him to find out if his thoughts are clear or muddled. Surely, the object of a university training is not so much the acquisition of knowledge as the development of the power to think. I believe we can strengthen scientific thinking by teaching scientific writing. If this is so, the teaching of scientific writing should not be, as it is at present, almost entirely neglected, but should be accorded a place at the very heart of a science curriculum. (745)

Becoming a competent scientific and technical writer will sharpen your ability to think clearly about scientific and technical topics and so will help ensure your professional success. The stereotype of the busy manager who lets an assistant write a report is outdated. As a professional, you can look forward to working at a word processor that will help you create an outline, file material for further reference, and point out some errors in spelling, punctuation, grammar, and style. But your writing will always be your own, and you will surely be judged by it.

WRITING ABOUT A SPECIALTY

The first step in writing clearly is to write about something you know very well. Because of the information explosion, it is difficult for most educated people to acquire in-depth knowledge in more than one or two fields. Therefore, we tend to become specialists. For instance, if a family-practice doctor discovers that a patient has developed an irregular heart rhythm, that doctor will send the patient to a cardiologist, a specialist, for further diagnosis because of the complexities involved in the treatment of such problems.

As we grow and learn, we all try to develop a specialty. A chemical engineer with a bachelor's degree might decide to return to graduate school to earn a master's degree in business administration. Or that same engineer might apply to be a member of a special research team. Either choice would lead to mastering a new specialty.

We can find technical writing in any discipline. In the following example from *Diet for a Small Planet,* Frances Moore Lappé describes why humans need protein; the description advances her claim that we should receive more of our protein from grains and less of it from meat.

Why Do We Need Protein Anyway?

Frances Moore Lappé

Given protein's importance to the body, perhaps it is not so surprising that a certain mystique grew up around it. We simply cannot live on fats and carbohydrates alone. Protein makes up about one-half of the nonwater components of our bodies. Just as cellulose provides the structural framework of a tree, protein provides the framework for animals. Skin, hair, nails, cartilage, tendons, muscles, and even the organic framework of bones are made up largely of fibrous proteins. Obviously, protein is needed for growth in children. Adults also need it to replace tissues that are continually breaking down and to build tissues, such as hair and nails, which are continually growing.

But talking about the body's need for "protein" is unscientific. What the body needs from food are the building blocks of protein—amino acids, specifically the eight that the body cannot manufacture itself, which are called "essential amino acids." Even more precisely, what the body actually requires are the carbon skeletons of these essential amino acids that the body cannot synthesize, although it can complete them by adding nitrogen, if the nitrogen is available. The body needs many more amino acids than just these eight essential ones. The body can, however, build the others *if* it has sufficient "loose" or extra nitrogen to build with. Thus, what is popularly referred to as the "protein" the body needs to eat are the eight essential amino acids and some extra nitrogen.

The body depends on protein for the myriad of reactions that we call "metabolism." Proteins such as insulin, which regulate metabolic processors, we call "hormones"; other proteins, catalysts of important metabolic reactions, we call "enzymes." In addition, hemoglobin, the critical oxygen-carrying molecule of the blood, is built from protein.

Not only is protein necessary to the basic chemical reactions of life, it is also necessary to maintain the body environment so that these reactions can take place. Protein in the blood helps to prevent excess alkalinity or acidity, maintaining the "body neutrality" essential to normal cellular metabolism. Protein in blood serum participates in regulating the body's water balance, the distribution of fluid on either side of the cell membrane.

Last, and of great importance, new protein synthesis is needed to form antibodies to fight bacterial and viral infections.

In this passage, Lappé uses excellent technical prose to present information about the function of protein in the body's metabolism—a specialty she has developed.

From Lappé, F. M., 1982. *Diet for a small planet.* New York: Ballantine, 168–69. Reprinted by permission of Frances Moore Lappé and her agents, Raines & Raines, 71 Park Avenue, New York, NY 10016.

She keeps to high standards by not bringing an unscientific bias to her material and by not introducing irrelevant, subjective attitudes.

The roots of the words "science" and "technology" reinforce the notion of writing objectively about something you know well: your specialty. "Science" comes from the Latin "scientia," derived from "sciens," the verb "to know" ("scire"). "Technology" comes from the Greek "tekhne," meaning art or skill. Literally, science is a function of pure knowing, and technology is a function of applied knowledge or skill.

When Lappé describes why we need protein, she is acting as a scientist presenting knowledge. When, later in her book, she describes how to apply this knowledge to improve human nutrition, she is using a technological approach by applying knowledge to a problem in order to recommend a certain course of action—that humans should derive more of their protein from grains than from meats. By presenting information and then describing its usefulness, Lappé is embodying one of the ideals of technical writing.

Writing Exercises

1. Revise the following piece of writing according to Woodford's standards of effective prose. Begin by making a list of his standards. Then read the passage below and circle words and phrases, or even entire sentences, that deviate from these standards. Finally, rewrite the entire passage.

> Computer systems have become an integral part of most organizations. The need to provide continuous, correct service is becoming more critical. However, decentralization of computing, inexperienced users, and larger, more complex systems make for operational environments that make it difficult to provide continuous, correct service. This document is intended for the computer system manager (or user) responsible for the specification, measurement, evaluation, selection, or management of a computer system.
>
> This report addresses the concepts and concerns associated with computer system reliability. Its main purpose is to assist system managers in acquiring a basic understanding of computer system reliability and to suggest actions and procedures which can help them establish and maintain a reliability program. The report presents discussions on quantifying reliability and assessing the quality of the computer system. Design and implementation techniques that may be used to improve the reliability of the system are also discussed. Emphasis is placed on understanding the need for reliability and the elements and activities that are involved in implementing a reliability program. (National Bureau of Standards 1987)

2. We all pride ourselves on having a specialty. Whether it is making fishing lures, taking a black-and-white portrait, understanding how and when a person becomes angry, fixing a lab specimen for microscopic analysis, or writing a computer program, each of us does something well as a result of long practice and intense interest. Think about something you do as well as or better than most others, the one area in which you have a well-developed skill or ability. Describe the activity (in 250–500 words), following as a model the straightforward, simple writing style used by F. M. Lappé.

Reference List

Anderson, P. V. 1985. What survey research tells us about writing at work. In *Writing in nonacademic settings,* ed. L. Odell and D. Goswami, 3–85. New York: Guildford Press.

Davis, R. M. 1978. How important is technical writing? A survey of the opinions of successful engineers. *Journal of Technical Writing and Communication* 8, no. 3.

Faigley, L., and T. P. Miller. October 1982. What we learn from writing on the job. *College English* 44, no. 6.

Keyes, L. C. January–February 1961. Profits in prose. *Harvard Business Review*:105–12.

National Bureau of Standards. 1987. Abstract for guidance on planning and implementing computer system reliability. Special publication 500–121.

Sharplin, W. S., T. P. Birdsong, and A. Sharplin. Winter 1986. Bridging the gap. *Technical Writing Teacher* 13, no. 1:29–32.

THE GOAL OF TECHNICAL WRITING: CONVEYING INFORMATION FOR A SPECIFIC PURPOSE

Facts and Their Context: Information, Audience, and Purpose

The Goal of Objectivity

In the broadest terms, technical writing concerns itself with the disciplines of science and technology, from the structure of DNA to the framework of a bridge. Whether it is an article in *Popular Mechanics* describing an innovative hydraulic jack for automobiles or one in *Scientific American* on black holes and the bending of light waves under their magnetic force, all good technical writing is based on objective knowledge—that is, knowledge whose truth and accuracy can be verified by others.

The goal of objective knowledge is to enlighten us about the nature of the world by describing that world with as little egotistical or subjective bias as possible using the measuring devices at our disposal. Conversely, the goal of subjective knowledge is to draw attention to the self, for a variety of purposes. Goleman (1985), a science writer for the *New York Times,* says the subjective self—if left undisciplined—tends to become a "totalitarian self" that desires to make the world into a mirror of itself. Yet, he notes, "the entire experimental enterprise in science is geared to counteract just such bias" (105). Furthering such experimental enterprise is the primary goal of the worldwide society of scientists and engineers—what Goleman calls "a mutually respectful worldwide family. It does not solve problems by blaming others but rather by undertaking objective analysis." Goleman believes that, at its best, "the scientific community functions . . . [to gather] information, with self-correcting mechanisms built in to guard against bias." As a scientist or engineer, you will become part of this community, and what you write will enrich that community's history.

Information

Goleman's ideal of objective knowledge is a hallowed article of faith among most scientists and engineers, but it must be enriched and tempered by an understanding of bias and the nature of the experimental environment.

Objective knowledge, of course, is composed of facts, and the main goal of scientific and technical writing is to present these facts. When imitators of Sergeant Joe Friday of "Dragnet" ask for "Just the facts, ma'am," they are presenting themselves as detectives and communicators of objective fact.

But facts are often more elusive than we like to admit. There are at least two problems in establishing facts as objective:

1. The amount of truth in a fact is often relative to the investigator's perspective.
2. Facts, often called data, are useful only in relation to the relative sophistication of the audience addressed.

Let's look at each of these problems more closely.

The Relativity of Facts

Most of us view the hard sciences and the engineering technologies as depositories of absolute, objective knowledge. In general, this faith is justified. The ability to perform the mathematical calculations needed to predict the trajectory of a rocket assumes both that scientists have correctly ascertained the force of gravity and that gravity will remain constant during the phases of ascent and the orbital insertion. Science and engineering progress on the basis of shared, objective knowledge.

The truth and accuracy of many alleged facts are not as easy to establish, however. Any insurance investigator will tell you, for instance, that there can be as many versions of an automobile accident as there are witnesses to report it. Facts, in such cases, are relative to the physical location of each observer of the accident as well as to how much the observer saw. Attaining a complete perspective on the accident requires piecing together all the partial pictures, correcting for observer error and bias, and coming up with a composite picture, which is as close to an objective account as we can achieve, short of having multiple cameras scanning the crash scene before and during the accident. Even scientifically trained automative engineers often disagree over what the length of a skidmark implies. Because the event cannot be repeated—and such repeatability is the hallmark of scientific knowledge—absolute certainty cannot be attained. In such a case, we can approximate objective truth, but never attain it.

Such relativity, partly based on insufficient data or an insufficiently exact analysis of such data, holds true also in the sciences and engineering. For instance, a Canadian scientist has recently found such a seemingly universal constant as the estimation of the speed of sound in air to be in error. As reported in the *New York Times* (Gleick 1986), George S. K. Wong, who is a senior research officer of the National Research Council of Canada, discovered that the accepted figure for the speed of sound is incorrect. "In the course of trying to calibrate microphones as accurately as possible . . . and with the help of refined instrumentation and what he called 'a year and a half of detective work,'" the article states, "he produced a figure that was slightly but significantly lower: 741.1 miles per hour, or 331.29 meters per second." The traditional figure is 741.5 miles per hour, or 331.45 meters per second.

Dr. Wong's new calibration reminded scientists and engineers that the speed of sound, like all other constants with the exception, apparently, of the speed of light, is relative to its medium—in this case, the type of air through which the sound is passing: "His figure," the article explains, "applies to dry air at 0 degrees centrigrade and at standard atmospheric pressure at sea level. Outside a laboratory, the vagaries of temperature, humidity, air pressure and even the composition of air can all raise or lower the effective speed of sound by many miles an hour." In the future, scientists and engineers using the speed of sound will need to account for all the variables of air in computing their results.

The Indeterminacy Principle

An extension of this problem can be found in the field of quantum mechanics, specifically in the area of subatomic particles. In investigating these particles, Werner Heisenberg, a twentieth-century German physicist, concluded that because of the disturbances created by measuring devices, the position of an electron and its momentum could not be determined simultaneously. Known commonly as Heisenberg's principle of indeterminacy or Heisenberg's uncertainty relations, this theory and its implications are hotly debated by physicists (Nagel 1961). Critics of the principle argue that *all* measurements and measuring techniques disturb the phenomenon being measured, which casts doubt on all standards of objectivity. Are researchers measuring the phenomenon or the phenomenon as it exists while being disturbed by the measuring device? This controversy has thus made all practicing scientists and engineers sensitive to the interrelationship of researcher and the phenomenon being analyzed.

Two examples will demonstrate how relative facts can be. The 1986 space shuttle Challenger tragedy taught us much about the management of engineering projects. It also taught us about testing, showing us that seals tested in one environment will not necessarily perform as predicted in a different environment. Although most of the testing of the seals took place under nonextreme temperature conditions and with the rocket or seal casket positioned horizontally, the seals were required to perform under conditions of extreme cold and in a very stressful vertical position, with strong gravitational pull. That is, the seals in the tests were essentially a different entity, in terms of practical performance, from the seals in the actual launch.

Or take the case of how many roles each of us plays every day. We act as a friend, a son or daughter, a student, maybe a lover or a flirt, even a competitive athlete. We act differently in these social roles: we are usually more relaxed around our friends and more nervous and serious around people in authority. So, if asked to give an objective definition of your self, you could certainly begin by saying that you have multiple selves, that the nature of your self is relative to the environment.

Information, then, defines: It presents a context for us to understand facts and fills a gap between the known and the unknown, like a bridge from a familiar to an unfamiliar shore. It enables us to learn, to find a relationship between two things that were not previously thought to be related. And it creates an insight, a new understanding: "Oh, I see now."

Finally, information enables the mind to create a new, more complex pattern of the world by associating previously unassociated elements. Such elements are pieces of raw data, waiting to become information. We can say, then, that information is patterned data.

Useful Information and Pattern Making

When we present facts, we need to make them useful information. An article appearing in a local newspaper attempted to present "Facts About Nicaragua" as a background piece on the question of whether the United States should send aid to the Contras. The article presented a long list of facts, under various headings (geography, history, people, economy, government). The history section read as follows:

> Nicaragua gained independence from Spain in 1821, then was briefly part of Mexico until formation of the United Provinces of Central America. It became an independent republic in 1838. U.S. Marines landed in 1910, and American troops often were present until 1933. In the 1930s, Anastasio Somoza started the family dynasty that exercised rightist rule until the Sandinista victory in July 1979 overthrew his son, also named Anastasio. ("Facts about Nicaragua" 1986)

Here we have a list of facts or data, which do not become information because no connection is made between each discrete piece of information. Because no connections are made, no pattern is formed—and information represents a set of facts woven into a pattern. For instance, we never find out

What political status Nicaragua had before gaining independence from Spain,

Why it then became part of Mexico,

How it became independent,

Why the U.S. Marines landed there in 1910,

Whether any Americans remained after 1933,

Who Anastasio Somoza was and what economic class he came from,

What "rightist rule" means, and

Who the Sandinistas are.

Because the paragraph makes no interconnections, we learn little about Nicaragua; we gain no perspective about the country's history. Therefore, when we present facts, they should have the same status as technical information. As they stand, the facts in this article are useless to an intelligent reader.

Readers are willing to assimilate new information, but they can be overwhelmed by too much material. Writers, therefore, need to be careful of how they present information and of how much they present. Too much new information, like too much of any new type of data, will overwhelm the readers; too little new information will bore them. Presenting just enough information to draw the audience into a new understanding, but at the same time not to overload it, requires a delicate balance, the skill for which comes only with experience.

The Technical Writing Triad

Authors of technical material must always be aware of three aspects of their writings:

1. The level of complexity of the information,
2. The relative sophistication of the intended audience, and
3. The purpose the information is to serve.

The accomplished technical writer sees these as mutually dependent variables: if one changes, the other two must be adjusted.

Much the same principle applies in the study of pipelines. If you connect a one-foot diameter line to a nine-inch diameter line and keep the initial pumping pressure constant, the velocity of the fluid will increase, along with the pressure on the pipeline and its seals. Similarly, if you write an interim laboratory report for your colleagues and then are asked to present the knowledge to all the employees in your company, you will usually reduce the level of complexity of the information because you are writing to a more diverse audience. Alternatively, if you take the same information and want to publish it in the leading journal in your field, you will have to follow a precise organizational format, present the material in its full complexity, and add several new sections, usually on background and bibliography.

Or, if you are writing to someone to explain how to change the oil in a car, your explanation will have to account for the person's previous mechanical background. Does he or she, for instance, know where the oil pan is? If not, you cannot just direct the person to unscrew the drain-plug bolt; you first have to orient that person to the car.

We will discuss audience and purpose more in the next chapter. At this point it is important to understand, however, that this triad is purposive:

Information + Audience ~ Purpose: to act
to understand
to contemplate

Your technical document will be more successful if you keep these three functions in mind while you are writing and revising. The success of any piece of technical writing is relative to its context of purpose, audience, and complexity of information.

Writing Exercises

1. Write a description (about 500 words) of a social event you have recently attended (a party, a concert, etc.) Make your description as objective and technical as possible. Then interview others who attended the same event, and use your interview notes to summarize their descriptions. Make this summary as objective as you can. Compare your version with that of your friends and note the discrepancies. Trying to make sense of the discrepancies, draw some conclusions about the "real," objective nature of the event: What was its primary purpose? What pattern of social interaction did it display?

Then, using your imagination, rewrite your description so that it functions for one of three rhetorical purposes: to encourage action, to further understanding, or to foster contemplation and reflection. Make sure to address it to a particular audience.

2. Do some research about Nicaragua and answer the implied questions asked above (the list of the pieces of information we never find out). Answer the questions in an objective and scientific way, and try to make bridges between relatively unknown information (that you find in your research) and commonly known facts. Write in the form of a "news analysis" that you might find in your local newspaper.

Reference List

Facts about Nicaragua. 16 March 1986. *Dominion Post* (Morgantown, W.V.), 1.
Gleick, J. 27 May 1986. Scientist corrects an old error on hallowed speed of sound. *New York Times,* 15, 17.
Goleman, D. 1985. *Vital lies, simple truths.* New York: Simon & Schuster.
Nagel, E. 1961. *The structure of science.* New York: Harcourt, Brace & World, 294–97.

ANCIENTS AND MODERNS: THE TRADITION OF SCIENCE AND TECHNOLOGY

Scientific and technical writing is as old as humanity's desire to understand and control its environment. When medieval monks wrote instructions in their monastic record books describing how to make and store wine or bake leavened bread, they were practicing technical writing. Likewise, when veterinarians send in reports of infectious or rabid animals that they have treated, they too are practicing technical writing. Technical writing is a culture's effort to collect useful information and use it, to enlarge the culture's realm of knowledge.

In oral cultures, traditional knowledge is passed on orally from parent to child; such knowledge is usually practical and based on firsthand experience. In literate societies, however, or in the global society currently being fashioned by computer technology, the sum of the available literature on any subject symbolically becomes the parent, and anyone who wants to learn, the child. Scientific and technical writing, then, is the repository of the accumulated knowledge of any culture. At its best, as we saw in the last section, it offers objective knowledge presented in a clear and precise writing style.

Science: The Rise of the Ideal of Objective Knowledge

To understand how easy it is to diverge from the goal of objectivity, the scientist or engineeer must understand the importance of objective knowledge and must realize how recently it has gained ascendency as an intellectual ideal. In the following selection, J. W. N. Sullivan (1886–1937) presents 21 points in his review of the history of Western science, revealing how the ideal of objectivity came about.

The Reason for Science

J. W. N. Sullivan

1. Science, like everything else that man has created, exists, of course, to gratify human needs and desires. The fact that it has been steadily pursued for so many centuries, that it has attracted an ever-wider extent of attention, and that it is now the dominant intellectual interest of mankind, shows that it appeals to a very powerful and persistent group of appetites. It is not difficult to say what these appetites are, at least in their main divisions. Science is valued

---

for its practical advantages, it is valued because it gratifies disinterested curiosity, and it is valued because it provides the contemplative imagination with objects of great aesthetic charm. This last consideration is of the least importance, so far as the layman is concerned, although it is probably the most important consideration of all to scientific men. It is quite obvious, on the other hand, that the bulk of mankind value science chiefly for the practical advantages it brings with it.

2. This conclusion is borne out by everything we know about the origin of science. Science seems to have come into existence merely for its bearings on practical life.

3. More than two thousand years before the beginning of the Christian era both the Babylonians and the Egyptians were in possession of systematic methods of measuring space and time. They had a rudimentary geometry and a rudimentary astronomy. This rudimentary science arose to meet the practical needs of an agricultural population. Their geometry, a purely empirical thing, resulted from the measurements made necessary by the problems of land surveying. The cultivation of crops, dependent on the seasons, made a calendar almost a necessity. The day, as a unit of time, was, of course, imposed by nature. The movement of the moon reckoned from one new moon to the next. Twelve of these months were taken to constitute a year, and the necessary adjustments were made from time to time by putting in extra months.

4. This degree of scientific knowledge was the bare minimum necessary for the regulation of practical affairs. But another of the great motives for scientific research, disinterested curiosity, would seem to have played some part. The Babylonian priests continued to observe the heavens long after their calendar had been established. They kept accurate records of the rising and setting of various heavenly bodies until, by the sixth century B.C., they were able to calculate in advance the relative positions of the sun and the moon, and so predict eclipses. The observations that were made during the centuries that elapsed before this stage of perfection was reached could not have served any obvious practical purpose. They must have been undertaken out of curiosity, in order to discover what regularities existed amongst the motions of the heavenly bodies. But, once this degree of scientific knowledge had been reached, it was turned to practical account. Not, it is true, to the practical purposes of agriculture and the like, but to the no less practical purpose of foretelling the future in human affairs. Astronomy, in fact, was made to serve the purposes of astrology. Indeed, astrology was regarded as the real justification of astronomical researches.

5. There is nothing reprehensible in all this. It would show a grave lack of the historical sense to sneer at these early astronomers as being "superstitious." It must be remembered that the scientific outlook was not yet born. Science is not created by the scientific outlook; it is scientific knowledge that creates the outlook. In the time of these early Babylonian and Egyptian astronomers there was too little scientific knowledge in the world to justify them in creating a new outlook to accommodate it. They already had a comprehensive world-outlook, an outlook based on their experience and on their reasoning about it. They fitted the new facts into their general outlook, just as we do today. It was not until many centuries later that scientific facts became so abundant and recalcitrant that they obviously could not be fitted into the old outlook. Even as late as the seventeenth century so great a scientific man as Kepler used his astronomical knowledge to make astrological predictions—a little with his tongue in his cheek, perhaps.

6. We can see that a rudimentary knowledge of space and time measurements was imposed by the necessities of everyday practical life. Another science of obvious practical importance is the science of medicine. Medicine, as we should expect, is one of the oldest of the sciences. But here the general Babylonian outlook on life put them at a marked disadvantage. Their experience of life had convinced them that the universe is governed by powers that are, on the whole, maleficent. It seemed to them that pain and disease could well be referred to the direct action of the gods. They therefore had recourse to sorcery and exorcism as the only way of dealing with the problem. Rational medicine made no progress whatever. Life in Egypt was more secure, was less liable to sudden storms and floods, and the universe appeared to the Egyptians as a less malignant affair. Their mythology shows the divine powers as being, for the most part, friendly to man. They practiced incantation in their treatment of

disease, but they also looked for other causes than the direct action of the gods. And their practice of embalming their dead gave them some knowledge of anatomy. Egyptian medicine reached a considerable degree of development.

7. In none of these discoveries does there seem to have been more than a trace of what is called the scientifc spirit. The only scientific problems that interested these ancient peoples were those that had a direct bearing on practical affairs. They seem to have shown little, if any, disinterested curiosity in the workings of nature. And they based no speculations on their scientific discoveries. These discoveries were incorporated into their religions and philosophical schemes and were interpreted in accordance with their religious and philosophical principles.

8. It is not until we reach the Greeks that we find science emerging as an autonomous activity. It is not until then, in fact, that we find anything that we can call the scientific spirit. Thales of Miletus (*c.* 580 B.C.), we are told, set out to answer the question, "Of what, and in what way is the world made?" Here we recognize the spirit, necessary to science although not peculiar to it, of disinterested curiosity. The Greeks appear to have been the first people with whom this feeling became a passion. They wanted to know—for the sake of knowing. All their predecessors, it seems, like so many of their successors, belonged to the type which asks, "What is the use of it?" It really seems as if the human consciousness, with the rise of the ancient Greeks, took a genuine leap forward. An unexampled freedom of the mind was born. This was a necessary condition for science to come into the world.

9. In another respect, also, the Greeks were unique. They seem to have been the first people with a thorough grasp of the nature of mathematical reasoning. The land-surveying formulae of the Egyptians gave rise, in the hands of the Greeks, to a deductive geometry. This was an immensely important step forward. Mathematical reasoning, the most powerful of man's intellectual instruments, was created. Overwhelmed by the almost magical power of this new instrument, the Greeks thought that in mathematics they had discovered the key to all things. To the Pythagoreans, in particular, number was the principle of all things. Everything, whether physical properties or moral qualities, was a manifestation of number.

10. This outlook has played a very large part in the development of science. Leonardo da Vinci's remark that a science is perfect in so far as it is mathematical has been very generally accepted by scientific men. If a complete mathematical description of the world could be given, it is felt that science would be complete. But is there any *a priori* reason to suppose that the universe *must* be the kind of thing that can be described mathematically? To Newton, at any rate, the attempt to describe nature mathematically was an adventure that might or might not be successful. And some modern men of science have been so astonished by the success of the adventure that they have been led to conclude that God must be a mathematician. On the other hand, there seems some reason to believe that any universe containing several objects can be brought within some sort of mathematical web, so that the mathematical character of the universe is a fact of no particular significance.

11. But whatever basis of truth there may be in the Pythagorean outlook it is certain that it greatly exaggerated the significance of mathematics. Nevertheless, a modified form of this outlook, after many fruitless centuries had elapsed, contributed very powerfully towards the origin and development of the modern scientific movement.

12. In the meantime the spirit of disinterested curiosity, and man's delight in this new and wonderful mathematical faculty, withered and died under the cold blight of the Roman Empire. The Romans were an essentially practical people, and they adopted the "What is the use of it?" attitude towards all abstract speculation. Such science as they had was borrowed from the Greeks, and they seem to have valued it solely for its practical applications in medicine, agriculture, architecture, and engineering. As a natural consequence of their obsession with practical affairs the Romans created nothing in science.

13. The ensuing centuries in Europe, up to the time of the Renaissance, also produced nothing in science. But this was not because the medievalists were exclusively absorbed by practical affairs. On the contrary, some of the greatest abstract thinkers the world has ever produced appeared at this time. But they had an outlook on life that made science unnecessary. Science could tell them nothing that they wanted to know, and they had no curiosity

about the sort of things science could tell them. The medievalist lived in an orderly universe. He knew the principles on which it was constructed, and he knew the meaning and purpose of everything in it. He knew the scheme of creation; he knew the end that every created thing was made to serve. He derived this information from two sources, reason and revelation. The highest discoveries of the human reason were embodied in the works of Aristotle; the Scriptures contained divine revelations on matters not accessible to reason. By synthesizing these two kinds of information everything worth knowing could be learned. This synthesis was accomplished, magnificently, by St. Thomas Aquinas.

14. The medievalist lived in a purposeful universe of which he himself was the centre. The reason why phenomena existed was to be found in their bearing on the eternal destiny of man. Nothing had any meaning except in so far as it fitted into this great logical scheme. In this atmosphere it is obvious that science would appear to be a trivial activity. It could be of no real importance, for the reason that it was concerned with merely secondary questions. *How* things happened was of no importance compared with the question of *why* they happened. Even Roger Bacon, the one man of his time who insisted on the experimental investigation of nature, agreed that the importance of this investigation was that it would assist in elucidating theology. It was only when faith in the all-pervading purposefulness of natural phenomena had faded that the scientific method of inquiry became important.

15. But although the scholastic outlook discouraged scientific inquiry, it furnished an essential element of the scientific outlook itself. This was the belief in nature as a rational whole. In the medievalist's universe, unlike that of the Babylonians and other early peoples, nothing was capricious or arbitrary. This belief, that "every detailed occurrence can be correlated with its antecedents in a perfectly definite manner, exemplifying general principles" is, as Whitehead says, the necessary basis for the whole scientific adventure. "Without this belief the incredible labours of scientists would be without hope." Yet this belief in universal order does not impose itself as an outcome of direct experience, as the very different conceptions prevalent in earlier times is sufficient to show. It may even be that this belief will utlimately prove to be unjustified. It may be, as Eddington has hinted, that the universe will turn out to be finally irrational. This would mean, presumably, that science would come to an end. This does not mean, of course, that the scientific knowledge so far obtained would be abandoned. As a set of working rules science would still be valid, for phenomena would presumably continue to occur in the same fashion as at present. But science would have reached a limit beyond which it could not go.

16. The development of science up to now, then, has assumed that nature is a rational whole, and this belief we owe, as a matter of history, to the great scholastic philosophers. Although, therefore, they achieved nothing, or practically nothing, in actual scientific discovery, they had a great deal to do with the formation of the modern scientific outlook.

17. That outlook comes to its first clear expression in Galileo. During the great intellectual ferment of the Renaissance a scientific genius of the first order appeared in the person of Leonardo da Vinci, but unfortunately he never published his scientific researches. What influence he may have had on the succeeding century could have been only indirect. And even Copernicus, immensely important though his work was, did not so completely manifest the scientific spirit as did Galileo. Copernicus was led to his assertion that the earth and the other planets went round the sun chiefly by considerations of mathematical harmony. The Copernican system was, regarded mathematically, a very much neater affair than the Ptolemaic system that it replaced. It was, however, open to objections that were at that time unanswerable. Also, it was in conflict with the general outlook of the time, which still regarded man as the centre of the universe. Nevertheless, its aesthetic charm, considered as a mathematical theory, was sufficient to secure it the enthusiastic acceptance of such rare spirits as Galileo and Kepler. They felt that so beautiful a thing must be true although, as Galileo admitted, it seemed to contradict the direct testimony of our senses.

18. Even Galileo himself was not the *perfect* scientific man. Perfection was reached only in the person of Isaac Newton. Galileo fell a little short of the possible by not fully realizing the necessity of confirming mathematical deductions by experiment. Fortunately, the objections of his opponents forced him to make test experiments.

19. This tendency to rest content with the mathematical deduction has always been characteristic of a certain type of scientific man, and was particularly noticeable at the beginning of the scientific movement. In the case of Kepler this tendency was supported by a whole philosophy. Kepler believed that the very reason for phenomena being as they are was that they fulfilled certain mathematical relations. By discovering these mathematical relations we seize upon the purpose that guided the Creator.

20. But although Kepler's philosophy led him into innumerable fantastic speculations, he was always stubbornly faithful to the facts. His anguish at finding that some wild and beautiful idea was not confirmed by observation was, as we know, sometimes very considerable, but he never hesitated to abandon it. He was spurred on, indeed, to look for an even more subtle and recondite harmony. And he succeeded in finding it. His three laws of planetary motion are not only of the first importance scientifically, they are also beautiful. And this quality of his imagination led him also to exceptionally beautiful ideas in the realm of pure mathematics. Kepler, more than any other man, conveys to us the breathless excitement that must have attended the opening of the great scientific movement. The poetry of science and its sense of unlimited adventure are conveyed by Kepler in the most magnificent prose that any scientific man has ever written.

21. When we come to Newton the sun is fully up. The scientific outlook has, in him, reached full consciousness. It would be fair to say that science, in the hands of Newton, has become a completely autonomous activity for, although Newton had a philosophy and a religion, they did not play any part in his science. The basis of science, according to Newton, was observation and experiment. From this basis mathematical deductions could be made. These deductions were then to be checked by further experiment. Thus science formed an independent and self-enclosed system, borrowing nothing, as it had done formerly, from metaphysics or theology. This outlook was not understood by Newton's contemporaries. It was, as it were, too austere for them. But it has become the dominating outlook of the scientific world.

Sullivan claims that Isaac Newton was the perfect scientist because he confined himself to observation and experiment, which are the bases for generating objective knowledge. If you duplicate any controlled experiment, your observations should be the same as those of other researchers. And as often as you run the experiment under the same conditions and controls, the results will be the same. The knowledge gained becomes the property of the scientific community, the members of which can communicate with each other because of their shared knowledge.

Sullivan also argues, in the first paragraph, that science and technology contribute to three realms of human endeavor: action, understanding, and contemplation (or appreciation). Most of the work in science and technology that industry and government support is oriented toward practical action: solving problems in the world. Some monies, of course, are reserved for pure research, which furthers understanding. But there is little left for contemplation.

Consider the space program, which was initiated by President Kennedy with the express practical purpose of putting "a man on the moon," in direct, competitive response to the progress of the Russian space program. We accomplished this practical task and gained knowledge, too. But in addition to the ripple technologies spawned by the space program, no one can underestimate the impact of being able to see, through an astronaut's camera eye from the porthole of a space craft, our glorious spot of earth, a blue-white orb hung against the void and cold blackness of space. The U.S. astronauts never cease to be amazed at the beauty, and several are moved to metaphysical contemplations broadcast back to Earth. Thus, the hard understanding and cool action of technology often result in inspired contemplation.

Engineering: The Rise of Technology

Sullivan's praise of science, nevertheless, must be balanced by an equal emphasis on technology—in this case, engineering. Indeed, when Sullivan refers to the paucity of Roman contributions to pure science, he fails to mention that empire's advances in engineering. As Richard Shelton Kirby and his colleagues point out, the Romans were unexcelled as engineers.

The Engineers

Richard Shelton Kirby, Sidney Withington, Arthur Burr Darling, and Frederick Gridley Kilgour

Whatever the causes, Rome rose to supremacy over the world from Scotland to Persia. And whatever their origins, Rome's engineers added to her power and her fame. When men turn from rural to urban living, they are likely to discover interest and aptitude in themselves for devising and constructing things, in short a flair for engineering. Or they will seek out those who have it. The need becomes insistent as men gather in crowds. There is heightened appreciation of individual concern in public welfare, not necessarily a moral sense of obligation to fellow men. There is realization that one is very much involved with others. Like the Sumerians, Babylonians, Egyptians, Minoans, and Greeks, the Romans responded to the stimulus of urban living. They greatly improved the buildings, communications, and utilities of their Republic. In fact, devotion to public works remained long after republican institutions of government had become mere ornaments to a regime of emperors.

Unlike preceding periods, the time of Rome was one of few discoveries or inventions. Romans did little theorizing, but they were skillful at learning and adapting the ideas and practices of others. As they conquered and absorbed their neighbors on the north in the fourth century B.C., they appropriated Etruscan knowledge of subsurface drains and building with arches and stone blocks. They took over Greek architectural forms, materials, tools, and methods, even the Greek engineers themselves, as they overran those neighbors in southern Italy and Sicily. In the middle of the third century B.C. virtually all that the early Romans knew about engineering came to them out of the civilizations of the eastern Mediterranean; for it is generally accepted from archaeological evidence that the Etruscans, though their records have not yet been deciphered, came to the northwestern shore of the Italian peninsula from Asia Minor and that they had close relationships with Egypt.

The engineers of Rome were developers rather than originators. They were practical men, as Frontinus boasted, expert in applying ideas, in extending and enlarging their public usefulness. To say this is not to assert that Romans thought only about the functional purposes of their work and had no appreciation of beauty. Their work was strong, solid, impressive, but it was balanced and well proportioned. It was pleasing, not wearisome, in its impressiveness. The word "grandeur," . . . though used until worn flat, . . . still defines, as no other word can, both the architecture of the Eternal City and the range of its engineers.

One historian has declared that the cities of the Roman Empire enjoyed systems of drainage and water supply, heated houses, paved streets, meat and fish markets, public baths, and other municipal conveniences that would compare favorably with modern equipment. Such generalizations can be dangerously misleading. They are relative to other factors and conditions which may not be presented to the reader at the same time. It would be a mistake to

---

From Kirby, R. S., S. Withington, A. B. Darling, and F. G. Kilgour. 1956. The engineers. In *Engineering in history,* 57–60. New York: McGraw-Hill. Reprinted by permission of the publisher.

conclude from this comparison, revealing though it is, that all Romans enjoyed freely all of the facilities of their cities. The comparison nonetheless does indicate clearly the extent and the efficiency of the Romans in applying their knowledge to the problems of their daily life. And in this they were, according to our conception of the term, exceptionally able engineers. Their public baths were a notable example of civic accomplishment.

The profession of the Roman *architectus,* master technician or engineer, was greatly respected. Appius Claudius Crassus, the censor, was acclaimed down through Roman history for his aqueduct begun in 313 B.C. and his great road of the following year. Julius Caesar praised the skill of the engineers who built his bridge across the Rhine. Vitruvius wrote for the approval of his patron Augustus a treatise on architecture and engineering that held the attention of builders far into medieval times. Agrippa, Minister to Augustus, was a noted engineer. The Emperor Claudius took a personal interest in public works. The historian Tacitus admired the genius of Nero's engineers, Severus and Celer. The Emperor Hadrian, himself an engineer, maintained a staff of experts, one of whom, Apollodorus of Damascus (ca. 98–ca. 117), built his great bridge across the Danube in record time, dwarfing the accomplishment of Caesar's men and setting an example for Charlemagne.

So keen was the interest in engineering and appreciation of its value in both military and civil affairs, that systematic training was encouraged at home and recruits for the service were sought abroad. This does not mean that institutes of technology such as we have today, or even anything like the universities of the Middle Ages, appeared in the Roman Empire. It does mean that such emperors as Trajan, Alexander Severus, Constantine, Julian, and Justinian searched for and financed the training of likely young men and that a system of apprenticeships directed by the state supplemented and improved upon the traditional method of handing down technical knowledge from father to son. Martial, sycophant to the Emperor Domitian and master of epigram, perhaps quite unconsciously drew attention to this wide interest in engineering. Certainly he paid oblique compliment to the profession with something less than his habitually light touch. "If the boy," he said, "seems to be of dull intellect, make him an auctioneer or architect." In spite of the poet's sarcasm, the engineer was a man of distinction at Rome whether he were native or foreigner, patrician or plebeian, master, freedman, or slave. The American reader must caution himself here against assuming that Roman slaves were inferior because they were slaves. Roman slaves, taken captive in war or purchased in the market, were often intellectually superior to their masters and, what is more significant, were recognized as being so.

Roman engineers set their profession firmly upon economic principles. They employed exact specifications and detailed contracts and they took into account varieties of materials and types of construction best suited to particular conditions and projects. Their public buildings, aqueducts, bridges, and roads show a sense of economy and efficiency just as the legal system of Rome reveals orderliness and analytical power. As Vitruvius expressed it, the engineers understood the "suitable disposal of supplies and the site." They knew how to make a "thrifty and wise control of expense in the works."

All ancient civilizations developed technologies to maintain the supply of water and to use water as a tool: they raised it out of wells, transported it to cities, and stored it in case of droughts; and they used it as routes of transport and harnessed its force to remove waste. The Romans surpassed most civilizations in their inventiveness. Kirby and his coauthors continue:

As the hydraulic engineers of Rome saw their problem, it was to obtain an ample supply of water under steady flow. This they could do with low dams, feeder canals, and settling basins at sources 1,000 feet above sea level; then it could be let down slowly through conduits underground most of the way to reservoirs in the city at an elevation of some 200 feet. The hills which sloped obliquely toward the Campagna afforded a steady decline for these aqueducts if the contours were followed, a few ridges tunneled, and ravines bridged. Thus the aqueducts of Rome came within 10 miles of the city. Then and then only were they carried upon high structures across the Campagna. These archways amounted to hardly more than an eighth of the total mileage in the water-supply system. For economy's sake, a newer aqueduct was often added to the substructure of an older one. The arches of Claudia, for example, carried the Anio Novus across the Campagna. The Tepula and the Julia were laid upon the arcade of the Marcia.

Rome's aqueducts with their great archways were its most impressive achievement in engineering. Frontinus, water commissioner of Rome in the first century A.D., wrote proudly of his aqueducts, "Will anybody compare the idle Pyramids, or those other useless though renowned works of the Greeks with these aqueducts, with these many indispensable structures?" They were copied everywhere in the Empire. Forty or fifty aspiring provincial cities had water systems on the Roman model. The aqueducts at Segovia in Spain, Athens, and Constantinople, are still in use. An outstanding example in Roman days was that which supplied Carthage. Its 7 miles of partly ruined towering arches are a feature of the African landscape, practically all there is left to mark the site of Rome's ancient foe. The engineers at Lyon, capital of Roman Gaul, used a combination of arcaded aqueduct and inverted siphon conduits. By lowering the grade more than half, they reduced both the height and the length of the arched structure. And then they carried the water down one slope across the Rhone bridge and up the other in eighteen 8-inch lead pipes laid side by side. These of course were many times stronger than a single pipe of the same thickness and total capacity.

Rome itself had 11 aqueducts, ranging in length from 10 to 60 miles. The first was constructed in 312 B.C. and the last in A.D. 226, five centuries later. Four of these carried nearly three-fourths of the supply. Herschel estimated that eight in the time of Frontinus, A.D. 97, delivered 220 million gallons daily, or from 110 to 120 gallons per capita. Today in New York and modern Rome the consumption per capita is about 130 gallons. There is no accurate way of comparing the relative apportionment then and now to industrial uses, sanitation, drinking, fire fighting, and waste. Some modern cities like Paris and a few American cities have dual supplies of water. Rome had three virtually independent services. Spring waters of the Marcia, Claudia, and Virgo aqueducts were piped from the distributing reservoirs first of all to the public fountains whence the majority of Romans daily carried home their supplies for drinking and other household purposes. But it should be remembered that the Romans, like modern Europeans, did not use water as freely as Americans do in their houses. The more turbid streams of the Anio flowed into the public baths, the fulling mills, and laundries where those white or gray woolen togas that distinguished the citizenry of Rome were washed. Overflows from all the reservoirs flushed the streets and the storm sewers.

There was relatively little water for private consumption. Some of the wealthier, either by paying outright, by bribing the inspectors, or by secretly tapping the mains, contrived to have service pipes into their houses. The authorities seem never to have thought of financing the water system by sale although they leased the right to charge fees for the use of public latrines and the right to cart away night soil (human waste) for manure. Water was free in Rome, often the gift of an emperor or a wealthy citizen who had appropriated to that purpose booty taken in war. The Aqua Marcia, 55 miles long, was built in 144 B.C. from the proceeds of the victories over Corinth and Carthage at a cost estimated as 9 million dollars in the purchasing power of modern currency. (66–68)

Balancing Sullivan's ideal of objective science is Kirby's admiration for the prowess of engineers: engineering technology can enrich our lives and make them comfortable. Consider, for instance, what life would be like without our major technological innovations, such as basic tools, which extend and empower the abilities of the human body (hitting, twisting, turning, etc); the devices to move, store, and control water; roads and bridges; techniques for mining, forging, and smelting; the mechanical and electrical clock, which has revealed to us how time can be used to measure other phenomena; movable type (the rise of printing) and its major related marvel, the development of interchangeable parts; the plow and the seed drill; the steam engine; the battery; electricity; optics and glasses; airplanes; and, of course, the computer.

Such marvels, often taken for granted, have developed from the dreams of humans to secure power and control over their environment. And many of these innovations, especially electricity, have been developed not by researchers who fully understood a scientific theory and were trying to apply it, but by inventive men and women who tinkered with something long enough until its properties began to unfold themselves.

The Synthesis of Science and Technology

Sullivan and Kirby describe and praise their own disciplines, one science and the other technology. Seeing the practical interrelationship between these disciplines enriches our understanding of both. Philip Sporn, one of the outstanding American electrical engineers of this century, has described the relationship between scientists and engineers in a series of lectures that he gave in 1963.

Foundations of Engineering

Philip Sporn

Science vs. Engineering

In appraising the relative roles of science and technology, one cannot help but be struck by the fact that for a long time basic scientific research was somewhat neglected, or at least de-emphasized, but is now in danger of overcorrection. The pendulum certainly seems to have swung too far towards science to the detriment of engineering and technology. The idea has become widespread that we must look almost exclusively to science if our society is to continue to progress—that engineering is a minor detail to be easily resolved once the scientist has completed his work.

I do not know what foundation there can be for the notion that science is today more important and perhaps much more important than technology.

The fact is that there has been no revolutionary change in the relationship between science and technology. It has been a fruitful relationship going back as much as 100 years. The development from Faraday's scientific experimentation to modern electric-power technology is one of the earliest, and perhaps one of the best exemplifications.

Science and technology are both important; each derives expanded scope, meaning, and significance from the other. Science may be said to represent an evolving body of systematic, experimentally verifiable knowledge regarding the relationships among the complex phenomena of the physical world. Scientists are concerned with improving man's understanding of his physical world and expanding the range of physical phenomena embraced by man's understanding. The technician, the technologist or the engineer, utilizing the knowledge made available by the scientist, develops the means for controlling man's physical environment and transforming the conditions of life.

The scientist usually works—but very seldom under the pressure of a timetable—in a field of his special interest, in which he has generally chosen to stake out a narrow sector for

---

From Sporn, P. 1964. *Foundations of engineering.* New York: Macmillan, 11–23. Reprinted by permission of Pergamon Press.

his own specialization. The engineer, on the other hand, while also operating within the area of his own competence, has to tackle a variety of problems, some of which may be new to him, but to which he has to apply his scientifically based knowledge and skill to produce workable and practical solutions; this work includes economics and involves both analysis and synthesis, generally within a rigid time limit. This is technology and engineering.

In their simplest concept, technology and engineering are applied science. But engineering is really a great deal more than this. It embraces all human experience with science, tools, methods, systems, and social organization that add leverage to man's effort and make possible much greater abundance than his unaided physical strength and skill alone could deliver.

Despite the close bond between science and technology, perhaps even because of it, there has developed a serious misunderstanding about the respective roles of science and technology that persists at almost all levels of our society. This has had an important impact on education and educational programs and could have severely adverse effects on our ability to cope with the difficult problems confronting our society and our future as a nation; it could adversely influence our security and our defense, and the defense and survival of the entire West.

Some Reasons for Confusion

This misunderstanding has become pervasive throughout our whole society—laymen, high-school students, faculty and counselors, technical-college students and faculty, and even among the scientists themselves.

Even such a distinguished analyst of science and scientists as Sir Charles P. Snow ascribes to scientists the exclusive possession of powers and skills which they simply do not possess. When Sir Charles says he wants "scientists active in all the levels of government" because they are trained in foresight, or that "scientists have it within them to know what a future-directed society feels like," or again that scientists "have been in certain respects just perceptibly more morally admirable than most other groups of intelligent men," he is ascribing qualities to scientists that are either totally absent in many cases or, if they exist, certainly do so to no greater degree than among other members of our society. This type of glorification of the scientists with the accompanying, if subtle, downgrading of the engineer, has resulted in a great deal of damage.

Scientists are not the only ones who have made claims for science that would be difficult to demonstrate. When the scholarly president of one of our great Midwestern universities says he believes "that science today has the magnificent power of creating conditions that will allow man, for the first time in recorded history, to master the created universe in such a way that human dignity can rise above the miserable conditions that reduce man to little better than an animal," he is claiming for science more than it can achieve alone.

All this is cause for concern because, if the present trend to elevate the importance of science at the expense of a proper emphasis on engineering continues, our rapid rate of technological development may be seriously impaired.

Engineers themselves may, in large measure, be responsible for this lack of proper understanding and appreciation of the role of engineering. Among other shortcomings engineers have failed to develop, or at least have failed to articulate a cohesive philosophy of engineering and its relationship to society as a whole. "Philosophy of Engineering" is not a subject that is frequently discussed, at least, in a brief search of the relevant literature, I could find nothing dealing directly with the subject. This is regrettable, and constitutes another demonstration of the lack of a generally accepted philosophy at the present time.

An Engineering Philosophy Needed

By philosophy I mean the postulation of a basic attitude and approach on the part of the engineer to the problems of engineering, and an orientation of engineers and engineering with respect to the society in which they function. The lack of a philosophy has contributed significantly to the declining influence of engineers in our society and, indeed, to the failure of engineers to recognize their responsibilities to participate adequately in the making of important decisions in our society that involve broad engineering, engineering-economic and social-economic judgment. A point has been reached where a preponderating percentage of the public at large does not think of engineers as even being capable of providing a constructive contribution to the broader problems of our society.

In 1930 Charles Piez, President of ASME, writing in the fiftieth anniversary issue of *Mechanical Engineering* at a time when severe economic depression and social unrest was beginning to scar our society, said the following:

> *Technologically the future is secure. We have learned the art of applying science to useful needs. What we still lack is that greater wisdom to wipe out the plague spots, to bring about an orderliness in the control of the vast forces we unleash, so that we may have progress without the waste in human and other material which has marred many of our past efforts. The problems involved are largely* outside of engineering in its narrow sense—*they are humanitarian, economic, and political problems. But before engineers can hope to realize the position that is rightfully theirs, they must arouse their consciences and recognize that they are members of society and not merely technicians.*

It seems that Charles Piez laid his finger on the nub of the problem of the engineer in the world of tomorrow. After describing reasonably comprehensively some of the problems with which the engineer has to be concerned in his work, he points out that this is not enough. The engineer who is incapable of grasping these problems or ignores them on the ground that they are outside engineering in its narrow sense cannot achieve his proper place in society. I maintain with Piez that engineering cannot be practiced as a professional discipline in a narrow sense; when it is practiced in a narrow sense it is not engineering.

This has been recognized by the great engineers who have influenced the course of engineering history for the past 250 years. Among the earliest was Abraham Darby, the great ironmaster, whose coke-smelting process for iron ore was started in January 1709; in retrospect, it is easily seen as one of the important milestones of the Industrial Revolution. When he took out a patent in April 1707, he described his objectives and his achievement in the preamble as

> *A new way of casting iron bellied potts, and other iron bellied ware in sand only, without loam or clay, by which iron potts, and other ware may be cast fine and with more ease and expedition, and may be afforded cheaper than they can be by the way commonly used, and in regard to their cheapness may be of great advantage to the poore of this our kingdome, who for the most part use such ware, and in all probability will prevent the merchants of England going to foreign markets for such ware, from whence great quantities are imported, and likewise may in time supply foreign markets with that manufacture of our own dominions.*

Clearly, Darby was concerned with a great deal more than engineering in its narrow sense; indeed, he was concerning himself with the humanitarian, economic, and political problems of his times. Somehow this concern on the part of engineers and technologists for problems that extend beyond the limits of particular techniques seems to have been lost. It badly needs to be recaptured.

. . . Dean Gordon Brown, of Massachusetts Institute of Technology, classifies engineers into three types: the "composer type," the "arranger type," and the "custodian type." According to Dean Brown, the composer type is the person who can pull together a lot of abstract, off hand, unrelated knowledge, who can make mathematical models, and can postulate conceptual theories, and can propose something that is pulled out of the whole body of pure or new science to perform a superior role. The arranger type is the person who is more design-oriented, who is much more concerned with economic considerations, and who wants to build and create, but is definitely concerned with making things work. And the custodian type is the human being who can put complicated systems together from drawings, instructions and specifications written by others, generally written by the arranger type of engineer.

This classification is novel, even ingenious. Although it helps take us partially toward a definition, it is not comprehensive enough, and I would like to extend and recast it.

Technician, Technologist, Engineer

Engineers today, and even more as they are likely to be in the world of tomorrow can best be classified into these three groupings: Technician, technologist, engineer.

I would like to define "technician" as a person trained in scientific and engineering principles who is capable of designing, creating new if he is inventive enough, modifying and improving if he is not, all sorts of new or improved particular devices or arrangements of known devices to serve as parts of large groups of devices or of systems. In this group belong the designers of motors, printing presses, shoemaking machinery, steam turbines, steam-turbine governing mechanisms, designers of motor cars, trucks, jet engines, and rocket motors. All of these persons are essentially technicians who, if they are to accomplish anything, have to specialize very highly and eventually become highly expert and completely versed in the techniques of one particular device or operation, and frequently with only part of that.

The "technologist" may have begun his engineering work as a technician. However, by education and self-development, he has brought himself along to the point where he has achieved a mastery of the discipline of a particular technology and has become a master—a technologist—in one of the many areas of technological activity. Illustrative of such areas, are rail transportation, steel production, aluminium production, synthetic fiber production, and electric-power production or transmission.

The "engineer" is one who is capable, in his own right, or by completely integrating and synthesizing the work of technicians and technologists in numerous branches, of bringing into being a major product or system having for its objective the production of something vital or necessary in human society. In contrast with the technologist in electric-power transmission, he is concerned with the whole electric power system. His concern extends beyond any particular technology, and he visualizes the social-economic or human needs and methods for satisfying them more economically and more efficiently than has been accomplished heretofore.

So much for the engineer. What about engineering?

What Is Engineering?

. . . A committee of the School of Engineering at the Massachusetts Institute of Technology issued a statement on engineering, from which I quote:

> *Engineering is a profession. Its members do creative work which results in things that people need or want. These things may be highways, submarines, interplanetary vehicles, antibiotics, or television. Science, on the other hand, is a search for knowledge. The science of mathematics extends abstract knowledge. The science of physics extends organized knowledge of the physical world. In each of these, consideration can be limited to a carefully isolated aspect of reality.*
>
> *The engineer must deal with reality in all its aspects. He must not only be competent to use the most classical and the most modern parts of science, but he must be able to devise and make a product which will be used by people. Moreover, he must assume professional responsibility insofar as the safety and well-being of people are affected by the thing he makes.*

While one cannot take too much issue with these able academicians, it seems to me this is too limited a concept of engineering—it is altogether too narrow and uninspiring.

In the lecture of Dean Gordon Brown to which I referred earlier, he also discussed engineering and said:

> *Engineering is not merely knowing and being knowledgeable, like a walking encyclopedia; engineering is not merely analysis; engineering is not merely the*

possession of the capacity to get elegant solutions to nonexistent engineering problems; engineering is practicing the art of the organized forcing of technological change, and this is something very different. That is why I am one who today is spending a great deal of his time attempting to find ways to inject this purposefulness of engineering into the experience of engineering undergraduate students. I support the argument that unless the prospective engineer is exposed during his undergraduate experience to the attitudes of engineering, he will never be properly motivated for his later training and professional career. Engineers operate at the interface between science and society. . . . When an engineer works at the frontier of his field his main function is to couple science—old science and new science—with his particular problem in order to build something and make it work. He must frequently make a commitment long before he has ever demonstrated the feasibility of the objective.

This says a great deal that is revealing about engineering, but I would go even further. The engineer must often go beyond the limits of science, or question judgment based on alleged existing science. He must frequently assert his own over-riding judgment and stake his reputation to go into areas beyond that which has been fully explored scientifically and indeed, may even contravene that which may be claimed to have been demonstrated incontrovertibly by the science of the day. To paraphrase the late Mr. Justice Holmes: Science and scientists have all been cocksure of things that were not so.

Sometimes, the engineer has to go beyond science and the scientist. We are all familiar with the great tragedy of the Tacoma Narrows Bridge, sometimes referred to as "the tragedy of Galloping Gerte." That bridge tragedy, brought about by failure to thoroughly grasp and evaluate the simple principle of dynamic stability, was not the first great tragedy in bridge failures. One of the most appalling of such tragedies occurred in England in 1879 when the great bridge over the Firth of Tay collapsed in a storm, carrying with it an entire passenger train with its complement of 75 passengers and crew of which there were no survivors. The designer of this bridge, Sir Thomas Bouch, was ruined by this disaster; a Court of Inquiry found that the bridge was defective both in design and construction. Most notably, insufficient allowance had been made for lateral wind pressure and this was chiefly responsible for the fall of the bridge.

When Sir Thomas was asked why his Tay Bridge was so much weaker in this respect than other tall viaducts which he had built previously, he replied that his ideas on the subject of wind pressure had been modified by advice given to him by the Astronomer Royal, Sir George Airey. Sir George had pronounced that: "the greatest wind pressure to which a plane surface like that of the bridge will be subjected in its whole extent is 10 lbs per square foot." Poor Sir Thomas had subordinated the lessons of his own practical experience to a piece of specious scientific poppycock, and this was proved when wind gages in the Firth of Forth a few years later recorded pressures up to 34 pounds per square foot.

Sir Eric Ashby succinctly expressed the outstanding characteristic of technology and engineering a few years ago when he said:

Technology is of the earth, earthy; it is susceptible to pressure from industry and government departments; its end an obligation to deliver the goods.

The engineer is the key figure in the material progress of the world. It is his engineering that makes a reality of the potential value of science by translating scientifc knowledge into tools, resources, energy, and labor to bring them into the service of man. Engineering goes a great deal beyond technical know-how, beyond the work of the technician, and even beyond the work of the technologist, skilled in the technology of a particular field.

To make contributions of this kind the engineer requires the imagination to visualize the needs of society and to appreciate what is possible as well as the technological and broad social understanding to bring his vision to reality.

Sullivan and Sporn offer different perspectives about science and technology. As a professional writing about science and a practicing engineer, respectively, they offer richly diverse ideas with which we can begin to consider the problems that technology and science pose.

Writing Exercises

1. J. W. N. Sullivan calls Isaac Newton the first perfect man of science, but he doesn't describe his contributions in detail. Do some research on Newton, and write a summary (of about 1,000 words with two good examples) of what you learn about Newton's commitment to empirical research.

Or, if you prefer an engineering topic, research the work of Leonardo da Vinci, and write a report describing some of his work with experimental designs—designs that his era lacked the technical know-how to construct.

2. Select an area of science and technology that you know well and describe it to show how it contributes to what Sullivan believes are the three needs of humans: to act, to understand, and to contemplate. Your piece should be about 1,000 words long, with about 250 words devoted to each of the three needs.

2

The Audiences for Technical Writing

AUDIENCE AND PURPOSE

The Range of Audiences

Anytime you say something or write something down, you assume that someone will hear it or read it. This observation, though seemingly trivial, reveals the basis for all human communication: we speak to be heard, and we write to be read. Consequently, we assume that a hearer or a reader exists; we assume that there is an audience.

Such an audience can be one person, a group of people, an organization, a professional society, a nation, or the world. This audience can even be yourself, as in the case of purely expressive discourse, such as diaries and other kinds of personal writing. But there is always an audience, and before we speak or write anything, we already assume that we know how to appeal to it.

For instance, when babies cry, they want to be fed or changed or have some attention given to them. The utterance—the cry—has a persuasive purpose and assumes an audience: the persuasive purpose is, "pay attention right now," and the infant assumes the audience is someone who will provide that attention. Likewise, when an engineer makes a presentation of a new design to a client, that engineer is assuming that his or her audience is intellectually advanced enough to understand the design and interested enough in the message to pay close attention.

The concept of audience affects every mode of communication, and it holds special power in the fields of science and technology because of the complexity of the material under discussion. For instance, many people do not know that telephones are electrified and that they form part of an electric circuit. Consider, then, how much the scientist and engineer would have to explain to this audience to describe how a telephone worked. Ineffective writing often results when writers assume that their audiences have the same amount of knowledge that they have.

Hans Selye, the pioneer in research on stress, discusses this problem:

Whenever you decide to write anything, be it only a letter or an entire encyclopedia, the first thing to ask is: Who should and will read this? There is no such thing as a perfectly written piece; at best it can be perfect only for a certain type of reader. It is a common mistake for beginners to send articles to the wrong journals or to adjust the general tone of their communication to a level quite different from that of the probable reader. The usual errors are to talk down to the audience; to be too far above their heads; to use a chatty narrative tone when dry conciseness is desirable, or vice versa. (334–35)

In the writing that you will be doing as a professional scientist or engineer, you will be addressing three different kinds of audiences (Bowen and Mazzeo 1979, 197).

Scientists and Engineers Writing to Others in Their Particular Discipline

Highly specialized and technical in nature, prose written for this audience can be sampled in journals such as the *American Journal of Physics, Nature* (British), *Science*, the *New England Journal of Medicine,* and the *Society of Petroleum Engineers,* as well as in books written to define the most current status of any discipline.

The language used at this level is usually highly specialized; much of it represents the jargon, or specialized vocabulary, of complex disciplines or of subdisciplines. This vocabulary is often created by professionals when they are breaking into new territory that cannot be described with conventional vocabulary; the new words they create are called "neologisms." The specialized vocabulary of science and technology is perfectly valid; it represents the odyssey into uncharted territory.

But when the jargon or specialized vocabulary first appears, it can prevent any but the current insiders from understanding the new knowledge, and thus it can become a code that excludes most readers. For instance, the vocabulary used by Stephen Nesbitt, the public affairs officer at Mission Control who described the events of the *Challenger* explosion, contains many jargon words. Watching his computer screen, but not a TV monitor, Nesbitt said seconds after the explosion: "Obviously a major malfunction. . . . We have no downlink. We have a report from the flight dynamics officer that the vehicle has exploded." (*New York Times,* 25 February 1986, 20). When Nesbitt uses "downlink," he is talking jargon. Although others at the consoles, specialists in NASA procedure, would understand him, the average listener has to construct for himself or herself what the word might mean (it refers to the lack of confirmation from tracking stations farther down the flight path).

As an undergraduate, you will probably not be writing professional articles designed for this audience. But you will be expected to be able to read material at this level and be able to understand and summarize it; for instance, the reading you do in some specialized journals and in conference proceedings will be at this level.

On the other hand, most of the writing you do at the undergraduate level—lab reports, feasibility studies, and senior research projects—will have as an audience experts in the field, those experts being your professors. This situation is challenging: you must address an expert audience while being only a novice yourself. The

---

solution is to present your knowledge and expertise directly and objectively, knowing that your audience will judge it partly on the basis of your previous academic experience. The selections of student writing in this text show student writers doing what they do best: displaying their knowledge through specific document formats.

Scientists and Engineers Writing for Professional Audiences

The audience for this level of writing is assumed to be college educated and to have had some formal studies in one or more of the sciences or in engineering or technology. This audience may or may not have had graduate courses in one of these areas. *Scientific American* represents the best tradition at this level. Readers of this journal will understand, for instance, the theory of physical stress and the difficulty of predicting how much stress will actually be applied to any one girder of an elevated walkway, but they will not necessarily know the latest mathematical or computer-modeling techniques used to predict such stress.

Many of your writing assignments as an undergraduate will give you experience at this level of writing, since you may be at this intellectual level when you write: you may have basic competence in a discipline, but you may not yet have a specialized understanding. Mastering this level of writing assumes that you have an understanding that transcends the level of your audience. Attaining this level, and then using it to write to an audience with less expertise, is a real challenge.

Scientists, Engineers, Lay Scientists, and Science Writers Who Write to the Generally Educated Layperson

Though this level is often criticized by sophisticated scientists and engineers as giving overly simplified notions to nonscientists, and though it is often referred to as "popularized" writing, books and articles written at this level are read and enjoyed by a wide audience.

One of the ironies of these three levels of prose and their respective audiences is that the prose at the lowest technical level is often the most interesting and the most readable of all scientific and technical writing; it demands that the writer has complete command of the discipline and can make that knowledge comprehensible to someone not as highly trained. It also assumes that the knowledge that science and technology accumulate should be understood by the public, which is the direct beneficiary of that knowledge.

Publications such as the *New York Times,* which includes a special section on science and technology in every Tuesday issue, books on ecology by Rachel Carson *(The Silent Spring),* the astounding series of books on astronomy and the natural world by Isaac Asimov and Carl Sagan, the profound essays in medicine by Lewis Thomas and Richard Selzer, and a whole world of other books and journals, including *Popular Mechanics,* fall into this category.

Hans Selye, quoted earlier, comments on the importance of being able to write for professional and lay audiences:

> Most scientists are so deeply involved in their work that they never write about science, except in the technical language of their specialty. I think this is a pity. Such outstanding

investigators as . . . Poincaré, Darwin, [and] Einstein . . . have written interesting and stimulating books about the general problems of the scientist in the form of diaries, biographies, or essays on the philosophy and psychology of research. . . . In the years to come, as Bertrand Russell put it:

> *Not only will men of science have to grapple with the sciences that deal with man but—and this is a far more difficult matter—they will have to persuade the world to listen to what they have discovered. If they cannot succeed in this difficult enterprise, man will destroy himself by his halfway cleverness.*

Of course, a scientist who has never written other than purely technical texts runs into a great many difficulties when he first tries to address a more general public. Here, even more than in other forms of writing, he must constantly keep in mind for whom he is writing. Well do I remember my difficulties in this respect when I wrote my first semitechnical book, *The Stress of Life.* In dictating the rough manuscript, I continually found myself wavering between baby talk, which would have been below the level of any individual likely to pick up a nonfiction book, and highly technical language meant only for specialists. (354–55)

Knowing how to write to each of these three audiences is learned over years of work. For instance, not until you are at the end of graduate school can you seriously consider writing to the technical audience; much of your professional training after your undergraduate years will be devoted just to this task. For now, you will be writing reports for professors who know more about the subject than you do and who will be evaluating your writing according to the demands of the assignment under question. Thus, the assignment and your professor's expectations about it are your most practical and important audience.

Audience and the Concept of Mutual Knowledge

To understand the academic audience and the three levels of audiences outlined above, you should keep in mind the *mutual knowledge* that you as a writer share with your audience. Meaningful communication occurs only when the knowledge being assumed is shared between writer and reader.

The teaching of writing is classified under the discipline of rhetoric, which is the study of the elements of spoken or written discourse; *rhetoric* comes from Greek, meaning "to speak." A rhetor (an old word) is someone who speaks, such as at a commencement, or, by extension, someone who writes. The concept of audience is related to rhetoric simply because we always speak or write to someone (even when we are mumbling we are talking to ourselves). The word *audience* comes from Latin and French, meaning "to hear." An audience can be someone we are talking or writing to, a gathering of spectators or listeners (as at a concert), the readers of an article, or, in its older sense, a formal meeting with a distinguished person (for instance, the Pope). You can see, therefore, how the concepts of rhetoric and audience go together.

To be a successful speaker or writer, you must make contact with your audience. To do this, you must understand what the audience's world view is and what its assumptions and opinions are likely to be. What you say must seem to them to be understandable and familiar.

For instance, say that a biology professor is giving a lecture on genetics to a freshman biology class. If the professor began the lecture by giving a full description of the organic chemistry of the bonds in the DNA molecule, he or she would lose the class simply because the students would not have the background in organic chemistry or biochemistry necessary to understand the material. The professor would have failed because he or she did not begin with knowledge the students already had. The chasm between the professor's level of knowledge and the students' would not have been bridged, and no useful information would have been communicated during the lecture. In simple terms, the professor and the students did not share enough mutual knowledge.

Gordon P. Thomas, an American rhetorician, defines mutual knowledge as "the knowledge, attitudes, and beliefs that a speaker or writer and the audience *knowingly* have in common" (1986, 582). Thomas calls this "world knowledge," which he defines as the sum of what the audience "already knows and believes about the world" (587); it allows the speakers or writers to "know how much and what sort of detail to include" in their writing (593).

A Case Study of Audience: The Cambridge City Council Confronts Genetic Research

A dramatic example of the potential gulf between speakers and their audiences can be found in the series of meetings that the City Council of Cambridge, Massachusetts, held between June 1976 and February 1977 on whether or not to allow recombinant DNA research to be conducted at Harvard, which is in the city of Cambridge.

Four Factors of the Drama

The drama of the case involved the members of the city council and engineers and scientists from Harvard and the Massachusetts Institute of Technology. The drama was heightened by four factors:

1. All involved believed that this was the first time any city council in the United States had considered regulating the scientific work being carried out in scientific laboratories housed on university campuses and funded by the federal government.
2. Understanding the intricacy of what was actually being done was difficult because of the complexity and newness of the biochemical engineering used and the inherent intellectual difficulty of the science itself.
3. Luminaries in the field, including Nobel laureates, disagreed about the dangers involved.
4. The public, and especially Cambridge's mayor, feared that an uncontrolled epidemic could start if genetically altered bacteria escaped from the lab.

The case, according to a series of articles in the *New York Times,* concerned Harvard's intended use of a new technique to study DNA. The technique, called recombinant DNA, as the paper explained, "involves isolating a tiny portion of animal or plant DNA [the basic genetic material] and then inserting it, for analysis, into a common laboratory bacterium. The fusion that results yields a previously unstudied type of organism" (8 July 1976).

Disagreement Among Experts

Aside from the challenging task of understanding enough of the science and technology to make a responsible decision, the city council was also confronted with experts who disagreed. Although Professor Matthew Meselson, chairman of the Harvard Biochemistry and Molecular Biology Department, assured the council members that Harvard was following not only the N.I.H. guidelines but also "stricter regulatory processes established by the university" (8 July 1976), Dr. George Wald, a Harvard biologist and Nobel laureate, expressed his concerns about "real potential hazards" in constructing an artificial, functional gene. Dr. Wald said: "Just one mutation has taken 5 to 10 million years. Here one has suddenly the technology that can interchange whole groups of genes. It is a fantastic breakthrough, but it contains real potential hazards. Understanding living organisms has been our first goal. It hasn't been part of the bargain to mess them up. . . . A living organism, self-producing, is forever." Dr. Wald also expressed concern that the experiments might run amok. "The question of potential hazards raises a question in which all of society is involved—it ceases to be a scientific issue and becomes a socio-economic issue" (29 August 1976).

This dramatic conflict made it difficult for the city council to acquire new and reliable technical knowledge, since even the experts disagreed. Other information made their task harder still:

1. Michael Wolfson testified that he had quit his job as a lab technician at M.I.T. after he observed "safety standards constantly violated, either by carelessness or accidents in the research lab" (8 July 1976).

2. Such experiments had been raising controversy in the wider scientific community. As a *Times* story noted, "In 1974, in an action rare in the history of science, a group of American biologists voluntarily renounced two types of genetic experiments they considered dangerous. In 1975, more than 100 scientists from 16 countries, meeting in California, proposed a voluntary deferral of potentially hazardous experiments" (17 January 1977).

3. An apocryphal story fueled existing fears that scientists often could not contain their own experiments. The story went that "the Harvard University biology building where the experiments are to take place . . . is infested with a particularly hardy variety of Egyptian ant that has resisted the most sophisticated efforts to wipe it out. The ants are believed to have escaped from a professor's briefcase years ago" (17 January 1977).

4. The mayor of Cambridge, Alfred Velucci, feared the professors might create a "Frankenstein germ." He declared, "We sure to God don't want another legionnaires' mystery" (17 January 1977).

The Growing Sophistication of an Uninitiated Audience

After seven months of hearings, the city council had become a sophisticated audience, having been partially educated in the new technology of genetic engineering. After 75 hours of testimony from 35 persons, the review board—a citizens' panel and a subcommittee of the city council—said that its members had "come to appreciate the brilliant scientific achievements made in molecular biology and genetics." It also decided, however, to extend the ban on the research, pending more study. Decisions to go ahead with potentially dangerous research, they said, should not be left "within the inner circles of the scientific establishments." Rather, "the social and ethical implications in genetic research must receive the broadest dialogue in our society. We citizens must insist that in the pursuit of knowledge appropriate safeguards be observed by institutions undertaking the research" (7 January 1977).

Finally, however, the citizens' review panel recommended that Harvard and M.I.T. be allowed to proceed with their experiments under the umbrella of careful safeguards. And on February 8, 1977, the Cambridge City Council, putting its faith in the conclusions of the panel, decided not to ban the genetic research (that is, not to ask the city health department to close the lab).

The city council, through the citizens' review panel, had become a sophisticated audience with an increased level of mutual knowledge shared with experts. By acquiring new knowledge, the council and the panel could speak to the scientists and engineers as near-equals and could make what they believed to be a responsible decision. The subtlety of that decision was to create a safeguard guaranteeing that genetic engineers would alter the new cells so they could not live outside of a laboratory environment in human hosts.

The attainment of mutual knowledge, in this case, occurred through an incremental and reciprocal process, in which the city council and the citizens' panel gained technical and scientific knowledge from the scientists and engineers and the scientists and engineers gained an understanding of the fears of the council and the citizens of the town. Both groups acquired knowledge that enabled them to adopt positions similar enough to allow a compromise agreement. The compromise could occur because the two groups came to share more and more of the same premises.

Gaining Mutual Knowledge Reduced Conflicts

This success story exemplifies how education about concerns and responsibilities, combined with a willingness to find a common ground, enables conflicting interests to solve their problem. The scientists and engineers adopted a socially responsible stand, having come to understand that they too are part of a social fabric; likewise, the citizens came to understand the intricacies of a problem that they had little comprehended before. Of course, both audiences were highly motivated. The citizens of Cambridge had an investment in supporting their great universities, and the scientists and engineers knew they had to live where they worked. Such realism enabled a rational decision.

A month after the decision, Steven Toulmin, a member of the Committee on Social Thought of the University of Chicago and a distinguished historian of science, gave an address on recombinant DNA research to the National Academy of Sciences in Washington. In it, he discussed the need for these two groups—the scientists and the public—to become each other's audience in a pursuit of mutual knowledge:

> *The need for improving the forums for negotiation between the scientific community and the larger public has been clear, at least, since*

Hiroshima; but the actual task of creating the necessary arrangements has remained in suspension over the intervening 30 years.

Now this task cannot be put off any longer. Where large public interests are at stake, no profession or industry can for long remain entirely self-regulating. In this respect, the recombinant DNA problem is as much a symbolic *as it is a* practical *issue.* (Toulmin 1977)

How an Engineer or a Scientist Writes to a General Audience

The attainment of mutual knowledge between an expert and a beginner, of course, is called learning. The following excerpt from Victor F. Weisskopf, an eminent twentieth-century physicist, shows how an expert can write to beginners and work with the knowledge that they already have, the knowledge they mutually share with the expert. In the preface to *Knowledge and Wonder: The Natural World as Man Knows It,* Weisskopf describes the origin of the book:

This book had its beginning in a series of lectures the author gave at the Buckingham School in Cambridge, Massachusetts, before an audience with no special grounding in science. The idea was to sketch our present scientific understanding of natural phenomena and to try to show the universality of that understanding and its human significance.

Now such an understanding runs into difficulties that are only too well known. Scientific knowledge is hard to communicate to the nonscientist; there is so much to be explained before one can come to the essential point. All too often the layman cannot see the forest, but only the trees. The difficulties, however, should not prevent, or even discourage, scientists from tackling the job in different ways. This book is one way of giving the uninitiated an idea of the greatest cultural achievement of our time. (1979, ix)

After setting the stage by describing the place of Earth in space and time, Weisskopf launches into a description of the two features of nature that are common to all objects: gravity and light. Here is how he describes gravity.

Knowledge and Wonder: The Natural World as Man Knows It

Victor F. Weisskopf

Gravity on Earth and in the Sky

Gravity is a well-known phenomenon here on Earth. All things around us, large or small, are attracted by the earth—they fall downward when they are not held up by some support. The attraction of every piece of matter by the earth is the best-known example of a force in nature. Still, tremendous effort and centuries of thinking were needed before mankind recognized that the motion of the moon around the earth and of the planets around the sun is based upon the same force. It long was thought that the laws governing heavenly bodies were different from those that held on Earth. The universality of the laws of nature, their validity for the whole universe, has been recognized only since the days of Isaac Newton.

The moon and the planets do not fall toward the earth nor toward the sun. How, then, could their motion be governed by the force of gravity? There is a big gap between our

---

From Weisskopf, V. F. 1979. *Knowledge and wonder: The natural world as man knows it.* 2d ed. Cambridge: The MIT Press, 41–48.

terrestrial experience of things falling toward the earth and the heavenly appearance of bodies orbiting around a center (moon around earth, planets around sun). The bridging of this gap was a decisive step toward the understanding of the universe. Let us see how it came about.

Imagine that we are at the top of a very high tower and throw a stone horizontally into space. [See Figure 2-1.] The stone's path will be bent down toward the earth because of gravity, and the stone will hit the ground at a certain distance away from the tower. The harder we throw the stone, the more gradual will be the bending of the path. We can imagine that the stone could be thrown with such vigor that the downward bend of its path would just equal the curvature of the earth's surface, which is, of course, the surface of a sphere. Then the stone would never reach the surface because whenever its path bent down, the surface of the earth would bend by the same amount. We have thrown the stone, as it were, beyond the horizon. If the air did not slow it down, our stone would circle the earth as a satellite. This is, of course, the principle of launching a rocket satellite. In a typical rocket firing, the first stage raises the satellite above the atmosphere, and then a second rocket explosion pushes it into a horizontal motion. The horizontal speed necessary for the bending to be equal to the earth's curvature is about five miles per second. Thus we see how the falling motion of an object can go over into an orbiting motion around the earth if the object receives a strong horizontal push.

Let us now look at the orbit of a body around a center of attraction in a different way. When a planet circles the sun, the attractive force of gravity keeps the orbit circular, just as a weight tied to the end of a string keeps moving on a circle if you whirl it around while holding the other end of the string. The attractive force counteracts the centrifugal force, which in a circular motion pushes things outward.

The centrifugal force (the drag on the string) is greater the more rounds the object makes per second. It is also greater for larger radii, and it is proportional to the mass of the object.

FIGURE 2-1

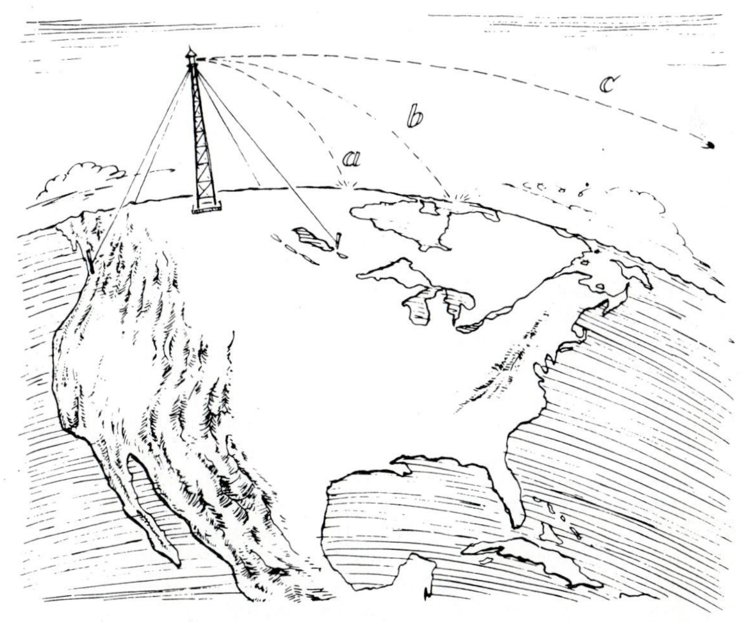

A stone thrown from a tower. Paths *a, b,* and *c* correspond to throws of increasing power. Throw *c* will never reach the earth.

In fact, we can easily calculate the centrifugal force on each planet, since we know its revolution time and its distance from the sun.

The centrifugal force is just balanced by the attractive force of gravity; hence whenever we calculate a centrifugal force in an orbit, we have determined the force of gravity. That is the way Newton measured the force of gravity of the sun upon the planets and of the planets upon their moons. He found that gravity follows a very simple law: The attraction between two bodies is proportional to the product of the masses and inversely proportional to the square of their distance. For example, the distance of Venus from the sun is 0.7 times the distance of the earth from the sun. The square of 0.7 is ½, and from this we conclude that the sun's attraction for Venus has to be twice as strong as the sun's attraction for the earth. Here we have calculated and measured a force far, far beyond direct human experience, a force in our sky.

In order to be sure that the force between the sun and the planets is a universal force that acts between any two masses, one must show that the same kind of attraction exists between two blocks of lead or any other two objects, and that this force also decreases with the square of the distance and is proportional to the product of the masses. Of course, the gravitational force between two lead blocks would be extremely small, since their mass is small compared to that of the heavenly bodies. If the blocks weigh one hundred pounds each, the force between them at a distance of one foot is about as small as the gravity force that the earth exerts upon one ten-thousandth of a gram. Still, it has been measured, and these measurements do bear out the general validity and universality of the law of gravity.

The Generality of the Law of Gravity

Newton's discovery of the law of gravity explained for us the orbits of the planets around the sun. But it also put an end to the old and cherished dream of many philosophers. The dream was to find a fundamental significance in the actual sizes of the orbits and the durations of the periods of the planets. One might have expected that the radii sizes of the planetary orbits would have simple relations; for example, the size should always be doubled from one planet to the next or exhibit some other simple numerical regularity. The Pythagorean philosophers, for instance, attributed special importance to the numerical ratios between heavenly orbits and considered them as the essence of their system. These relationships were the embodiment of a "harmony of the spheres"; they were supposed to reflect an inherent symmetry of the heavenly world as contrasted to the earthly world, which is full of disorder and without any symmetry. The harmonious interplay of the various celestial motions was supposed to produce a music whose chords were audible to the intellectual ear, a manifestation of the divine order of the universe. Even Johannes Kepler, whose analysis of planetary motions led to the discovery of the law of gravitation, tried hard to explain the observed sizes of the orbits by inventing a universe of regular solids—the sphere, cube, tetrahedron, etc.—one inscribed in the next, and each determining the size of one of the orbits by virtue of some deep, fundamental, all-embracing principle. [See Figure 2-2.]

With Newton all these ideas turned out to be illusions. The fundamental principle underlying planetary motion is the law of gravitational attraction. It determines the orbits of the planets only insofar as it requires them to be circles or ellipses with the sun in the center of the circles or in one of the foci of each ellipse, and establishes a special relationship between the radius (or major axis of the ellipse) and the period of revolution. But the principle does not prescribe any special size or radius. In fact, the actual size of an orbit depends on the conditions at the beginning, when the solar system was formed, and on the subsequent perturbations upon the orbits. For example, if initially the earth had received a different speed, it would have circled on a different orbit. Furthermore, if another star should pass near our solar system, all planetary orbits would be changed, and the relationships between their sizes and periods would be quite different after the encounter.

We can see from this that the orbit sizes, as observed today, are of no great significance. They could just as well be quite different without violating any law of physics. The fundamental law of gravity determines only the general character of the phenomenon. It admits a

FIGURE 2-2
The Kepler Device

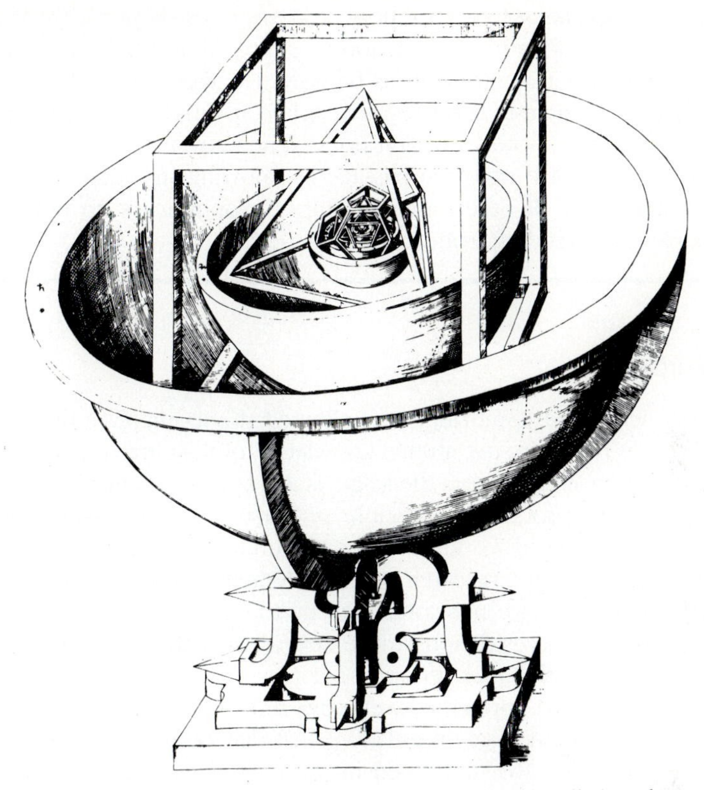

Kepler's model of the universe, showing how he thought all the planets were positioned with relation to certain geometric shapes. Redrawn from *Mysterium Cosmographicum* (1597, edition of 1620).

continuous variety of realizations. The sizes of the present orbits in the solar system were determined by the circumstances and special conditions prevailing during the formation of the solar system, or to the influence of passing stars, but there is nothing fundamental in their present magnitudes. We expect the planets of another star to circle on quite different orbits, even if the star is very similar to our sun in its size and constitution.

Because of its universality, the force of gravity reaches beyond the solar system and even beyond our galaxy. The stars within each galaxy attract each other by virtue of gravity, and each galaxy exerts gravitational forces upon other galaxies; hence the motions of the stars and also the motions of the galaxies are regulated by their mutual attraction. We don't know enough yet about these motions because they are very hard to observe. We would have to solve a very difficult problem of mathematical analysis if we were to find out what motions an assembly of fifty billion stars would perform under the influence of mutual gravitational attraction. There are very good indications, however, that the same principle governs the motions of the stars. The stars seem to circle around the center of the galaxy in much the same way the planets circle around the sun.

Are the motions of the galaxies also determined by gravity forces? Here we come to an unsolved problem of astronomy. We don't know much about it except for the striking motion of the galaxies away from each other—the expanding universe. This motion obviously cannot come from gravity; there must be some other fundamental, but as yet unknown, explanation connected with the Big Bang explosion, which may have been the beginning of the universe.

Notice that, in speaking to high school students, Weisskopf uses words that are generally familiar, develops his explanation by simple illustrations, and defines new or difficult terms (such as centrifugal) as he goes. He also uses familiar experiences (throwing a stone, launching a rocket) to explain unfamiliar concepts. By moving from what his audience knows to what it doesn't, he allows them to learn, to acquire the knowledge he has. He ends the passage by showing his audience that there is something that he and all other contemporary astrophysicists do not understand: why the galaxies are receding from each other.

Writing Exercises

1. The last act of the Cambridge City Council drama involves the arrival of an audience that does not share the mutual knowledge of the citizens and the scientists: this audience, or new character on the stage, is a group of commercial concerns that uses gene-splicing to produce marketable products. In the original debate, both citizens and scientists were working under the umbrella of government standards that controlled gene-splicing research at federally financed universities.

Companies, however, are not controlled by these standards and fear that local municipalities will enact restrictions more severe than those listed in the federal guidelines. The *New York Times* reported (31 May 1981) that some companies are concerned that biohazard committees, empowered by communities to oversee research, "may impose unnecessary regulations that may be unduly expensive to comply [with]," because they are "untrained in the intricacies of genetic engineering." Arguing that gene-splicing has been done successfully for a number of years with no accidents, one industry specialist said, as the *Times* reports, that "the issue of safety from the scientific standpoint is totally dead. . . . But the public perception of this runs a few years behind."

Once again, a gulf may be forming, with the stereotype of the concerned but naive citizen on one side and the industrial spokesperson, out for profit and willing to cut corners on safety, on the other. The naive citizen and the media are accused of holding up progress: "Scientists working in the field scoff at the possibility of something dangerous happening at a genetic facility. They point to widely accepted National Institutes of Health procedures that require the use of specially weakened bacteria that cannot survive outside the test tube. Mark Skaletsky, a Biogen vice president, dismisses the fearful scenario surrounding genetic research as 'media hype' " (*New York Times,* 23 May 1982).

Obviously, the last chapter has not been written. The three sides now find themselves far apart in the attempt to attain mutual knowledge. In an interview in *Technology Review,* David Baltimore, a professor of microbiology at M.I.T. and a winner of the 1975 Nobel prize for work that helped to clarify the interaction between cancer and tumor viruses, tries to bridge the gulf between citizen, scientist, and businessperson by providing information that he believes the media often ignores. Dr. Baltimore (1986, 46) argues,

a. Many genetically altered products, such as livestock viral vaccine and frost-free bacteria, are created merely by deleting a gene, a process that occurs continually in nature. Such deletions seldom produce an organism that has a selective advantage for survival.

b. The breeding of dogs and the selective modification of plants represent a much greater manipulation of genes than does anything currently being done in genetic engineering. We accept these because they have helped us survive and because there have been no catastrophic results.

c. The real danger is not introducing a new organism into an environment but removing one. The gypsy moth, for instance, was introduced from Europe and has no natural enemies in the United States; hence it consumes and destroys a significant percentage of the foliage on the East Coast.

d. Since the mid-1970s, biologists have shown themselves to be professionally responsible by regulating their own genetic research, by establishing standards for safe research, and by working with local groups, including the Cambridge City Council.

By providing such information, Baltimore is attempting to increase the level of mutual knowledge and thus to decrease the irrational fears of the general public. His arguments have the potential to allow the three sides to reach a new harmony.

Given what you know about this controversy, write a 1,000-word analysis of Baltimore's argument. Do you agree with Baltimore, or do you disagree? Give *specific* reasons for your position. To buttress your argument, do some research about what has happened to genetic research in the last 10 years. The *New York Times Index* is a good place to start for citations.

2. In some of the best journals in your major, find an article and rewrite it for a high school audience, following Weisskopf's model. Try to imagine what you knew as a high school student and write to that level.

You might want to use Weisskopf's method of identifying the basic concept, discovering why it is hard to understand, using familiar terms and experiences to explain it, and then discussing the usefulness or practicality of the knowledge attained.

Reference List

David Baltimore: Setting the record straight on biotechnology. October 1986. Interview. *Technology Review* 89, no. 7:38–46.

Bowen, M. E., and J. A. Mazzeo. 1979. *Writing about science.* New York: Oxford University Press.

Thomas, G. P. October 1986. Mutual knowledge: A theoretical basis for analyzing audience. *College English* 48, no. 6.

Toulmin, S. 12 March 1977. DNA and the public interest. *New York Times.*

Weisskopf, V. F. 1979. *Knowledge and wonder: The natural world as man knows it.* 2d ed. Cambridge: The MIT Press.

3

The Writing Process

THINKING AND PREWRITING

Professional writers do not create a final draft at their first sitting. None is so smart and so well organized that he or she can magically produce a draft in almost perfectly edited form. The written product—the letter, the report, the journal article—emerges from a disciplined *process,* which involves gathering information, thinking, writing, revising, editing, and, most importantly, rewriting.

The writer often organizes, composes, and revises at roughly the same time. Writing does not occur in separate and discrete stages: successful writers work backward and forward through their drafts, picking up and elaborating on undeveloped ideas and condensing or revising others. The more experienced writers become, the more efficiently they perform this process, and the better results they achieve.

Because professionals spend at least 20 percent of their time—the equivalent of one day per week—writing, it is necessary for students to learn to write as efficiently and as articulately as possible.

Preparations for Writing

Writing involves much more than just putting pen to paper. The writer needs to plan a schedule for the entire project, find a time and a place where he or she can work undisturbed every day, make decisions about content, find a colleague to read and assess the final draft, and create a file to serve as a record of the progress of all the writing projects. We will look at each of these in turn.

First, when you are given the writing project, schedule what you will do, day by day, between the time the paper is assigned to the time it is due. If you wait until the last minute to write a report, you deny yourself the opportunity to write thoughtfully, reflect, and revise your work (i.e., your product will have the substance of an early draft). Your schedule should include time to define what you wish to discuss in the paper, to do all necessary research, to write *at least* two drafts (this does *not* include the final copy), and to have the drafts and final copy typed into and printed from a word processor (or typed on a typewriter if a word processor is unavailable) and *proofread*. Make sure to allot sufficient time after each draft to distance yourself from the work and to have a fellow student or colleague read and evaluate the final draft.

If your work contains illustrations, don't forget to allow time for the artist or graphic designer to draw and insert the artwork. Finally, allow some time in the schedule—perhaps a couple of days—to accommodate any problems that might arise.

Second, once you have laid out the plan for the project, designate a place where you can write undisturbed and where you can keep all of the material you need. This includes reference books and papers, writing equipment, and any additional items you find helpful.

As Dorothea Brande notes, "Writing calls on unused muscles and involves solitude and immobility" (1981, 71). Students usually have a desk or work area, but finding solitude can be difficult in the workplace. Coordinating projects with coworkers, being available to answer the telephone, being ready to work in the lab or to go out in the field, and just being responsive to colleagues in a friendly and sociable way are everyday requirements. Despite the difficulties, however, you must find the space and time to do your writing. If you do not have a private desk, perhaps you could find a solitary work area, or you could go to your company library or ask your coworker to take phone messages while you are writing. The effort in securing whatever time and place you can find will pay off in a written document that expresses clarity of thought.

A more realistic way to describe the act of finding solitude might be to think of putting blinders on: You need to be able to *focus* and work on a single project at a time. In the practical environment of college or an office, concentrating on one task is not easy to do, given that you might be working on several projects at one time, all with different and often overlapping deadlines. Finding the time to focus and give all your energy to your writing for a short period of time, however, will definitely pay off in higher quality work.

Although much writing itself is usually done in solitude, other parts of the writing process—such as brainstorming, developing and organizing ideas, collecting relevant data, revising, and editing—can be a collaborative endeavor. Just as many major research projects in science and technology are now accomplished by research groups, instead of by an individual working alone, so is the writing associated with such projects becoming a group undertaking. The collaborative authoring of books, articles, and reports is commonplace, as attested by the list of authors on many publications.

In such collaborative writing, one of the authors is often the primary writer, with the other authors contributing to other parts of the writing process, such as reviewing and editing. If you eventually publish in the scientific and technical literature, most probably you will take part in such collaborative writing practice. But even in these groups, the primary author, when constructing the first draft, usually must exercise the discipline and rigor of concentrated thought.

Third, every piece of writing, technical or otherwise, in order to convey its message successfully, relies on a strong foundation. Other features may change or be relative to your topic or situation, but the four cornerstones listed below are vital to the effectiveness of your document. When planning your writing project, you must make decisions about the following:

1. Define your *audience.* To whom are you writing? If you are writing to specialists in your own field, your piece will be different from one whose audience includes professionals from several fields.

2. Define your *purpose.* What is your document intended to do? Traditional technical and scientific writing usually has one of three purposes: to inform, to persuade, or to motivate to action. Your tone and presentation will be a function of your purpose.

3. Follow a specified *format and length.* Most scientific and technical writing follows well-established formats, each of which has a predetermined style and length. Letters, for instance, are usually shorter than reports and have a specific format. Articles in scientific and technical journals also follow specified formats.

4. Collect all necessary *data.* You should assemble all the relevant data before you begin to write the first draft. The data provide the basis for the organizing ideas you generate.

The requirements of audience, purpose, format, and data will shape your writing. They form the structure of your document.

Fourth, arrange to have a reliable friend or colleague read a completed draft. A fresh reader can often point out some deficiencies in your writing or logic that you wouldn't see because you are too close to the work.

While your colleague is reading your draft, leave your copy unread in your desk. When you take it out for review in two or three days, when your colleague has returned the copy with his or her comments, you will be able to look at it with a fresh perspective and so will be able to see more of its problems.

Finally, keep a record of how each of your papers was evaluated so you can improve your performance the next time. Compare your work with that of your fellow students or coworkers for quality and completeness. Keep a file of this material so you can review it before you begin or while you are revising your next report.

When you have met each of these preconditions—a reasonable time schedule, a suitable time and place for work arranged, the bounds of your paper defined, a colleague lined up in the wings to review your draft, and some reinforcement from your last writing performance—you are ready to begin to write.

Flow Chart: Creating and Communicating

The process of writing can be graphically represented by a flowchart, which uses the idea of loops, from computer programming, to represent the way experienced writers write in a recursive, nonlinear way. They don't march through the writing process step by mechanical step, as an overly simplistic flowchart might imply. The chart in Figure 3–1 (page 44) shows that the writing process has two basic steps: creating and communicating.

Creating

The writer composes a first draft, assembling material and working toward an organizing idea for the data and the most effective sequence in which to present that data. Linda Flower (1985) has called the writing in this stage "writer-based prose," simply because the writer is writing and discovering at the same time. This enables writers to create and organize their ideas.

Several techniques will enhance your creativity. We list three here:

Brainstorming

On a blank sheet of paper, write the four cornerstones (audience, purpose, format and length, and data) at the top so they continue to form a boundary for your thinking. List the major data, results, or material you are working with. Then write down any possible interpretations of the data. List even interpretations you never think you will use—that is, don't edit out ideas during this process. Your sheet of paper will now look like a jumble of scratchings.

Grouping

Draw lines between ideas that look as if they might become part of a longer chain of meaning. Try to assemble your ideas into two or three distinct groups, and give each group a tentative title.

Fitting in Data

Take each piece of data or information and fit it into one of these groups. You are now forming tentative organizing ideas. After you have worked at it for a while, put into a separate group all the data that don't easily fit into one of the categories you have listed. Does this new group have an organizing idea, or is it just a group of material you should put on hold and use another time? If the former, create one or more new categories.

The most important thing during this stage is to remain flexible. Don't rush to get a fixed form. Play with the material for a while.

Communicating

After organizing the available information, a writer is able to formulate an organizing idea and to create an outline for sequencing the data and information. The writer then uses the outline to create the first draft. This sequence of sketchy paragraphs gives the reader enough information at every stage to understand the new material being presented. Flower (1985) calls the type of writing in the first draft "reader-based prose" because it is designed formally for the reader, not informally for the writer. One major quality of reader-based prose is that it can stand alone; all of the material needed to understand the paper is in the document itself, and a complete context is given for why the material is being presented and for what the author hopes the document will achieve.

FIGURE 3-1
The Process of Scientific and Technical Writing

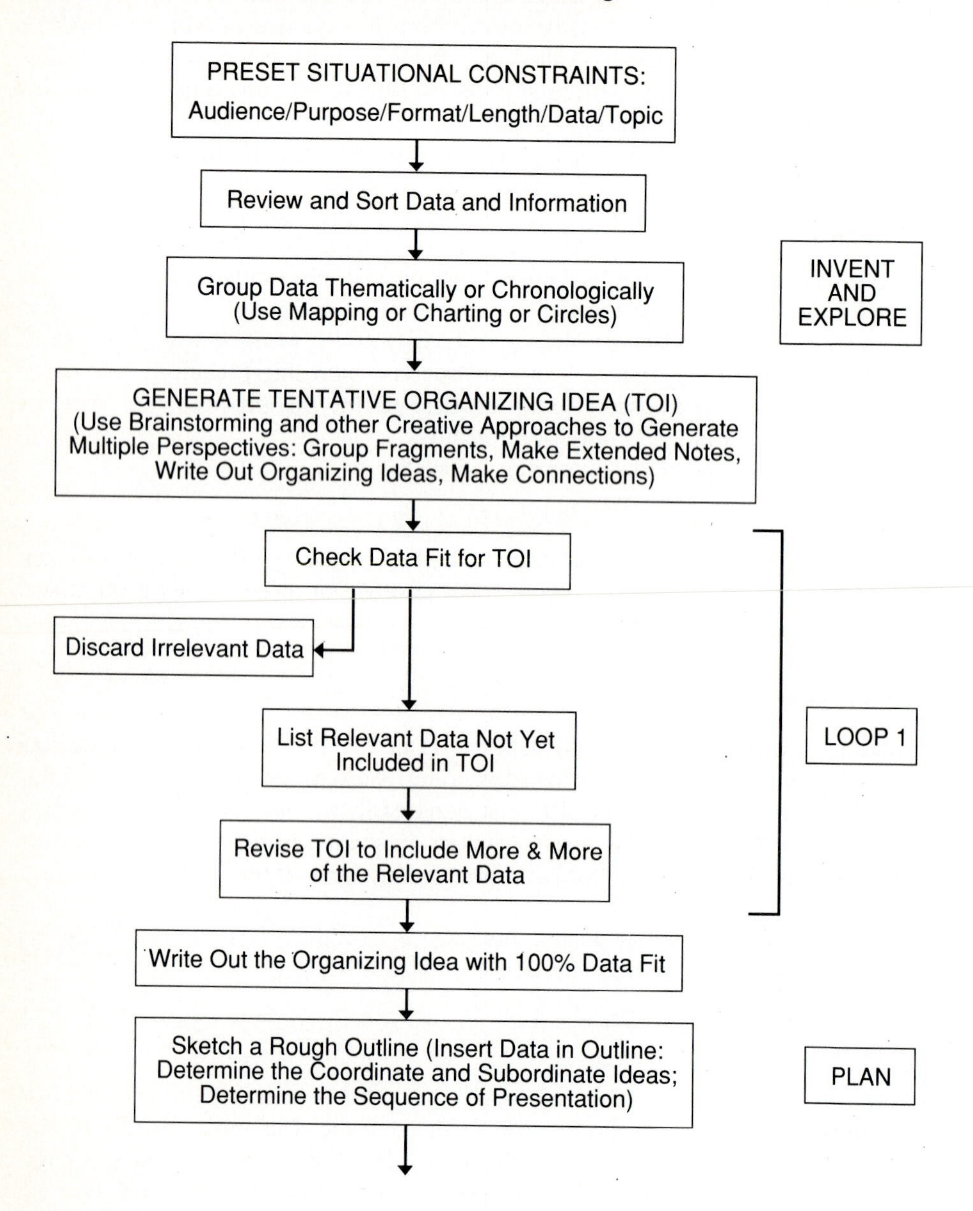

Several techniques will help you create and revise the rough draft. As you work, try the following:

1. Keep a one-page outline handy to remind you of the structure. Then, when you add or delete material, it will fit in the overall organization.

2. Add headings, subheadings, and transitional sentences so your reader can eas-

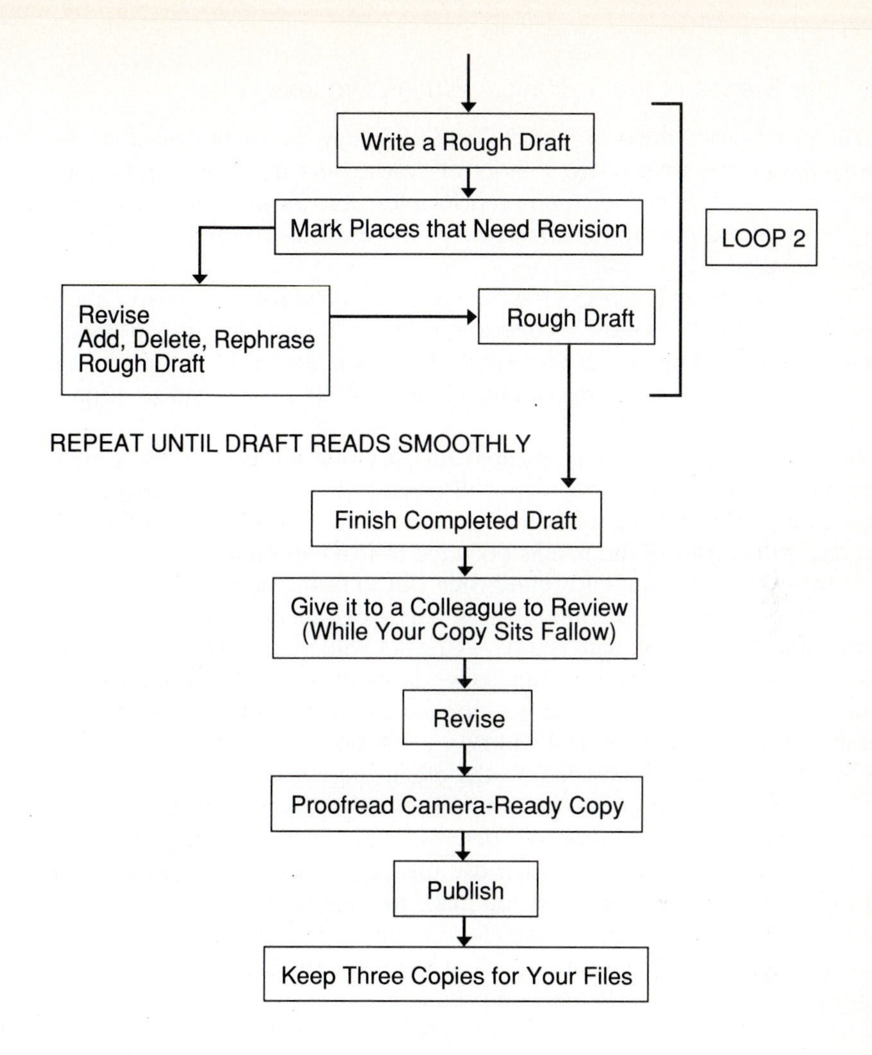

ily follow your train of thought. Try to include one of these three kinds of markers after each group of three or four paragraphs.

3. The examples you use must demonstrate what you want them to. Be sure they don't have more complex implications that, if fleshed out, might invalidate them as proofs. For instance, despite its extraordinary degree of technical innovation and the awe with which it was received, don't use the example of the machine gun in an argument designed to prove how technology has improved human life.

4. Read the first paragraph, the last paragraph, and one key paragraph in the middle. Does the document seem to be going anywhere? Is the progression of ideas logical?

If you present your writer-based prose as a finished document, your audience will seldom read it with pleasure since it represents your process of thought (incomplete as it might be) instead of a finished, clearly reasoned document. Inexperienced writers often hand in writer-based prose; experienced writers never do.

The Four Stages of the Technical Writing Process

The process of writing in science and technology can be broken down into four stages: *summarize* what other authors or researchers have said about your topic; *synthesize* their ideas to find common principles; *describe and analyze* the reliability of these ideas and principles; *contribute* your views and research in order to assert a new synthesis.

The flowchart in Figure 3–1 is based on this fourfold analysis of the writing process. You should begin by collecting your data, grouping it, and generating a theme, a Tentative Organizing Idea (TOI). Using the graphic techniques of mapping, charting, or putting data in overlapping circles will help you organize disparate data more easily.

Note: If you are writing a research report (what those in industry often call a "technical status report"), your data will consist of summaries of important recent research. If you are writing a laboratory experiment, as in a regular college exercise, your data will consist of the results obtained in that experiment.

In Loop 1, you continue to refine your TOI to bring more and more data into it. Your goal, of course, is to come up with a TOI that includes, or accommodates, all the important or relevant data you have to work with. Almost by its nature, your first TOI will not include all of the data; some data will be irrelevant and can therefore be discarded. Continue to refine your organizing idea until it includes all the relevant data. At this point, your TOI will have a 100 percent data fit—that is, all the data will fit into or be accommodated by, the organizing idea.

Once your organizing idea accommodates all the significant data, sketch out an organizational plan—an outline—to determine in what sequence you can best present your information. You can then use this organizational plan to select the data you will include in your report. If you make the mistake of including nonsignificant data, your report can become encyclopedic instead of purposive; many technical documents are marred by including all of the research data.

Loop 2 represents the actual composing and revising stage. Write, edit, and revise until the document reads smoothly from beginning to end. Then revise on the basis of your reader's comments, and proofread the final copy.

This process approach is based on the belief that writing and revising stimulate critical thinking processes that are never fully activated unless such a disciplined, open-ended, exploratory process takes place, as Peter Woodford argued in Chapter 1. The act of writing, both the physical and the intellectual process, stimulate optimal mental functioning.

Finally, from this flowchart you can infer the difference between the experienced and inexperienced writer. The inexperienced writer will quickly pass over the intellectual challenge demanded by Loops 1 and 2, will seek a fast closure on the writing task, and will be content to let difficult but relevant data wait for another day. Experienced writers, by contrast, will work energetically but slowly through the two loops and will use them to allow ideas to ferment and grow, thus tapping their best thoughts. These writers will be willing to try out several organizational strategies and outlines until they find the one that groups all the relevant data most effectively.

Reference List

Brande, D. 1981. *Becoming a writer.* Los Angeles: J. P. Tarcher.
Flower, L. 1985. *Problem-solving strategies for writing.* 2d ed. San Diego: Harcourt Brace Jovanovich.

CASE STUDY OF CRITICAL RESEARCH SKILLS AND THE COMPOSING PROCESS

To demonstrate how the writing process works, we have chosen the following case study, which concerns one of the most spectacular and deadly structural collapses in the history of American engineering. The study is presented at length so you can get a feel for how writing, research, and rewriting go together to form a continuous process of creative thinking. It describes the steps you might actually go through to write an engineering report, and its length indicates the rigors of such a process.

Imagine that you are employed as a research engineer for a civil engineering construction firm that has won a contract to build a hotel in your area. The hotel's owner wants the lobby to have very high ceilings, a great atrium, and an elevated walkway along one side of the atrium. The architect and your company's chief operating officer well remember the collapse of a similar walkway at the Kansas City Hyatt Regency Hotel in 1981. The collapse killed 114 people and injured almost 200 others, making it the worst structural collapse in the history of American engineering.

Wanting to avoid any possibility of such a problem in their walkway, they ask you to prepare a research report on the tragedy that describes for them the engineering problem responsible for the collapse. Since preliminary work needs to begin immediately, they give you one week to prepare the summary report.

Notice that, as is usual in technical writing tasks, the audience, purpose, format, length, and topic are preset. You are writing to the two supervisors of the project (the audience), to give them information on a design failure so they do not make the same mistake (the topic and the purpose). The format and length will follow those of a normal informational report: a summary of the information (with relevant references) in fewer than 10 pages, with an executive summary up front.

Doing Research to Create a Database: Collapse of the Hyatt-Regency Walkway

Your first step is to collect information, which you will then pare down for your report. You begin by searching through the *Applied Science and Technology Index* (or *ASTI*) for relevant articles in professional journals. To use *ASTI,* you need to find the correct subject category; in this case, it's "Walkways." And you need to check that subject heading for 1981 through to the time when the story has been fully covered. By doing this, you would find the following references (in *ASTI* reference style):

1981
Column-supported walk to replace Hyatt bridges. il *Eng N* 207:14 Aug 20 '81
Hotel disaster triggers probes. il diags *Eng N* 207:10–11 Jl 23 '81
Hyatt walkway design switched. il diags *Eng N* 207:11–13 Jl 30 '81
Kansas City, common sense, and simple communication. [editorial] W. F. Wagner, Jr. *Archit Rec* 169:7 Mid-Ag '81
Walkway probes continue. *End N* 207:16 Ag 6 '81
1982
Collapse of the Kansas City Hyatt Regency walkways. E. O. Pfrang and R. Marshall. il diags *Civil Eng* 52:65–8 Jl '82
Connection cited in Hyatt collapse. il diag *Eng N* 208:10–12 Mr 4 '82
Kansas City tragedy: there is not always strength in numbers. il diags *Tech R* 85:29–30 Ag/S '82
NBS reproduces Hyatt walkways. il *Eng N* 207:12–13 O 22 '81
[University Center, Flint, Mich.] il plans *Archit Rec* 169:18–23 S '81
Welding clear in collapse of Hyatt walkways. J. R. Birchfield. diags *Weld Design and Fabr* 55:64–7 Je '82
1983
Collapse of the Hyatt Regency Walkways; implications. G. A. Leonards. *Civ Eng-ASCE* 53:6 Mr '83
Hyatt-Regency walkway collapse; design alternates. G. F. W. Hauck bibl diags *J Struct Eng* 109:1226–34 My '83
1984
Experts testify GCE not responsible for walkway collapse. J. Jacquet. *Consult Eng* 63:18 O '84
Hyatt hearing traces design chain. il diag *Eng News-Rec* 213:12–13 Jl 26 '84
Hyatt hotel engineers cited for 'negligence.' *Eng News-Rec* 212:14 F 9 '84
Hyatt Regency walkway collapse: design alternates [discussion of 109:1226–34 My '83]. G. F. W. Hauck bibl diags *J Struct Eng* 110:934–6 Ap '84
Judge bars Hyatt tests. *Eng News-Rec* 213:31–2 S 20 '84 Weld aided collapse, witness says. *Eng News-Rec* 213:12 S 13 '84
1985–1986
Hyatt engineers found 'guilty' of negligence. il *Eng News-Rec* 215:10–11 N 21 '85; Discussion. 216:9–10+ Ja 23 '86

Reviewing the Database

Your next step is to scan the titles of these articles looking for any kind of pattern. By doing this, you find out that

The collapse was investigated by the National Bureau of Standards, a government agency.

The original design was switched sometime during the construction process.

The architectural plans are available.

An article on the problem has appeared in *Technology Review* (a major journal from M.I.T. on the applications of technology).

The welding was originally found to be sufficient but was later questioned.

The project engineers were first cited for negligence and then found guilty.

From Applied Science and Technology Index. 1981–82.

There has been an ongoing discussion of the problem in the *Engineering News-Record,* an industry standard.

Then, at the nearest library, you borrow or photocopy these articles so you can work with them at your desk. As you scan through them chronologically, you discover two additional sources:

1. The report from the National Bureau of Standards is available. Named *Investigation of the Kansas City Hyatt Regency Walkways Collapse* NBSIR 82-2465, February 1982, it can be obtained for $10 from the Superintendent of Documents, U.S. Government Printing Office, Washington, D.C. 20402 (request Document #003-003-02397-3). You order it, but expect it won't arrive before your report is due. You will send it on when it arrives.

2. The author of the article in *Technology Review* is Henry Petroski, a professor at Duke University who writes widely on issues in technology. You decide to check *Books in Print* to see if he has written anything relevant to your report. You find that he has written *To Engineer is Human: The Role of Failure in Successful Design* (New York: St. Martin's Press, 1985). The title sounds right for your topic. You find the book and discover a chapter on the Hyatt Regency tragedy. The chapter states that a Pulitzer Prize–winning story appeared on the tragedy in the *Kansas City Star,* so you send to the *Star* for a photocopy of the article.

Organizing and Partitioning the Database

The articles and Petroski's book are now on your desk, and you begin to read them in chronological order. You find that the articles, on the basis of their titles and the first few paragraphs, break into two sections: (1) articles describing and analyzing the collapse of the walkways from a purely structural point of view (this descriptive analysis ends, essentially, with the publication of the report by the National Bureau of Standards—Pfrang and Marshall, principal investigators—and the summary overview by Petroski); and (2) articles describing the prosecution of the president and vice-president of the engineering firm involved and letters-to-the-editor discussing other facets of the case.

Generating the Initial Organizing Idea

The material in the section on thinking and prewriting easily arranges itself into your initial organizing idea, which is basically a summary of the NBS report and Petroski's descriptive comments about it. You decide to base the summary on four illustrations:

1. The hanger rod–box beam as designed and as built (Figure 3–2a)
2. A rendering of the stringers, box beam, and floor (Figure 3–2b)
3. A drawing of the atrium walkway area (Figure 3–2c)
4. A rendering of the box beam and hanger rod after separation (Figure 3–2d)

Your summary would highlight three main concerns as well as the material from the conclusions of the NBS report and from Petroski:

FIGURE 3-2a
The Original and Revised Versions of the Hanger Rod–Box Beam

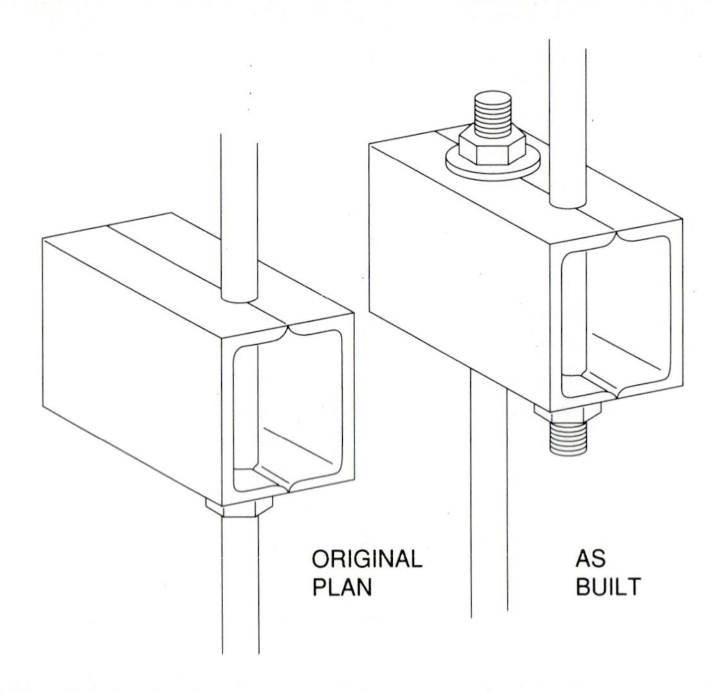

1. The original architectural renderings showed a nut and washer under the box beam, but no threading of the hanger rod below that beam. Since the architects had not provided a way for the nut to be threaded, the design was not feasible (though it was theoretically possible, given that a company could be found to supply such a threaded beam of extreme length).

2. The design change solved the problem of the threading but doubled the stress on the joint.

3. As originally designed, the hanger rod–box beam assembly (the continuous rod design) did not meet Kansas City building codes. When the walkways collapsed, they met only 53 percent of the code requirements.

Writing a First Draft

At this point, you could write a straightforward research report using the illustrations to develop the three main concerns. Your report would make the following conclusions and recommendations:

1. Construction should adhere to local and national building codes.
2. Architects and structural engineers should consult with each other about any change in design.
3. Unique or innovative designs should be thoroughly tested.

Although your report would be competent and professional and would have followed your first organizing idea, your supervisors might look at your conclusions and say, "So what else is new. These policies are a matter of normal, professional

FIGURE 3-2b
Arrangement of the Stringers, Revised Hanger Rods, and Floor

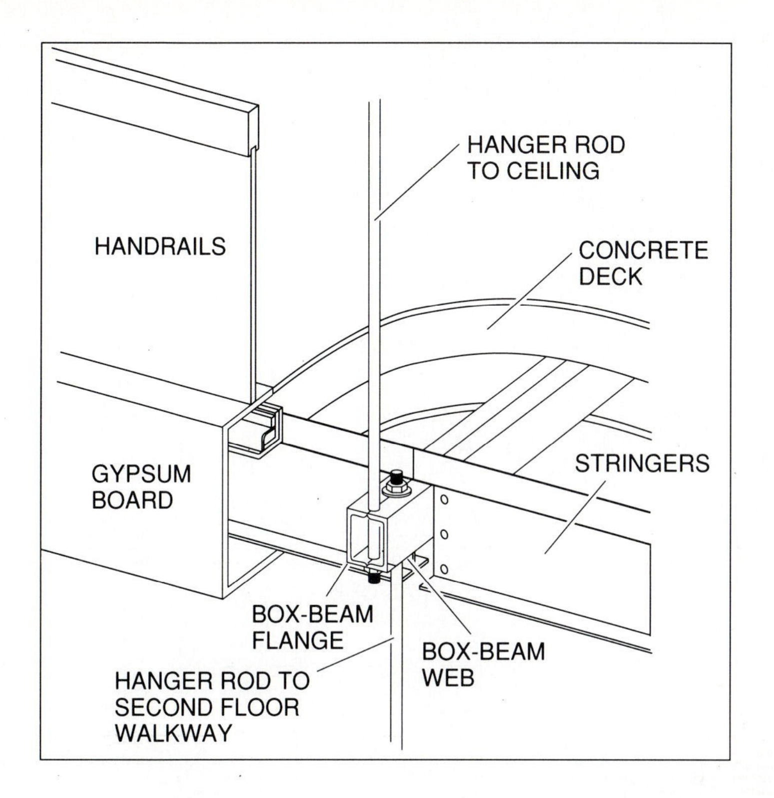

routine." And they might also wonder if you thought these conclusions would be helpful.

Including Anomalous Data

The problem with your report, based on your initial organizing idea, is not that it is too conservative but that it is incomplete. It does not consider the entire environment in which engineering and structural decisions are made and implemented. You needed to push ahead and read the articles from 1984 to 1986, which provide information that is anomalous with the purely technical information of the NBS report (such reports, by their nature, are not designed to assess blame; their goal is to analyze only the mechanical nature of the engineering problem).

For instance,

1. The NBS tests did not reconstruct what actually happened. When the hanger rod and box beam were stressed in the lab, they were stressed only until the hanger rod pulled through the bottom hole in the weld of the box beam. In the accident, it was the nut and washer at the end of the hanger rod pulling through the upper hole of the double-channel box beam that allowed the chain-reaction collapse to take place.

FIGURE 3-2c
A View of the Walkways

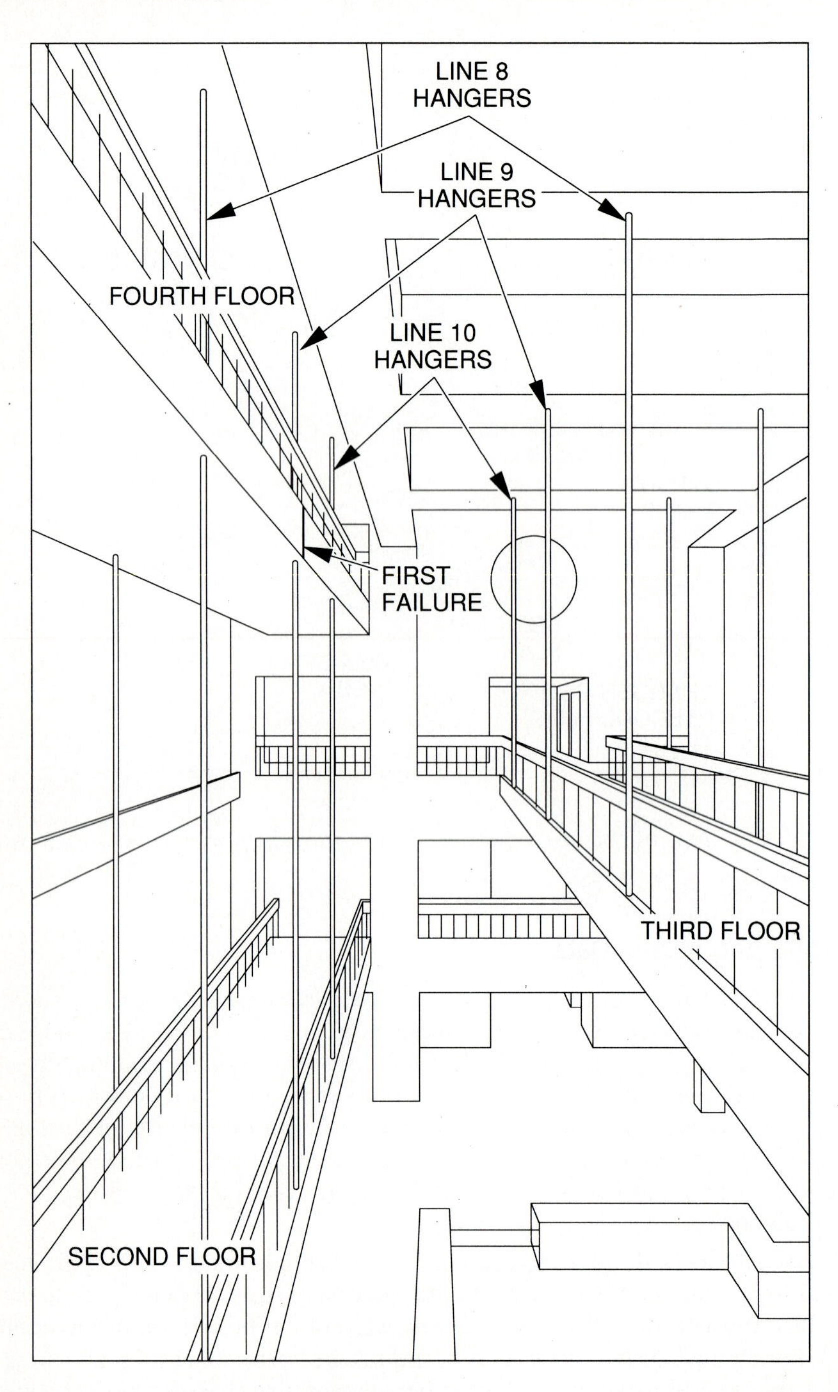

FIGURE 3-2d
The Failed Box Beams

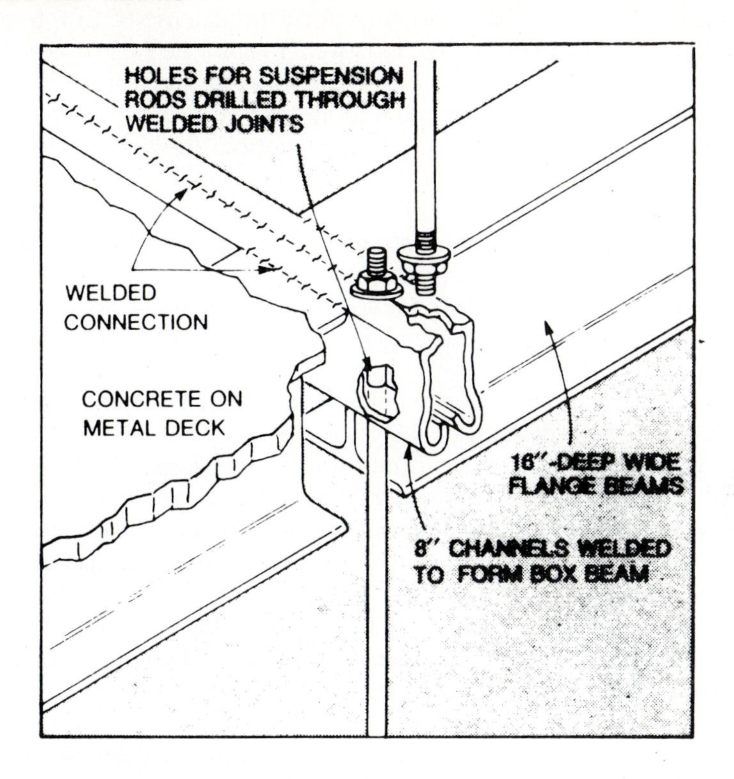

2. Jack D. Gillum, the chief engineer for the design and construction of the Hyatt Hotel, and Daniel M. Duncan, the project engineer (both of GCE International, Inc., of St. Louis) were charged with "gross negligence" in the design and analysis of the project. The two engineers were charged with failing "to perform calculations to determine the load capacity of the bridge rods and connections." Had they done so, continued the complaint, "they would have discovered that the connections were so grossly inadequate that they not only lacked the capacity to bear the design load intended or to comply with the Kansas City building code, but also were grossly inadequate to support even the weight of the bridges themselves" (*Engineering News-Record,* 9 February 1984, 14).

The charges also stated that after the design of the hanger rods was changed, no one checked to determine if the new design would be adequate; and that the analysis requested by the hotel's developer-owner after the atrium roof collapsed during construction was never performed.

3. An experienced steel detailer would have known, by experience and "eyeball" judgment, that the box beam was a lightweight channel (a thin-webbed channel box) that should have been replaced with a single, thicker walled tube, at a substantial savings (*Journal of Structural Engineering,* April 1984).

4. GCE subcontracted the steel fabrication to Havens Steel Company of St. Louis. Being overworked and not having time to complete the project, Havens

subcontracted the steel detailing to WRW Engineering of Kansas City. GCE never knew of this subcontract (*Engineering News-Record,* 26 July 1984).

5. A consulting engineer hired to strengthen the defense of GCE testified that Havens Steel used only a partial-penetration weld to fabricate the box beams, instead of a full-penetration weld (as required by the drawings), and that they failed to use web stiffeners (steel plates welded to the inside of box beams to keep the webs of the beams from buckling) as specified by the American Institute of Steel Construction (*Engineering News-Record,* 20 September 1984, 31).

6. This engineer, Lev Zetlin, said that his conclusions disagreed with those of the NBS report: workmanship and materials *were* involved in the collapse. Zetlin also criticized the NBS report because the tests did not force the hanger-rod through the upper connection on the box beam and because the tests "only showed that a deficient structure was indeed deficient" (*Engineering News-Record,* 13 September 1984, 12).

7. Gillum and Duncan were found guilty of gross negligence for abdicating their responsibility to review all shop drawings of the critical steel-to-steel connections that failed. Most damaging, and most influential in the decision, was Duncan's admission that "the key drawings that contained the final connection details were never reviewed by an engineer. The Gillum technician who performed that last check admitted in testimony that he did not know how to do the calculations required" (*Engineering News-Record,* 21 November 1985, 11).

8. Passionate letters to the editor of the *Engineering News-Record* (23 January 1986, 9, 10+) followed the announcement of the judge's decision. Among the opinions expressed were the following:

a. Although the engineers were derelict in allowing a design change to be made without carefully checking it, the steel fabricator is ultimately responsible for a construction design that is consistent with the structural design.
b. Fabricators should never be allowed or required to design construction details. The fabricator and the detailer must follow the shop drawings that have been approved by the engineering firm.
c. A retired ironworker claimed that the "flimsiness of the toe-to-toe channels should have screamed for reinforcement." He said that the foreman, the superintendent, the ironworkers, or others on the project should have objected. "Some people apparently didn't give a damn, some made more money by not doing the extra work and some, who held the final responsibility, did not take their duty seriously enough."
d. Fabricators are not expected to have the knowledge of stress that engineers do. When they make shop drawings, they mechanically follow the latest edition of the manual of the American Institute of Steel Construction.
e. The hanger rod–box beam connection was not a standard connection covered in the manuals and therefore should have been specifically checked by the engineers.
f. Engineers must be responsible for all aspects of the construction. To do this, they must be paid adequately, as they often are not now.

Revising the Initial Organizing Idea

This additional information indicates that a complete description of the failure must go beyond seeing it as only a problem in engineering. It forces you to adopt a systems approach. That is, the architect, structural engineer, steel fabricator, shop drawer, ironworker, construction superintendent, foreman, and laborer—in short, everyone from the white-collar professional to the experienced construction worker—must work as a team, with communication channels open and with workers able to ask questions and offer constructive criticism on the basis of their extensive experience. Responsibility cannot be compartmentalized and, thereby, avoided.

Generating the Second Organizing Idea and the Final Report

This approach, based on information not available in your first set of readings, leads you to your second organizing idea. The walkways collapsed not only because of initial mistakes in the design (the single rod was not threaded from the bottom and thus the connection could not be made) but also because changes made in the original design were not properly tested and analyzed and because experienced steel fabricators and ironworkers did not object to what was obviously an inadequate design.

Your report would recommend that a new process of team review be established within your company that would require teamwork. This would add time to a schedule, but it also would help avert potential disasters. You would, in this report, show your superiors that you can grasp the concept of a problem, not just its details.

Creativity and Stimulation

Although the term "creativity" is commonly associated more with the arts than with science and technology (we talk of the "creative arts," for instance), truly creative thinking forms the foundation for scientific discovery and technological innovation and progress. And creativity is especially important in writing scientific and technical reports and in leading a rich, imaginative life as a scientist or engineer.

An article in *Science News* (Weisburd 1987) reports on a conference about creativity held at the Smithsonian Institution. David N. Perkins, codirector of Harvard University's Project Zero study of cognitive skills in the sciences and humanities, believes there are six traits common to creative people:

1. A desire to find and frame unusual problems, as well as to solve them.
2. An ability to create metaphors (unexpected connections between previously unrelated phenomena) and, consequently, an ability to challenge traditional assumptions.
3. The ability to be objective, "to temper the energy of creation with testing and judgment"; the ability to come up with many ideas, and to recognize and discard the bad ones.
4. The desire to take risks and to "live on the edge of uncertainty"; and the ability to tolerate, even enjoy, ambiguity.
5. The ability to be stirred by intrinsic motivations—"money, grades, recognition

and awards are relatively unimportant." "At any early age," reflects Linus Pauling, "I had a strong curiosity about the nature of the world. I don't think I was ambitious."

6. A drive to uncover the aesthetic, to reduce chaos, and to find beauty, to find order and simplicity (as in the creation of the Periodic Table of the Elements and many discoveries in modern physics).

Writing Exercises

1. Review the case study on the walkway. How many of the characteristics of creativity (listed above) can you find in the process of accumulating the data, reviewing the data, creating an initial organizing idea, and creating a second organizing idea? Write an analysis of this case study as a creativity exercise.

2. Investigate other disasters, of both natural and human origin, such as the collapse of a section of bridge on the Connecticut Turnpike, the explosion of the shuttle Challenger, the 1985 flood in southwestern Pennsylvania and northern West Virginia, the 1987 earthquake in the Los Angeles area, the nuclear accident at Three-Mile Island, the continuing series of U.S. Army helicopter training-flight crashes, the depletion of the shrimp and crab nesting areas off the northeastern coast line, the destruction caused by acid rain, and the deterioration of the ozone layer above Antarctica.

Find articles about your research area by consulting these and similar sources:

The *New York Times Index*

The *Applied Science and Technology Index*

The index to *Scientific American*

The index to *Technology Review*

The index to *Science News*

The index to *Engineering News-Record.*

Apply one or more of the six traits of creativity listed above to try to come up with a tentative solution to one of these types of problems.

3. Use one of the six traits of creativity to solve, in an unusual or unexpected way, one of the lab or research projects you are currently working on in class.

Reference List

Hyatt engineers found guilty of negligence. 21 November 1985. *Engineering News-Record* 215: 11.

Hyatt hearing traces design chain. 26 July 1984. *Engineering News-Record* 213: 13.

Hyatt Hotel engineers cited for "negligence." 9 February 1984. *Engineering News-Record* 212.

Hyatt Regency walkway collapse: Design alternates. April 1984. *Journal of Structural Engineering* 110, no. 4: 934–36.

Weld aided collapse, witness says. 13 September 1984. *Engineering News-Record* 213: 12.

Weisburd, S. 7 November 1987. The spark: Personal testimonies of creativity. *Science News* 132: 298–300.

Part Two

Thinking and Writing in Science and Technology

4

The Approaches of Scientific and Technical Literature

The literature of technology and science presents an extraordinarily varied and fascinating picture. It ranges from a professional geologist describing how continents form by using the theory of plate tectonics, to an engineer writing a series of descriptive specifications for the foundation of a hydroelectric generating plant, to a biochemist announcing a new research direction for the cure of a communicable disease (such as AIDS), to a research physicist meditating on the philosophical implications of space folding over on itself.

This vast literature of science and technology can be divided into four categories or types of works:

1. Traditional science and technology,
2. Innovative or revolutionary science and technology,
3. Exploration in science and technology, and
4. Reflecting about science and technology.

Let's look at each of them.

TRADITIONAL SCIENCE AND TECHNOLOGY

This group includes the majority of the literature and most of the writing that you will do as a student and as a professional engineer or scientist. This literature is based on scientific theories that have been empirically and mathematically verified and that most practicing scientists and engineers use to describe the world and to predict its behavior. The traditional form attempts to explain these theories, both in themselves and in the ways they can be used to elucidate the behavior of the physical world. Although its main purpose is to describe information gathered by state-of-the-art technologies, this literature also reports new research that attempts to complete or fill out the implications of an established theory (often called a "paradigm" or a "research program").

The literature of traditional science and technology changes, of course, as the knowledge base in science and technology broadens. For instance, a report about

DNA (deoxyribonucleic acid, the genetic building block) written after Watson and Crick's discovery would be different from one written before it. Likewise, Lord Rutherford's description of the atom as having a nucleus whose charge was different from the electrons circling it was an improvement over J. J. Thompson's earlier description, which asserted that the positive charge dominated the atom's entire environment; and Rutherford's view is now being replaced with a description of the atom as a bounded field of vibrational force. In contemporary technology, if an analysis of computer programming ignored the innovation of parallel processing, it would be out of date as soon as it was written. Descriptions in the literature of traditional science and technology are relative to the current state of knowledge.

Despite this historical relativity, writing under the banner of traditional science and technology carries enormous authoritative weight. The following are typical examples of writing in this group:

A description of new information or research that fills in the gaps of a traditional scientific or technological paradigm—as when crystallographers discover and describe the seemingly crystalline structure of common clay

A description in a research journal of the solution of a scientific or technological puzzle or special problem, using traditional methods of analysis—as when psychiatrists discover why the drug Ritalin, which is a stimulant, actually calms down some overactive children

A description of the workings of an internal combustion engine

A description of how to replace a gasket on a pressurized oil line, written for a service engineer or a textbook audience

A description of the classification system for a biological genus or a pharmacological grouping

A report updating laboratory research

The definition of a theory written for a textbook or an encyclopedia

A solution to a given problem (a feasibility or a technical status report)

Learning to write these kinds of descriptions, definitions, summaries, and technical status reports is a must for becoming a successful scientist or engineer. Chapter 5 introduces you to the rhetorical building blocks of writing within the framework of traditional science and technology.

INNOVATIVE OR REVOLUTIONARY SCIENCE AND TECHNOLOGY

History reveals that there are times in science and technology when a well-established world view held by scientists, engineers, and even the general public, is discarded and replaced by another. Sometimes these revolutions are dramatic, changing the way ordinary people as well as scientists and engineers look at the world. Among these dramatic innovations are the Copernican/Galilean revolution, which replaced the Ptolemaic earth-centered universe with a sun-centered one; the Newtonian revolution, which laid the foundation for classical physics and mechanics by describing the relationship among force, mass, and acceleration, applying the inverse square law to astronomical problems, and developing the calculus; the Darwinian revolution, which argued that homo sapiens evolved from lower species by a process of natural selection based on competition and survival of the fittest group, and that this evolution was nonteleological, having no preordained purpose; the atomic revolution (attributable to J. J. Thompson and Lord Rutherford), which discarded the age-old belief in the indivisibility of the atom and replaced it with a more accurate description of a nuclear structure; and the Einsteinian revolution, which exploded the Newtonian idea of the independent status of space and time and introduced us to the mysteries of electromagnetic energy. These revolutions usually occur when aggregates of observed anomalies force scientists and engineers to abandon a traditional interpretation and search for a more encompassing one.

The literature describing these innovations, discoveries, and revolutions is often the most exciting and most thought-provoking of all the published materials in science and technology. Its excitement derives largely from the fact that many revolutionary discoveries come about by extraordinary means. Albert Einstein, for instance, reported that he first realized the speed of light must be constant when he was dreaming about riding on a beam of light. It struck him that if he were to measure the light beam's speed relative to himself it would be static (Burke 1985, 303). This revolution, like many others, changed our vision of humankind's place in the cosmos.

Other revolutions have been no less important, despite the fact that they have not led to a revolution in popular ideologies. Among these are the following:

- The development of writing, and the ability to mine and work with metals
- The development of cultivation, and the domestication of animals
- Engineering advances enabling people to control the flow of water, sail the oceans, and chart the skies
- The development of engineering marvels, like the clock, the steam engine, and the internal combustion engine
- The discovery of the relationship between electricity and magnetism, leading to the study of electromagnetism and our technological revolution
- Georg Mendel's discovery of the method of transmission of hereditary traits
- The discovery of X-rays and radioactivity
- The discovery of DNA

This list could be extended almost indefinitely. Chapter 7 introduces you to some of the literature of scientific and technological revolutions.

The literature of both traditional science and technology, as well as of innovation, discovery, and revolution, uses the basic forms of exposition. Technical knowledge, whether traditional or innovative, is expressed through description, narration (especially that of a process), explanation, analysis, and persuasion. The readings and the writing assignments, therefore, will illustrate these forms.

In the other two groups of scientific and technical prose, these expository forms are less distinct. They cannot be as neatly isolated as those used for writing in the traditional or innovative environments. Therefore, these latter two groups will be organized by other categories.

EXPLORATION IN SCIENCE AND TECHNOLOGY

This group could easily be included under traditional science and technology, especially because Thomas Kuhn, a distinguished philosopher and historian of science, believes that elaborate strategies for problem-solving and research methods for exploring the limits of a research program or theoretical paradigm (a world view) are aspects of normal or traditional science. But creating a special category is valuable because it provides a valid place for the kind of intellectual activities most scientists and engineers would consider to be at the limits of their disciplines.

The working out of conundrums, exploratory playfulness, and the tinkering that often accompanies or foreshadows ingenious technological invention all fall within this group. Here are some examples of the kinds of literature to expect:

An architect who bases a scale model on a dream, or a chemist who has a dream elucidating the structure of a compound under study

A psychologist who has subjects play a game to test the limits of one kind of human behavior

An inventor who creates a useful piece of equipment even though not fully understanding the physics implied in the invention

The computer specialist who teaches a computer to play tic-tac-toe in order to see if it can "learn" about no-win situations, as in the movie *War Games.*

This kind of literature often leads to the progress of traditional science and technology, but its methods and procedures—its protocols—differ from more straightforward approaches. Chapter 8 introduces you to the literature of this category by presenting a case study of an extraordinary invention.

REFLECTING ABOUT SCIENCE AND TECHNOLOGY

Scientists and engineers, and even lay people who are students of these disciplines, often reflect about issues relating to science and technology. For instance, E. F. Schumacher, formerly chairman of the British Board of Coal and an engineer, wrote the book *Small Is Beautiful,* in which he argues that technology should be kept closer to the human scale in order to prevent humans from being alienated from technology and from themselves as users of technologies they don't understand. Richard Selzer, a surgeon, has written eloquently about the technique and the art of surgery. Many scientists write memoirs and informal essays in which they discuss the excitement of their discipline and reflect on philosophical and metaphysical implications of what they have learned.

Though you may not ever publish material of this kind, it is useful for you to experiment with writing it so you get the feel for the way scientists and engineers think when they are letting themselves roam beyond the traditional borders of their disciplines.

Chapter 9 introduces you to the literature of this category by exploring the following topics: writing as a humanizing endeavor, applying scientific concepts to human life, speculating on the impact of a piece of technology, and evaluating technology from the point of view of the human scale.

Writing Exercises

Your instructor will assign an appropriate length for each of these exercises.

1. Scan *Scientific American, Science, Nature, Discover, Omni, Technology Review, New England Journal of Medicine,* and (if you are majoring in engineering) a primary engineering journal in your field. List several articles for each of the four major kinds of writing in science and technology. Write summaries of several of these and report on them in class.

2. *Traditional Science and Technology:* Describe a technique you know how to perform very well (such as gapping new spark plugs). Mention the tools that are needed, and give a sequenced set of instructions for performing the process. Then describe one innovation that would make the technique simpler and more efficient to perform or that would give a more precise result. Invent the innovation, if necessary.

3. *Revolutionary or Innovative Science and Technology:* Pretend that you are writing part of a science fiction story. Following the schema for scientific and technical revolutions that Thomas Kuhn describes in Chapter 7 (pages 195–98)—namely, stable paradigm, anomaly(ies), new hypothesis, testing, new paradigm—write a description of a strange occurrence in your research station and say how that occurrence, if verified, would require overturning some basic principle in physics, chemistry, mathematics, or engineering.

4. *Exploration in Science and Technology:* Many of us built model airplanes when we were young. In doing this, we thought of a model as a small-scale replica of the real thing, but without the ability to fly. When scientists and engineers model something, they usually construct a small-scale version to test the design before a full-scale prototype is built.

By building a model, or by modeling a process, and then submitting that model to a test, scientists and engineers can improve the design. Much of this is now being done through mathematical models and with the aid of computers. Automakers, for instance, construct two kinds of models: small-scale replicas and computer models. With these they can test drivability, wind resistance, and other structural features before the car goes into production.

Describe a model or a sketch for something that hasn't been built yet but that would be worthwhile and practical if it were. In your description, say what need the new device will satisfy.

5. *Reflecting About Science and Technology:* Humans have been called the toolmaking animals. Select several common hand tools (e.g., a hammer, a pair of pliers,

a screwdriver, and a saw), and describe how they extend the natural physical powers of the arm and hand. Speculate about how and why these tools might have been invented and developed, and say what everyday life would be like without them.

Reference List

Burke, J. 1985. *The Day the Universe Changed.* Boston: Little, Brown.

5

Traditional Science and Technology, I: Description

The chapters that discuss the four categories of literature offer a rich anthology of excellent and often classic examples of writings in science and technology. All the readings are of high quality, both intellectually and stylistically. Some also have historical interest, chronicling discoveries or innovations by pioneering scientists and engineers.

These readings provide examples of the kind of prose that practicing engineers and scientists actually write. Consequently, they give you experience with the primary sources of the disciplines of science and engineering, in contrast to the textbooks you are most often exposed to as an undergraduate.

Chapters 5 and 6 present the full range of technical and scientific exposition. Chapter 5 covers description, explaining a process, classification, definition, and reporting trials and experiments. Chapter 6 continues the traditional theme, discussing explanation, analysis, and persuasion, and providing the format of an academic article.

Description is the most basic form of scientific and technical communication, mainly because such communication attempts to convey objective knowledge about what is known. As such, description includes a large range of forms, namely static objects or organisms; objects or organisms as they change through time; classification structures or hierarchies, including simple definition and comparison and contrast, which by their very nature are based on classifications; the reporting of experiments; the workings of a process or mechanism, including descriptive instructions for performing technical tasks; extended definitions of abstract concepts; and exemplification. These forms make up the majority of the writing in traditional science and technology. They are the basis both for description and for more complex documents whose main purpose could be analytic or persuasive.

Rhetoricians have classified or organized this vast literature of scientific and technical description in different ways. As Richard Young and his colleagues (1970) argue, it is intriguing to be able to describe things from three perspectives: the thing in itself, unchanging, static, and unrelated to its surroundings; the thing as it changes through time (dynamic or temporal); and the thing as it relates to its surroundings or to the class of objects like itself. Thus we can describe an object as static, as changing through time, and as related to its environment; doing this enables us to see the object's richness and complexity. The third class of description, for instance, allows us to create major classifications of natural phenomena, such as the classification of the chemical elements, and then to define each class by comparing and contrasting it with other elements. Thus, classification, definition, comparison and contrast, and description itself become interrelated.

Not coincidentally, modern physics also describes matter at the subatomic level in the same triadic way: matter can be interpreted as a particle (static), as a wave (dynamic and temporal), or as a field (environmental and structural). In each type of description, we learn something about matter that cannot be learned from the other types. And combining all three descriptors gives us a richer picture of matter than we get from any one description alone. That is, to shift perspectives enables us to see all the facets of any object.

By reading examples of these forms and then trying them out for yourself, you will become familiar with the various formats of the literature of science and technology.

Reference List

Young, R., et al. 1970. *Rhetoric: Discovery and change.* New York: Harcourt Brace Jovanovich.

STATIC DESCRIPTION

Robert Hooke (1635–1703), a British physicist and inventor, worked as a laboratory assistant for Robert Boyle (1627–1691), who formulated Boyle's Law (at a fixed temperature, the pressure of a confined gas varies inversely with its volume). In 1663, Hooke became a Fellow of the Royal Society and curator of its experiments, serving as its secretary from 1667 to 1682. In 1665, he was appointed Professor of Geometry at Gresham College, London. Hooke formulated what came to be known as Hooke's Law (1678): the deformation of an elastic body is proportional to the deforming force. For instance, the force required to stretch a string 2 inches is twice that required to stretch it 1 inch. Subsequent researchers have found that this law is valid only for small deformations (those of less than 1 percent).

The following selection is from Hooke's *Micrographia,* which he published in 1665. Hooke's observations of common objects under a microscope brought him great fame and renown, for he was the first human on record to have seen such things up close. His colleagues at the Royal Society were amazed at his findings.

The illustrations of his observations are early examples of the high standards necessary in scientific and technical description.

Hooke's language sounds somewhat archaic to us, which is natural given its date of composition. Yet its precision and exactness perfectly mirror the magnifying power of his microscope. Hooke's punctuation also differs from ours, as does the punctuation in the later example from Benjamin Franklin (the most notable difference is the use of the colon where we would use a period). Style of punctuation, of course, changes with the times.

Observation 53: *Of a Flea*

Robert Hooke

The strength and beauty of this small creature, had it no other relation at all to man, would deserve a description.

For its strength, the *Microscope* is able to make no greater discoveries of it than the naked eye, but onely the curious contrivance of its leggs and joints, for the exerting that strength, is very plainly manifested, such as no other creature, I have yet observ'd, has any thing like it; for the joints of it are so adapted, that he can, as 'twere, fold them short one within another, and suddenly stretch, or spring them out to their whole length, that is, of the fore-leggs, the part *A* [see Figure 5–1] lies within *B,* and *B* within *C,* parallel to, or side by side each other; but the parts of the two next, lie quite contrary, that is, *D* without *E,* and *E* without *F,* but parallel also; but the parts of the hinder leggs, *G, H,* and *I,* bend one within another, like the parts of a double jointed Ruler, or like the foot, legg and thigh of a man; these six leggs he clitches up altogether, and when he leaps, springs them all out, and thereby exerts his whole strength at once.

But, as for the beauty of it, the *Microscope* manifest it to be all over adorn'd with a curiously polish'd suit of sable Armour, neatly jointed, and beset with multitudes of sharp pinns, shaped almost like a Porcupine's Quills, or bright conical steel-bodkins; the head is on either side beautify'd with a quick and round black eye *K,* behind each of which also appears a small cavity, *L,* in which he seems to move to and fro a certain thin film beset with many small transparent hairs, which probably may be his ears; in the forepart of his head, between the two fore-leggs, he has two small long jointed feelers, or rather smellers, *MM* which have

FIGURE 5-1

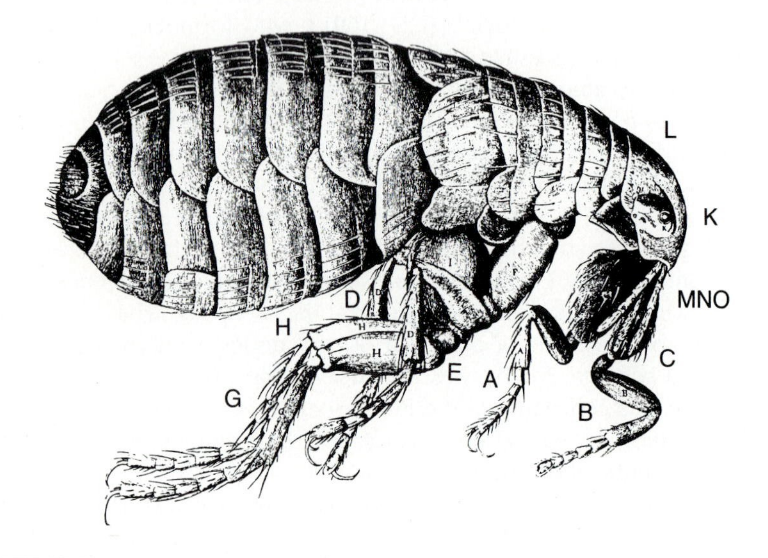

From Hooke, R. 1665. Observation 53: Of a flea. In his *Micrographia: Or some physiological descriptions of minute bodies by magnifying glasses with observations and inquiries thereupon.*

four joints, and are hairy, like those of several other creatures; between these it has a small *proboscis*, or *probe*, *NNO*, that seems to consist of a tube *NN*, and a tongue or sucker *O*, which I have perceiv'd him to slip in and out. Besides these, it has also two chaps or biters *PP*, which are somewhat like those of an Ant, but I could not perceive them tooth'd; these were shaped very like the blades of a pair of round top'd Scizers, and were opened and shut just after the same manner; with these instruments does this little busie Creature bite and pierce the skin, and suck out the blood of an Animal, leaving the skin inflamed with a small round red spot. These parts are very difficult to be discovered, because, for the most part, they lye covered between the fore-legs. There are many other particulars, which, being more obvious, and affording no great matter of information, I shall pass by, and refer the Reader to the Figure.

Benjamin Franklin (1706–1790)—entrepreneur who was independently wealthy by the age of 42, author, printer, diplomat, philosopher, scientist, and inventor—is America's premiere Renaissance figure, rivaled only by Thomas Jefferson. Franklin's most enduring scientific discovery, that lightning is a form of electricity, enhanced his fame in Europe and America. His most elegant engineering invention is the Pennsylvania fireplace, now called the Franklin stove.

In introducing his innovative fireplace—what we would call a wood stove—Franklin pointed out that inhabitants of the northern colonies kept fires for seven or eight months of the year and that firewood was becoming more expensive. Here is his discourse on the function of his invention and the physics of fire, heat, air, and smoke.

An Account of the Newly Invented Pennsylvanian Fire-Place

Benjamin Franklin

The fire being made at *A*, the flame and smoke will ascend and strike the top *T*, which will thereby receive a considerable heat. The smoke finding no passage unwards, turns over the top of the air-box, and descends between it and the back plate to the holes at *B* in the bottom plate, heating, as it passes, both plates of the air-box and the said back plate; the front plate, bottom and side plates are also all heated at the same time. The smoke proceeds in the passage that leads it under and behind the false back, and so rises into the chimney [Figure 5–2]. The air of the room, warmed behind the back plate, and by the sides, front and top plates, becoming specifically lighter than the other air in the room, is obliged to rise; but the closure over the fire-place hindering it from going up the chimney, it is forced out into the room, rises by the mantle-piece to the ceiling and spreads all over the top of the room, whence being crowded down gradually by the stream of newly warmed air that follows and rises above it, the whole room becomes in a short time equally warmed.

At the same time the air, warmed under the bottom plate and in the air-box, rises, and comes out of the holes in the side-plates, very swiftly if the door of the room be shut, and joins its current with the stream before mentioned rising from the side, back and top plates.

The air that enters the room through the air-box is fresh, though warm; and computing the swiftness off its motion with the areas of the holes, it is found that near 10 barrels of fresh air are hourly introduced by the air-box; and by this means the air in the room is continually changed, and kept at the same time sweet and warm.

It is to be observed that the entering air will not be warm at first lighting the fire, but heats gradually as the fire increases.

From Franklin, B. 1744. *An account of the newly invented Pennsylvanian fire-place.* Philadelphia: B. Franklin.

FIGURE 5-2
***Profile of the Chimney and* Fire-Place**

M The Mantle-piece or Breast of the Chimney.
C The Funnel.
B The false Back & Closing.
E True Back of the Chimney.
T Top of the Fire-place.
F The Front of it.
A The Place where the Fire is made.
D The Air-Box.
K The Hole in the Side-plate, thro' which the warm'd Air is discharg'd out of the Air-Box into the Room.
H The Hollow fill'd with fresh Air, entring at the Passage *I*, and ascending into the Air-Box thro- the Air-hole in the Bottom-plate neat
G The Parition in the Hollow to keep the Air and Smoke apart.
P The Passage under the false Back and Par tof the Hearth for the Smoke.
↗↗↗↗↗ The Course of the Smoke.

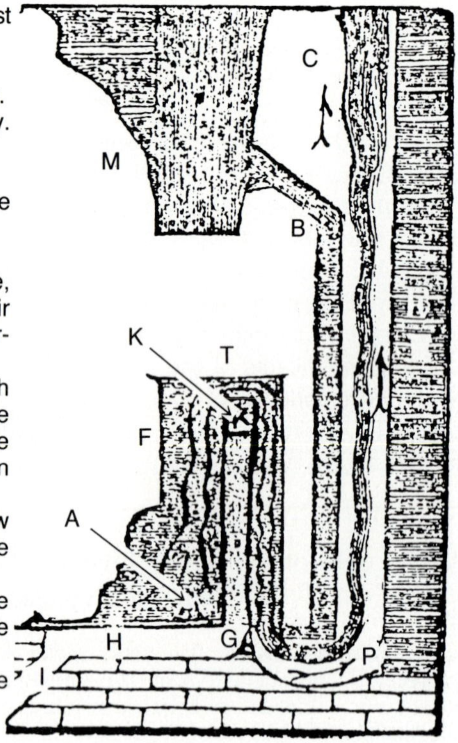

A square opening for a trap-door should be left in the closing of the chimney, for the sweeper to go up: The door may be made of slate or tin, and commonly kept close shut, but so placed as that turning up against the back of the chimney when open, it closes the vacancy behind the false back, and shoots the soot that falls in sweeping, out upon the hearth. This trap-door is a very convenient thing.

In rooms where much smoking of tobacco is used, it is also convenient to have a small hole about five or six inches square, cut near the ceiling through into the funnel: This hole must have a shutter, by which it may be closed or opened at pleasure. When open, there will be a strong draft of air through it into the chimney, which will presently carry off a cloud of smoke, and keep the room clear: If the room be too hot likewise, it will carry off as much of the warm air as you please, and then you may stop it entirely, or in part, as you think fit. By this means it is that the tobacco-smoke does not descend among the heads of the company near the fire, as it must do before it can get into common chimneys.

Francis Harry Compton Crick (1916–   ), a British biophysicist, and James Dewey Watson (1928–   ), an American geneticist and biophysicist, shared the 1962 Nobel Prize for Physiology or Medicine for their discovery of the structure of deoxyribonucleic acid (DNA), the chemical structure in chromosomes responsible for controlling heredity. While working together in the Cavendish Laboratory at Cambridge University, they used X-ray diffraction to discover that DNA was composed of a double-helix (spiral)—intertwined strands of sugar-phosphate, bridged horizontally by flat organic bases. They reported their discovery in the journal *Nature.* The article, reprinted below, is one of the most famous in modern biology and biophysics. They also wrote a related article, "General Implications of the Structure of Deoxyribonucleic Acid."

J. D. Watson, head of the Laboratory of Quantitative Biology at Cold Spring Harbor, Long Island, New York, has published *Molecular Biology of the Gene* (1965), *The Double Helix* (1968), and *The DNA Story* (with John Tooze, 1981). F. H. C. Crick has published *Of Molecules and Men* (1966), which describes the revolution in microbiology, and *Life Itself* (1981), in which he argues the possibility that life on Earth may have arisen from bacteria from outside our solar system.

Molecular Structure of Nucleic Acids

James D. Watson and Francis H. C. Crick

A Structure for Deoxyribose Nucleic Acid

We wish to suggest a structure for the salt of deoxyribose nucleic acid (D.N.A.). This structure has novel features which are of considerable biological interest.

A structure for nucleic acid has already been proposed by Pauling and Corey.[1] They kindly made their manuscript available to us in advance of publication. Their model consists of three intertwined chains, with the phosphates near the fibre axis, and the bases on the outside. In our opinion, this structure is unsatisfactory for two reasons: (1) We believe that the material which gives the X-ray diagrams is the salt, not the free acid. Without the acidic hydrogen atoms it is not clear what forces would hold the structure together, especially as the negatively charged phosphates near the axis will repel each other. (2) Some of the van der Waals distances appear to be too small.

Another three-chain structure has also been suggested by Fraser. . . . In his model the phosphates are on the outside and the bases on the inside, linked together by hydrogen bonds. This structure as described is rather ill-defined, and for this reason we shall not comment on it.

We wish to put forward a radically different structure for the salt of deoxyribose nucleic acid. This structure has two helical chains each coiled round the same axis [see Figure 5–3]. We have made the usual chemical assumptions, namely, that each chain consists of phosphate diester groups joining β-D-deoxyribofuranose residues with 3′,5′ linkages. The two chains (but not their bases) are related by a dyad perpendicular to the fibre axis. Both chains follow right-handed helices, but owing to the dyad the sequences of the atoms in the two chains run in opposite directions. Each chain loosely resembles Furberg's[2] model No. 1; that is, the bases are on the inside of the helix and the phosphates on the outside. The configuration of the sugar and the atoms near it is close to Furberg's "standard configuration," the sugar being roughly perpendicular to the attached base. There is a residue on each chain every 3·4 A, in the z-direction. We have assumed an angle of 36° between adjacent residues in the same chain, so that the structure repeats after 10 residues on each chain, that is, after 34 A. The

From Watson, J. D., and F. H. C. Crick. 25 April 1953. Molecular structure of nucleic acids. *Nature* 171, no. 4356:737–38. Reprinted by permission from *Nature.*

FIGURE 5-3
The Double Helix

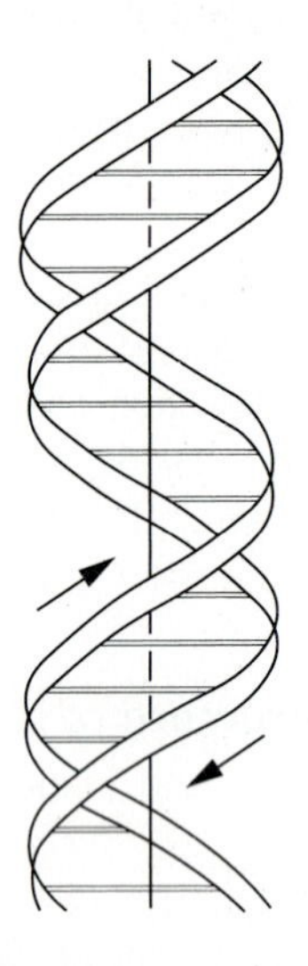

This figure is purely diagrammatic. The two ribbons symbolize the two phosphate–sugar chains, and the horizontal rods the pairs of bases holding the chains together. The vertical line marks the fibre axis.

distance of a phosphorus atom from the fibre axis is 10 A. As the phosphates are on the outside, cations have easy access to them.

The structure is an open one, and its water content is rather high. At lower water contents we would expect the bases to tilt so that the structure could become more compact.

The novel feature of the structure is the manner in which the two chains are held together by the purine and pyrimidine bases. The planes of the bases are perpendicular to the fibre axis. They are joined together in pairs, a single base from one chain being hydrogen-bonded to a single base from the other chain, so that the two lie side by side with identical z-co-ordinates. One of the pair must be a purine and the other a pyrimidine for bonding to occur. The hydrogen bonds are made as follows: purine position 1 to pyrimidine position 1; purine position 6 to pyrimidine position 6.

If it is assumed that the bases only occur in the structure in the most plausible tautomeric forms (that is, with the keto rather than the enol configurations) it is found that only specific pairs of bases can bond together. These pairs are: adenine (purine) with thymine (pyrimidine), and guanine (purine) with cytosine (pyrimidine).

In other words, if an adenine forms one member of a pair, on either chain, then on these assumptions the other member must be thymine; similarly for guanine and cytosine. The sequence of bases on a single chain does not appear to be restricted in any way. However, if only specific pairs of bases can be formed, it follows that if the sequence of bases on one chain is given, then the sequence on the other chain is automatically determined.

It has been found experimentally[3,4] that the ratio of the amounts of adenine to thymine, and the ratio of guanine to cytosine, are always very close to unity for deoxyribose nucleic acid.

It is probably impossible to build this structure with a ribose sugar in place of the deoxyribose, as the extra oxygen atom would make too close a van der Waals contact.

The previously published X-ray data[5,6] on deoxyribose nucleic acid are insufficient for a rigorous test of our structure. So far as we can tell, it is roughly compatible with the experimental data, but it must be regarded as unproved until it has been checked against more exact results. Some of these are given in the following communications. We were not aware of the details of the results presented there when we devised our structure, which rests mainly though not entirely on published experimental data and stereo-chemical arguments.

It has not escaped our notice that the specific pairing we have postulated immediately suggests a possible copying mechanism for the genetic material.

Full details of the structure, including the conditions assumed in building it, together with a set of co-ordinates for the atoms, will be published elsewhere.

We are much indebted to Dr. Jerry Donohue for constant advice and criticism, especially on interatomic distances. We have also been stimulated by a knowledge of the general nature of the unpublished experimental results and ideas of Dr. M. H. F. Wilkins, Dr. R. E. Franklin and their co-workers at King's College, London. One of us (J.D.W.) has been aided by a fellowship from the National Foundation for Infantile Paralysis.

---

[1]Pauling, L., and Corey, R. B., *Nature,* 171, 346 (1953); *Proc. U.S. Nat. Acad. Sci.,* 39, 84 (1953).

[2]Furberg, S., *Acta Chem. Scand.,* 6, 634 (1952).

[3]Chargaff, E., for references see Zamenhof, S., Brawerman, G., and Chargaff, E., *Biochim. et Biophys. Acta,* 9, 402 (1952).

[4]Wyatt, G. R., *J. Gen. Physiol.,* 36, 201 (1952).

[5]Astbury, W. T., Symp. Soc. Exp. Biol. 1, Nucleic Acid, 66 (Camb. Univ. Press, 1947).

[6]Wilkins, M. H. F., and Randall, J. T., *Biochim. et Biophys. Acta,* 10, 192 (1953).

Writing Exercises

Your instructor will designate the length and describe an appropriate audience for each of these assignments.

1. With Hooke's work as an example, use a magnifying glass to observe an everyday object. Write a description of the magnified object, showing your readers a feature of the object that they would not normally have known about.

This assignment is based on an idea that will be discussed at length later in this text: what you perceive is a function of the mechanism you use to make the observation. The world looks differently depending on whether you use your eye or a magnifying glass.

2. Following Franklin's model, write a description of a machine or physical object in such a way that you draw attention to an interesting feature of the object that usually goes unnoticed. (That is, there's no point in describing something that everyone already knows about.) If appropriate, include a functional description, as Franklin does, as part of your overall description. Franklin fuses functional and general description.

3. After reading the article by Watson and Crick, check the sections and the references that you do not understand. Look up each of them in a source book on chemistry or physics in order to understand them. Rewrite the article so that a person with a limited background in physics and chemistry could understand it (maybe someone with only freshman college physics and chemistry). Add a glossary of difficult words and phrases.

TEMPORAL OR DYNAMIC DESCRIPTION

Instructions for Performing a Process

Technical and scientific writing often involves a specialist providing instructions so a nonspecialist can reproduce a process or an experiment. The goal in writing such a set of instructions, like those in a cookbook, is for the nonspecialist to be able to perform the process solely on the basis of reading those instructions. In such a case, the written word takes on crucial importance, as you can see in the following examples.

With a droll, tongue-in-cheek tone, this next selection offers instructions for performing three arcane skills. Though the tone is light, the passage exhibits many of the characteristics of excellent technical writing, especially in the way it embeds information within the instructions themselves.

How To (in Three Parts)

William McKibben

One question that occurred to us last week was: Why is there unhappiness? One possibility, we decided, was that people are unhappy because they lack esoteric knowledge. A man who can prune his own hedge might be unhappy, but a man who can grow topiary shrubbery in the shapes of the various kings of rhythm and blues chatters at dinner. To change a light bulb is to grow depressed, but to throw the hammer is to die, in the good sense. Most people possessing arcane skills acquired them from their parents. But how, if you lack that advantage, can you gain such skills? Publishers, fearful of lawsuits from people attempting to perform odd feats, blackball many books that might help. We, too, must warn you not to try these stunts on either public or private property or for an hour after eating. Anyway, it's not really the experience you're after—it's the jargon.

Part I, Fly a Blimp:

(1) Right off the bat, let's be clear about aerobatic maneuvers. There is a sign just above the instrument panel that says "AEROBATIC MANEUVERS ARE NOT APPROVED."

(2) As the blimp leaves its mast, "weigh it off." That technical term means let it hang in the air. If it doesn't move up or down, it weighs nothing, which is perfect. If it sinks, toss some of the twenty-eight-pound sacks of lead shot overboard.

(3) If you fly over an open-air pop concert, such as the Mr. Mister concert in Daytona Beach, do not be alarmed if the music actually vibrates the "bag," or "envelope." "We stayed calm," says Mike Quinn, captain of the Fuji blimp.

(4) There are no ailerons, because they are not needed.

(5) The trick is to keep the same gas pressure inside the bag at all times. One and a half inches of water pressure is about right. But as the blimp rises the helium will expand, since the ambient air pressure is dropping. Rather than vent helium, you release air from two "ballonets," or internal air sacks, inside the main envelope. As you descend, you can refill the ballonets.

(6) Don't worry about leaks. Even if someone shoots the blimp with a rifle (as happened during the Fuji blimp's trip to Los Angeles for the Olympics), nothing bad happens. It sometimes takes weeks before anyone even notices the hole. Really.

From McKibben, W. 22 September 1986. How to (in three parts). *The New Yorker,* 32–33.

(7) Be watching for helicopters, not gawking out the window. It just looks like the world anyway, though smaller. On the Fuji blimp, we did fly over a parking lot full of maybe a hundred school buses, and they looked a little like pencils in a pencil box, and Manhattan from about three hundred feet above the tip of Rockaway Beach looked like spiky mold spores on a culture dish, but otherwise watch out for helicopters.

(8) On the Fuji blimp, the sign on one side advertises film and the sign on the other side plugs audio- and video-tape, so, depending on what Fuji is pushing, it makes either clockwise or counterclockwise circles, but if you climb above two thousand feet no one will be able to read your slogan anyway.

Part II, Train and Manage Birds of Prey:

(1) You can take a bird straight from the nest to train—that is, an "eyas." But he won't fly very stylishly, said Steve Herman, an ornithologist at The Evergreen State College, in Olympia, Washington, and a thirty-year veteran of falconry. It might be better to wait for the bird to learn from experience during his first year how to swoop down on unsuspecting wildlife, and then take him. Such hawks are known as "passage birds."

(2) Taming a bird of prey—also known as a "raptor"—is called "manning." You don't really train the birds—you simply get them to the point where you can call them back.

(3) "Keep him hungry," recommends Professor Herman, whom we found soaking in an open-air hot spring on an isolated antelope refuge in eastern Oregon. "Hunger is the only bond. Falconers get sentimental about birds, but birds do not get sentimental about falconers. If you feed a falcon too well, he gets 'high' and he'll just fly away."

(4) Beef heart is a good thing to feed him, and, for a treat, pigeon.

(5) If you're flying a golden eagle, don't have him land on your fist. Use a perch. Professor Herman knows a high-school-football coach from Albuquerque who got several bones in his left hand broken by a golden eagle.

(6) If you are stranded on the Utah-Colorado border with your hawk and no money and you meet a pigeon racer, bet him your hawk can catch his pigeon. When he was nineteen, Professor Herman won a bus ticket back to Oakland in this fashion, although he maybe pulled a feather or two out of the pigeon before releasing it. *Maybe.*

(7) If a hawk dives for a rodent, say, and plows into a snowbank, he will be leery of snow for a long time.

(8) Does all this make you queasy? If you plan to train and manage birds of prey, you must not be a wuss.

(9) A warning: Falcons, hawks, and eagles do not, says Professor Herman, "take disappointment very well." If they miss a few pieces of prey, they might just give up. "They know, unfortunately, that you'll feed them eventually."

(10) What is the point? Professor Herman said (rather beautifully, we thought), "You don't really *do* anything. You merely witness what the birds do in the wild anyway—the flight of the predator after the prey. But to witness that is a great privilege."

Part III, Mine Garnet:

(1) First, you need to find some garnet. One method of locating garnet, exemplified by Henry Hudson Barton, is to work as a jeweller and then marry into the Baeder family, of Philadelphia—prominent sandpaper kings—and then suffer a nervous breakdown and go to the North Creek area of the Adirondacks and take a hike and come across a huge garnet deposit (a cliff-face that looks as if it had a skin condition) and realize that, because garnet has a hardness of 8 on the Mohs' scale (only diamond, at 10, and corundum, at 9, are tougher), it could be used for sandpaper. This particular method gave birth, in the eighteen-seventies, to Barton Mines, still family-owned and operated and still the world's leading garnet mine. We toured part of it recently with Randy Rapple, a geologist and salesman, and his two-year-old daughter, Rebecca, who has a knack for spotting garnet.

(2) There are two extraction methods:

(A) Traditional

1. "Hand cob" all summer; that is, pick away with a hammer and chisel, freeing chunks of the red ore. (Garnet is only rarely of gem quality in this country, by the way.) This works best on football-size deposits.

2. Stash the garnet ore in caves.

3. When winter comes, pack it in burlap bags, take it down the mountain in a sleigh, and ship it to your father-in-law in Philadelphia.

(B) Modern

1. Drill the rock.

2. Load it with dynamite: Kaboom!

3. Use a "drop ball" to "down-size" the chunks.

4. Send it to the crushing plant, where the "jaw crushers" and "cone crushers" will crush it into small pieces. The key thing to remember here is that any piece larger than three-quarters of an inch won't be doing you any good, and will have to be recrushed.

(3) Once the ore has been crushed, you must separate the garnet from the rock. There are two ways to do this. The first makes use of the fact that garnet, with a specific gravity of 4, is denser than the surrounding rock, whose specific gravity is 3.2 or less. So you make a Dyna Whirl Pool (D.W.P.), using a liquid with a specific gravity of 3.5, and everything that sinks is garnet. The second technique is proprietary information. Perhaps you could figure out a technique of your own.

(4) Markets! Only forty per cent of Barton Mines' business still involves sandpaper. Barton sells most of the rest of its multimillion-pound annual output to color-TV manufacturers to use in grinding their screens. Do not make the mistake of trying to sell to black-and-white-television makers (who still dominate the Third World market). Because there is only one beam of electrons hitting the rear of the screen in the black-and-white set, it doesn't make nearly as much difference how finely the faceplate is polished.

(5) You will have competition. There are, besides the marvellous Barton operation (headquartered on top of a marvellous mountain), garnet mines in Maine, Idaho, India, and Australia. You could, if you have followed the directions closely and discovered garnet of a quality comparable to Mr. Barton's, point out, "Saying 'garnet' is like saying 'Cadillac.' You still have to specify 'Coupe de Ville' or 'Sedan de Ville' or 'Fleetwood Brougham.'" You could point out that some of your competitors sell to the sandblast market. This is a very cutting thing to say in the garnet world.

Changing the oil on a car is one of the normal maintenance procedures that most car owners can do themselves. Changing your own oil and filter can usually save you $10–15 over the charge at your neighborhood garage. Here is a set of instructions from Volkswagen for performing this process.

A Time-Honored Tradition: The Do-It-Yourself Oil Change

Mechanically minded Americans have been changing the oil and filters in their cars since the days of Barney Oldfield. A necessity back then, the time-honored tradition continues today on driveways and in garages across the country.

Volkswagen owners who are considering changing their car's oil for the first time, or perform this operation on a regular basis already, will find the following guidelines helpful in accomplishing the job safely and quickly.

From A time-honored tradition: The do-it-yourself oil change. Fall/Winter 1985. *Parts and advice Volkswagen* 10, no. 2:12. Reprinted by permission of Volkswagen of America, Inc., Corporate Parts Division.

Whether you are an accomplished Do-It-Yourselfer, or are going to have your first experience in home auto maintenance, it is always advisable to read your Volkswagen Owner's Manual before starting any repair procedure. The owner's manual contains detailed information, cautions and photographs that are specifically written for your model and make.

Our example of an oil and filter change applies to water-cooled gas engine Volkswagens whose model years are between 1975 and 1985.

After reviewing your owner's manual, the first consideration is the purchase of any necessary parts, tools and, of course, the oil itself. Your owner's manual specifies the grade and type of oil best suited to your Volkswagen, as well as the number of quarts you will need to fill your engine's oil reservoir. In addition to purchasing oil, you can get a Genuine Volkswagen oil filter and the correct oil drain plug gasket from your local Volkswagen Dealer Parts Department.

You will need a few basic hand tools to perform an oil and filter change:

Universal Oil Filter Wrench
13 MM Wrench
Oil Pour Spout/Funnel
Oil Drain Pan (6-quart capacity)

It's a good idea to have a supply of rags on hand, to wipe oil residue from parts and tools. Park your Volkswagen in your work area, setting the parking brake. Always use a level work area, for safety reasons as well as allowing proper oil drainage.

The oil should be drained from your engine while the engine is still warm (not hot). Turn off the engine, and place the oil drain pan under the drain plug. Remove the oil drain plug using the 13 MM wrench, and allow the oil to drain into the collection pan [Figure 5–4a]. You should have in mind where you are going to dispose of the used oil. Please be considerate of your environment.

Once the oil flow has diminished to about one drop per minute, reinstall the oil drain plug, using the new drain plug gasket. Do not overtighten the plug—a medium/hard pull on the wrench will seat the gasket properly.

FIGURE 5-4

(a)

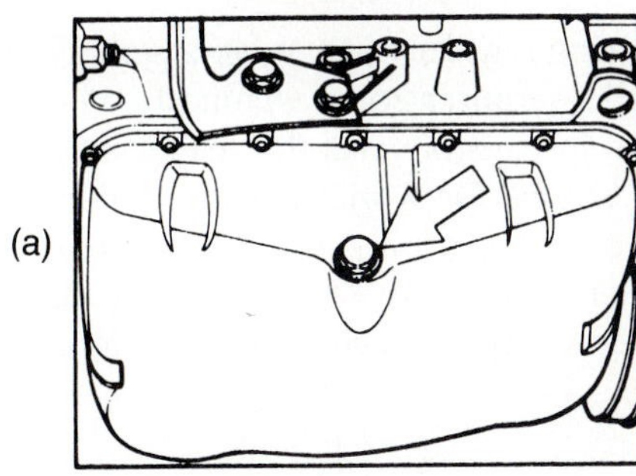

(b)

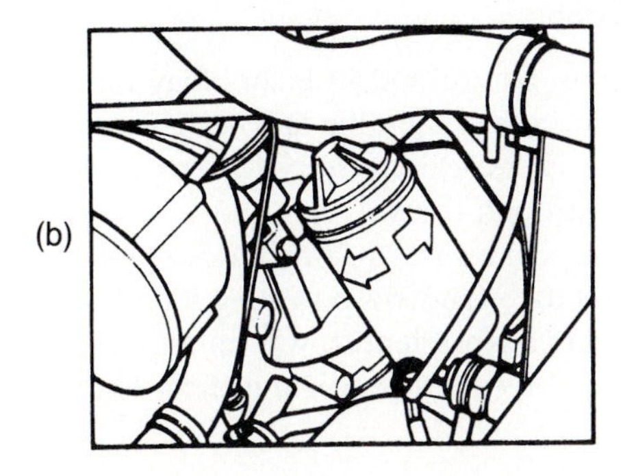

Next step is to change the oil filter. Again, it is important to consult your owner's manual for information specific to your model and make, as well as safety precautions.

The oil filter is located on a flange on the front side of the engine. To remove the filter, place your drip pan beneath the filter location. Then, using the oil filter wrench, turn the filter clockwise [as viewed from the top of the engine—see Figure 5–4b] until it is off the flange. If your car is equipped with an automatic transmission, or a turbo-diesel engine, the filter is attached to an oil cooler. If your car has an oil cooler, be careful that the oil cooler does not turn on the flange as you loosen the oil filter. Your owner's manual contains exact instructions on how to change an oil filter on a vehicle equipped with an oil cooler.

After removing the old filter, lightly coat the new filter gasket with a thin film of engine oil. Do not use grease. Wipe the surface of the flange on the engine clean. Screw the filter onto the flange by hand until the filter gasket is in firm contact with the flange all-around. For tightening procedure, refer to the instructions on the Genuine Volkswagen replacement filter.

Fill the engine with an appropriate amount of engine oil. Your owner's manual lists the correct amount of oil necessary for your model and make. Start the engine and allow it to idle until the oil warning light is out. Then run the engine at various speeds for three to five minutes, and check for any leaks.

Shut off the engine, wait a few minutes, then check the dipstick to see if you have correctly filled the oil reservoir.

After you have properly disposed of the used filter and oil, and cleaned up, don't forget to write the date and mileage of this oil change in your warranty and maintenance manual.

That's all there is to it. And, when performed at proper intervals—along with checking the engine oil level each time you fill your gas tank, this simple procedure will go a long way in maintaining your Volkswagen's quality performance and operating efficiencies.

Writing Exercises

1. Following the oil-change instructions, write a set of instructions for performing an everyday process that requires some technical competence. Include relevant explanatory material in the instructions that the reader may need to know.

2. Write a set of instructions for working with a piece of technical equipment, one that you know about firsthand from your own scientific or technical field.

3. Following the model of the *New Yorker,* write a set of instructions for performing a process that requires specialized knowledge to understand. Use humor if you think it's appropriate.

Description of a Process

Describing processes in nature and in technology further the understanding of such processes within the entire scientific and technological community. Such descriptions are part of the rhetorical category of temporal/dynamic description because they show change through time.

In this first example, Alexander Petrunkevitch describes a bizarre process in which the tarantula spider allows the *Pepsis* wasp to sting it and use its body for egg laying. Mr. Petrunkevitch, a Russian-born scientist, was an emeritus professor of zoology at Yale University when he published this article in *Scientific American.*

The Spider and the Wasp

Alexander Petrunkevitch

To hold its own in the struggle for existence, every species of animal must have a regular source of food, and if it happens to live on other animals, its survival may be very delicately balanced. The hunter cannot exist without the hunted; if the latter should perish from the earth, the former would, too. When the hunted also prey on some of the hunters, the matter may become complicated.

This is nowhere better illustrated than in the insect world. Think of the complexity of a situation such as the following: There is a certain wasp, *Pimpla inquisitor,* whose larvae feed on the larvae of the tussock moth. *Pimpla* larvae in turn serve as food for the larvae of a second wasp, and the latter in their turn nourish still a third wasp. What subtle balance between fertility and mortality must exist in the case of each of these four species to prevent the extinction of all of them! An excess of mortality over fertility in a single member of the group would ultimately wipe out all four.

This is not a unique case. The two great orders of insects, Hymenoptera and Diptera, are full of such examples of interrelationship. And the spiders (which are not insects but members of a separate order of arthropods) also are killers and victims of insects.

The picture is complicated by the fact that those species which are carnivorous in the larval stage have to be provided with animal food by a vegetarian mother. The survival of the young depends on the mother's correct choice of a food which she does not eat herself.

In the feeding and safeguarding of their progeny the insects and spiders exhibit some interesting analogies to reasoning and some crass examples of blind instinct. The case I propose to describe here is that of the tarantula spiders and their arch-enemy, the digger wasps of the genus Pepsis. It is a classic example of what looks like intelligence pitted against instinct—a strange situation in which the victim, though fully able to defend itself, submits unwittingly to its destruction.

Most tarantulas live in the Tropics, but several species occur in the temperate zone and a few are common in the southern U.S. Some varieties are large and have powerful fangs with which they can inflict a deep wound. These formidable looking spiders do not, however, attack man; you can hold one in your hand, if you are gentle, without being bitten. Their bite is dangerous only to insects and small mammals such as mice; for a man it is no worse than a hornet's sting.

Tarantulas customarily live in deep cylindrical burrows, from which they emerge at dusk and into which they retire at dawn. Mature males wander about after dark in search of females and occasionally stray into houses. After mating, the male dies in a few weeks, but a female lives much longer and can mate several years in succession. In a Paris museum is a tropical specimen which is said to have been living in captivity for 25 years.

A fertilized female tarantula lays from 200 to 400 eggs at a time; thus it is possible for a single tarantula to produce several thousand young. She takes no care of them beyond weaving a cocoon of silk to enclose the eggs. After they hatch, the young walk away, find convenient places in which to dig their burrows and spend the rest of their lives in solitude. Tarantulas feed mostly on insects and millipedes. Once their appetite is appeased, they digest the food for several days before eating again. Their sight is poor, being limited to sensing a change in the intensity of light and to the perception of moving objects. They apparently have little or no sense of hearing, for a hungry tarantula will pay no attention to a loudly chirping cricket placed in its cage unless the insect happens to touch one of its legs.

But all spiders, and especially hairy ones, have an extremely delicate sense of touch. Laboratory experiments prove that tarantulas can distinguish three types of touch: pressure

---

From Petrunkevitch, A. August 1952. The Spider and the Wasp. *Scientific American,* 20–23.

against the body wall, stroking of the body hair and riffling of certain very fine hairs on the legs called trichobothria. Pressure against the body, by a finger or the end of a pencil, causes the tarantula to move off slowly for a short distance. The touch excites no defensive response unless the approach is from above where the spider can see the motion, in which case it rises on its hind legs, lifts its front legs, opens its fangs and holds this threatening posture as long as the object continues to move. When the motion stops, the spider drops back to the ground, remains quiet for a few seconds and then moves slowly away.

The entire body of a tarantula, especially its legs, is thickly clothed with hair. Some of it is short and woolly, some long and stiff. Touching this body hair produces one of two distinct reactions. When the spider is hungry, it responds with an immediate and swift attack. At the touch of a cricket's antennae the tarantula seizes the insect so swiftly that a motion picture taken at the rate of 64 frames per second shows only the result and not the process of capture. But when the spider is not hungry, the stimulation of its hairs merely causes it to shake the touched limb. An insect can walk under its hairy belly unharmed.

The trichobothria, very fine hairs growing from disklike membranes on the legs, were once thought to be the spider's hearing organs, but we now know that they have nothing to do with sound. They are sensitive only to air movement. A light breeze makes them vibrate slowly without disturbing the common hair. When one blows gently on the trichobothria, the tarantula reacts with a quick jerk of its four front legs. If the front and hind legs are stimulated at the same time, the spider makes a sudden jump. This reaction is quite independent of the state of its appetite.

These three tactile responses—to pressure on the body wall, to moving of the common hair and to flexing of the trichobothria—are so different from one another that there is no possibility of confusing them. They serve the tarantula adequately for most of its needs and enable it to avoid most annoyances and dangers. But they fail the spider completely when it meets its deadly enemy, the digger wasp Pepsis. [See Figures 5–5a to 5–5d.]

These solitary wasps are beautiful and formidable creatures. Most species are either a deep shiny blue all over, or deep blue with rusty wings. The largest have a wing span of

FIGURE 5-5
Death of the Spider

a

b

In Figure 5–5a, the wasp digs a grave, occasionally looking out. The spider stands with its legs extended after raising its body so the wasp could pass under it.
In Figure 5–5b, the wasp stings the spider, which falls on its back.

about four inches. They live on nectar. When excited, they give off a pungent odor—a warning that they are ready to attack. The sting is much worse than that of a bee or common wasp, and the pain and swelling last longer. In the adult stage the wasp lives only a few months. The female produces but a few eggs, one at a time at intervals of two or three days. For each egg the mother must provide one adult tarantula, alive but paralyzed. The tarantula must be of the correct species to nourish the larva. The mother wasp attaches the egg to the paralyzed spider's abdomen. Upon hatching from the egg, the larva is many hundreds of times smaller than its living but helpless victim. It eats no other food and drinks no water. By the time it has finished its single gargantuan meal and become ready for wasphood, nothing remains of the tarantula but its indigestible chitinous skeleton.

The mother wasp goes tarantula-hunting when the egg in her ovary is almost ready to be laid. Flying low over the ground late on a sunny afternoon, the wasp looks for its victim or for the mouth of a tarantula burrow, a round hole edged by a bit of silk. The sex of the spider makes no difference, but the mother is highly discriminating as to species. Each species of Pepsis requires a certain species of tarantula, and the wasp will not attack the wrong species. In a cage with a tarantula which is not its normal prey the wasp avoids the spider, and is usually killed by it in the night.

Yet when a wasp finds the correct species, it is the other way about. To identify the species the wasp apparently must explore the spider with her antennae. The tarantula shows an amazing tolerance to this exploration. The wasp crawls under it and walks over it without evoking any hostile response. The molestation is so great and so persistent that the tarantula often rises on all eight legs, as if it were on stilts. It may stand this way for several minutes. Meanwhile the wasp, having satisfied itself that the victim is of the right species, moves off a few inches to dig the spider's grave. Working vigorously with legs and jaws, it excavates a hole 8 to 10 inches deep with a diameter slightly larger than the spider's girth. Now and again the wasp pops out of the hole to make sure that the spider is still there.

When the grave is finished, the wasp returns to the tarantula to complete her ghastly enterprise. First she feels it all over once more with her antennae. Then her behavior becomes more aggressive. She bends her abdomen, protruding her sting, and searches for the soft membrane at the point where the spider's leg joins its body—the only spot where she can penetrate the horny skeleton. From time to time, as the exasperated spider slowly shifts ground, the wasp turns on her back and slides along with the aid of her wings, trying to get under the tarantula for a shot at the vital spot. During all this maneuvering, which can last for several minutes, the tarantula makes no move to save itself. Finally the wasp corners it against some obstruction and grasps one of its legs in her powerful jaws. Now at last the harassed spider tries a desperate but vain defense. The two contestants roll over and over on the ground. It is a terrifying sight and the outcome is always the same. The wasp finally manages to thrust

In Figure 5–5c, the wasp licks a drop of blood from the wound.
In Figure 5–5d, the spider lies in its grave with the egg of the wasp on its abdomen.

her sting into the soft spot and holds it there for a few seconds while she pumps in the poison. Almost immediately the tarantula falls paralyzed on its back. Its legs stop twitching; its heart stops beating. Yet it is not dead, as is shown by the fact that if taken from the wasp it can be restored to some sensitivity by being kept in a moist chamber for several months.

After paralyzing the tarantula, the wasp cleans herself by dragging her body along the ground and rubbing her feet, sucks the drop of blood oozing from the wound in the spider's abdomen, then grabs a leg of the flabby, helpless animal in her jaws and drags it down to the bottom of the grave. She stays there for many minutes, sometimes for several hours, and what she does all that time in the dark we do not know. Eventually she lays her egg and attaches it to the side of the spider's abdomen with a sticky secretion. Then she emerges, fills the grave with soil carried bit by bit in her jaws, and finally tramples the ground all around to hide any trace of the grave from prowlers. Then she flies way, leaving her descendant safely started in life.

In all this the behavior of the wasp evidently is qualitatively different from that of the spider. The wasp acts like an intelligent animal. This is not to say that instinct plays no part or that she reasons as man does. But her actions are to the point; they are not automatic and can be modified to fit the situation. We do not know for certain how she identifes the tarantula—probably it is by some olfactory or chemo-tactile sense—but she does it purposefully and does not blindly tackle a wrong species.

On the other hand, the tarantula's behavior shows only confusion. Evidently the wasp's pawing gives it no pleasure, for it tries to move away. That the wasp is not simulating sexual stimulation is certain, because male and female tarantulas react in the same way to its advances. That the spider is not anesthetized by some odorless secretion is easily shown by blowing lightly at the tarantula and making it jump suddenly. What, then, makes the tarantula behave as stupidly as it does?

No clear, simple answer is available. Possibly the stimulation by the wasp's antennae is masked by a heavier pressure on the spider's body, so that it reacts as when prodded by a pencil. But the explanation may be much more complex. Initiative in attack is not in the nature of tarantulas; most species fight only when cornered so that escape is impossible. Their inherited patterns of behavior apparently prompt them to avoid problems rather than attack them. For example, spiders always weave their webs in three dimensions, and when a spider finds that there is insufficient space to attach certain threads in the third dimension, it leaves the place and seeks another, instead of finishing the web in a single plane. This urge to escape seems to arise under all circumstances, in all phases of life and to take the place of reasoning. For a spider to change the pattern of its web is as impossible as for an inexperienced man to build a bridge across a chasm obstructing his way.

In a way the instrinctive urge to escape is not only easier but often more efficient than reasoning. The tarantula does exactly what is most efficient in all cases except in an encounter with a ruthless and determined attacker dependent for the existence of her own species on killing as many tarantulas as she can lay eggs. Perhaps in this case the spider follows its usual pattern of trying to escape, instead of seizing and killing the wasp, because it is not aware of its danger. In any case, the survival of the tarantula species as a whole is protected by the fact that the spider is much more fertile than the wasp.

The process by which chromosomes duplicate themselves forms the basis for cell reproduction and the passing down of hereditary traits. Evelyn Fox Keller, a physicist, biologist, and historian of science, describes mitosis and meiosis, the two kinds of division that cells undergo. Her description comes from her fascinating book, *A Feeling for the Organism,* which chronicles the work of Barbara McClintock, a biologist who has spent her life working on the chromosomal structure of maize and who won the Nobel Prize for Physiology or Medicine in 1983 for her discovery of mobile genes in plant chromosomes that can alter future generations of the plants. Keller is Professor of Mathematics and Humanities at Northeastern University.

Mitosis and Meiosis

Evelyn Fox Keller

Chromosomes are found in the cell's nucleus. They first become visible as long slender threads in the early stages of nuclear division. Under suitable conditions, these "threads" can be then seen to double along the greater part of their length. Each chromosome thus consists of two half-chromosomes, or chromatids, that are held together at or near a region of the chromosome called the centromere—a structure that, later in the sequence, appears to regulate the movement of the entire chromosome. Throughout the first stage of division, prophase, the chromosomes become progressively shorter and thicker. Cells undergo two kinds

FIGURE 5-6
Mitosis in a Nucleus

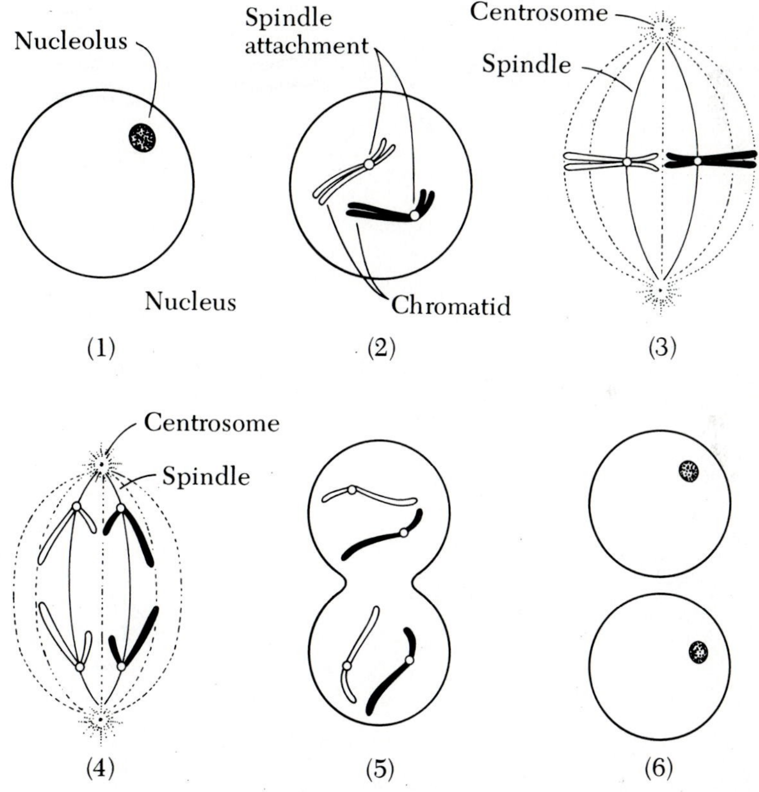

There are usually many pairs in each nucleus. (Cytoplasm is not shown.) (1) Resting stage. (2) Prophase, each chromosome already double (consisting of two chromatids). (3) Metaphase. (4) Anaphase. (5) Telophase. (6) The two daughter nuclei in resting stage.

---

of nuclear division—mitosis and meiosis—which, although they begin similarly, proceed in very different fashions.

In mitosis [see Figure 5–6], the usual process by which cells duplicate themselves, prophase ends with the dissolution of the membrane that normally separates the nucleus from the cytoplasm (see the drawing below). In the next stage, metaphase, a spindlelike structure of fibers appears, radiating from opposite poles in the nucleus; the chromosomes attach themselves to these fibers, lining up in the middle of the spindle. In due course, the centromeres divide, thus completing the division of the chromosomes. In the following stage, anaphase, the two new chromosomes, led by the centromeres, move apart along the spindle toward opposite poles. Finally, in telophase, the nuclear membrane re-forms around each set of

FIGURE 5-7
Meiosis

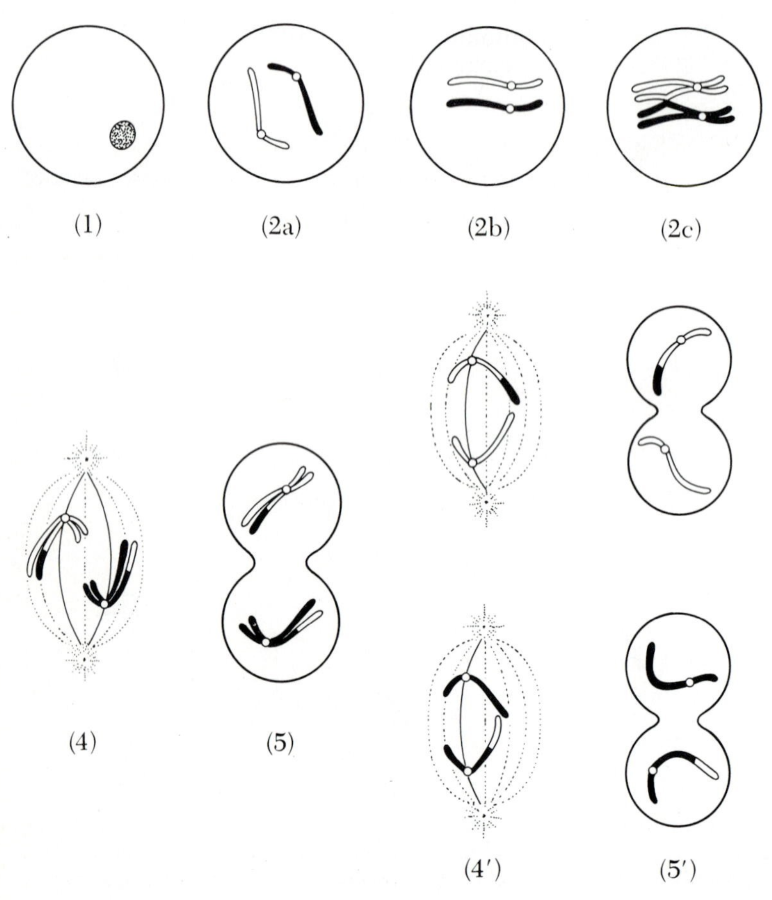

Stages are numbered to correspond to those of mitosis shown in the previous figure. Prophase in three stages: (2a) Appearance of chromosomes, (2b) pairing, (2c) separation of chromatids with chiasma formation. (4) and (5) First anaphase and telophase, two chromatids having mixed constitution due to chiasma. (4′) and (5′) Second anaphase and telophase. Four nuclei result from the whole process of meiosis.

daughter chromosomes, resulting in two complete nuclei, each with the same number of chromosomes as in the original nucleus.

Sexually reproducing organisms require, in addition to a mechanism for duplicating normal cells, another mechanism for the production of gamete cells, which contain not a full double complement of chromosomes, but a single (haploid) complement. This mechanism is meiosis, and it differs from mitosis in several crucial respects. The full or diploid complement of chromosomes in a normal cell consists of two sets of homologous chromosomes, one set from each parent. In meiosis, each chromosome, instead of dividing, lines up side by side with its complementary chromosome [see Figure 5–7]. This process of conjugation (or synapsis) occurs in the middle of prophase, in substages known as zygotene and pachytene. Later, the chromosomes in each pair pull a little way apart again (a substage called diplotene), but cross points (or chiasmata) can be seen at which pairs of chromosomes can exchange chromatids (crossing over). (It is above all the occurrence of crossing over in meiosis that makes this form of nuclear division of primary importance to geneticists.) At the end of prophase, a spindle appears, and the two complementary chromosomes proceed to separate along the fibers of the spindle, continuing as in mitosis, with one further exception. Following the first round of division is a second round of division in which, this time, the chromosomes do replicate themselves by longitudinal splitting. Four sets of nuclei result, each with a haploid set of chromosomes.

The miraculous coordination with which these processes are orchestrated makes the study of chromosomal dynamics of interest in and of itself.

Writing Exercise

Write a description of a natural, scientific, or technical process. Use an illustration to enrich your description. Your instructor will designate an appropriate length and audience for this assignment.

STRUCTURAL OR FIELD-RELATED DESCRIPTION

You have seen how static description describes what an object is like in itself, without analyzing its relationship to its environment, and how temporal or dynamic description shows us how the object changes through time, as if its nature could be understood only by viewing it as not static. These seemingly opposite modes of description, when put together, give us a fuller description of the object than we would have with either one alone; multiple perspectives enrich our vision of nature.

A third such perspective is the structural or field-related mode, in which we describe an object as it exists in relation to similar objects or to its environment. For instance, a structural description would include classification and definition, both of which describe an object as related to a set of similar objects. Biologists describe organisms that swim in the sea as cold blooded or as warm blooded; thus, sharks are cold blooded, and dolphins are warm blooded—or mammals. From this, all "fish" or "swimming creatures" can be grouped.

Once a classification system is established, objects can be defined: a dolphin belongs to the genus called "Delphinidae," with the common species we know called "Delphinus delphis." Once an organism is defined according to its genus and species, its characteristics can be listed and described. (The opposite sequence is used to identify the organism: we use the characteristics to find what genus and species it belongs to. In this case, the rhetorical process of description and the biological process of identification are reverse images of each other.)

A field-related description assumes that a naturally occurring phenomenon cannot be understood except in terms of its environment. For instance, most psychologists believe that emotions don't exist in themselves; rather, they are complex responses to the environment. Thus, it is difficult to define love unless you describe it as a set of responses to a particular person or concept. It is difficult, if not impossible, to think what love would be without a "loved object" in the environment.

The following examples present various approaches to classification and definition.

Classification

Making classifications of natural phenomena expresses one of the major purposes of scientists and engineers: to discover order in nature by finding similarities and differences within the seemingly endless variety of objects, both natural and fabricated. So, for instance, archeologists will classify arrowheads historically and structurally in order to discover the evolution of stone technology within any one tribe or area.

For a classification to be successful, each class in the general category must be distinct. Since the goal of identification is to determine how much an object can change and still remain in its class, classes cannot overlap. When they do overlap, the classification loses its organizing power.

For example, an Associated Press story (July 5, 1986) described a poll taken by the President's Commission on Americans Outdoors, the object of which was to determine why people do what they do outdoors. The pollsters found that the reasons fall into five personality styles: fitness driven, unstressed and unmotivated, getaway actives, health-conscious sociables, and excitement-seeking competitives. The conclusions drawn from such a classification will not be compelling because the categories are not discrete; they overlap. For instance, a fitness-driven person might also be excitement-seeking competitive, and so on. This classification system doesn't help us sort out the world: it just gives a list of different styles, more than one of which any individual might exhibit at different times. The following passage demonstrates a classifiction based on discrete categories.

The Private Well System

Basic Types of Wells

Water wells are classified according to the method used to build them. There are four common types: *dug, bored, driven and drilled.* [See Figure 5–8.] Dug, bored and driven wells are rarely more than fifty feet deep. Drilled wells, however, can be 1,000 feet or more in depth. Here are important facts about each well:

From *Water Systems Handbook: A Complete Text on Private Water Systems—Their Design, Operation and Maintenance.* 9th ed. Chicago: Water Systems Council, 9–10. Reprinted by permission.

FIGURE 5-8

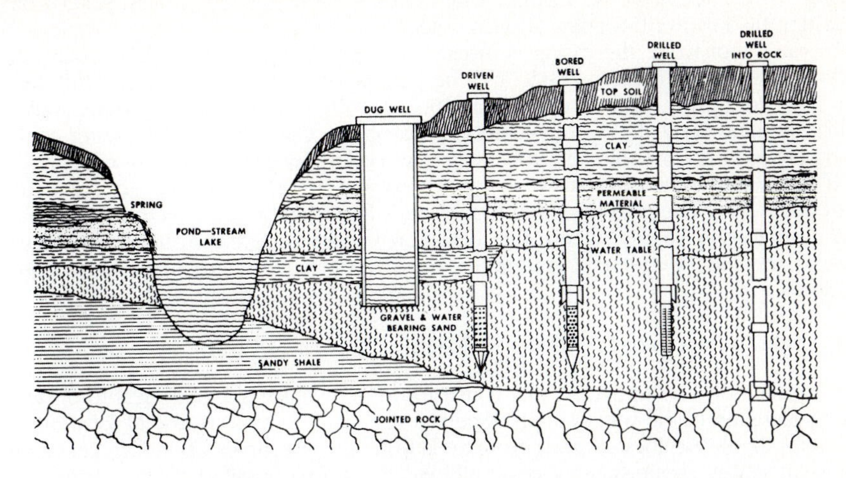

Typical cross section of underground strata, showing various types of well construction. Note that drilled well penetrating into rock formation does not require casing below bottom level of sand-and-gravel formation.

Dug wells are simply made by excavating a hole several feet in diameter to a fairly shallow depth. The circular hole is then lined with rocks, brick, wood or concrete pipe to prevent cave-in. Because depth is limited to 25–30 feet, a dug well is practical only when the water table is near the surface of the earth. And for this very reason, there is a strong likelihood of contamination by surface drainage and/or subsurface seepage. Because a high percentage of such wells are polluted, the use of dug wells should be avoided if at all possible. In many areas dug wells are prohibited by state law.

Bored wells are constructed by means of an earth auger which bores a cylindrical hole into the earth. . . . After water is reached, the well is cased with tile, steel pipe or other suitable material. The lower part of the well should be provided with a screen which keeps sand and other material from entering the water being pumped.

Like dug wells, bored wells are subjected to contamination unless the casing is sealed with cement grout to a depth of at least 15 feet below ground surface.

Driven wells consist of a series of tightly coupled pipe lengths fitted with a *well point* at the lower end. The well point is forced through the ground by a series of blows on the pipe above it. When the point reaches the water table, water flows into the pipe through screened openings on the well point.

Driven wells are most practical when the water table is not deeper than 50 to 60 feet. And obviously, the underground strata must be porous and free from rocks in order to permit driving of the well point. When these conditions are present, the driven well may be the simplest and cheapest means of developing a ground water supply.

Drilled wells are used when you need greater depth, volume, or diameter than a driven well can provide. They are also necessary when the subsurface is too hard to permit a well point to be driven through it.

Wells are commonly drilled by one of two methods. The first is known as the *cable-tool* or *percussion* method. . . . It involves a machine that alternately lifts and drops a long chisel-shaped bit within the borehole. The hole is deepened as the bit cuts and breaks rock or other materials in the subsurface. In the second method, *rotary* well drilling, a rotating bit is fixed to the lower end of a steel pipe, known as the *drill pipe*. . . . The rotary action of this bit "chews" into the rock or other earth materials. Air or water is used to transport the "cuttings" or chips to the surface.

The drilled hole is lined with steel or plastic pipe called the *well casing.* This serves two purposes: It (1) keeps the well from caving in and (2) protects the ground water from contamination by downward-seeping surface water.

In cable tool work, the casing is installed as the drilling proceeds. With rotary drilling, however, the hole is made oversize and the pipe is set in the borehole *after* full depth has been reached. In this latter case, the space between the borehole and the outer walls of the well casing must be sealed with a cement or clay grout. Otherwise, surface water could seep down between the borehole and the casing. This could cause severe pollution.

If the water-bearing formation is loose sand and gravel, the casing extends to the full depth of the well. The bottom end of the casing is fitted with a *well screen* to keep sand from entering the water supply. Where the formation is hard rock, the casing need extend only down to the top of the rock stratum in most cases. Screens are not usually needed in rock wells.

Good Location Important to Prevent Pollution

It's clear from the foregoing that construction of a well has a great deal to do with preventing pollution of the water supply. [See Figure 5–9.] The same is true of *location.* Generally, wells should be located away from and upstream of sources of surface contamination, such as barnyards, septic systems, cesspools, etc. If that isn't possible, it's doubly important to make sure that the well is: (a) driven to a depth sufficient to assure a *pure* ground water supply and (b) properly constructed to prevent surface-water contamination.

Construction of the well is the responsibility of a well drilling contractor. Some pump dealers do their own drilling. Others subcontract this work or take over the job after the driller has finished building the well.

FIGURE 5-9

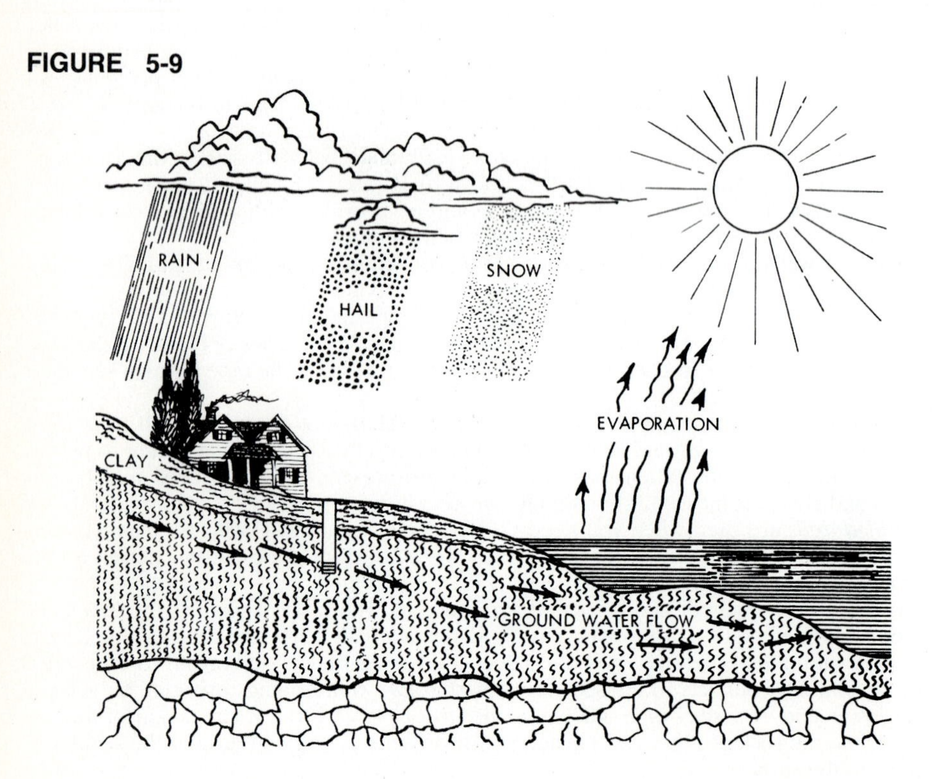

Points to Check With the Well Driller

If someone else drills the well in which you set pumps, it's important to make sure the job is done properly. Here are some of the main points to check:

Make sure the driller is a reputable, established contractor. There are a certain number of fly-by-nights in any business.

Require a written contract, spelling out the diameter of the well, the materials to be used, and the cost per foot for drilling the well. (Drillers can rarely predict the exact depth to which they'll have to go. Hence the practice of quoting price on a per-foot basis.)

See that the driller carries full insurance coverage.

Make sure that the well is constructed and sealed according to sanitary practices and local or state codes. It's especially important that the casing be installed so as to prevent contamination from surface seepage.

Upon completion of the installation, see that the well is cleared of all loose materials and is pumped clean of sand. It should then be sterilized or disinfected by means of chlorination.

Insist that the driller furnish a "log" of the well. This is a drilling record showing the depth of the well, the strata penetrated, and the depth at which water was found. This record, together with test data showing the pumping rate and pumping level, should be given to the owner. All of this information is needed to select and size the pump for efficient adequate operation.

Finally, remember that a private water system can never be any better than the well you start with.

Writing Exercises

1. Construct a classification scheme for the objects in your home or apartment. Write a description of your new classification system and contrast it with a more conventional system. For instance, you could classify objects in terms of their practical life expectancy and then correlate that with their cost and the amount of pleasure or comfort derived from each. In introducing the classification, say why it is innovative and what purpose it will serve.

2. Look up a major classification system in your field. Do some historical research on the classification to determine who first formulated it and how it has changed over time. Write an analysis of the classification system from a historical perspective and show how its changes mirror the changes in the discipline of which it is a part. (Geology and psychiatry would be particularly good subjects.)

Definition

The act of defining and explaining terms and abstract concepts is essential to the progress of science and technology. Civil engineers, for instance, must understand the difference between the terms "concrete" and "asphalt" and must be able to understand and apply the concept of stress in order to build roads; chemists must understand the difference between charged and uncharged particles and between systems in equilibrium and systems in a steady state.

Definition limits and thus constructs the meaning of any one term so that we all understand it in the same way. Such understanding is the basis of objective, shared knowledge.

As a scientist or engineer, you must be proficient at defining and explaining physical objects and their properties, processes or series of events, and concepts and laws. Let's look at these one at a time.

The examples come from the *McGraw-Hill Dictionary of Scientific and Technical Terms,* an excellent source of basic knowledge, which is supplemented by the 15-volume *McGraw-Hill Encyclopedia of Science and Technology.*

Defining Physical Objects

The *The McGraw-Hill Dictionary of Scientific and Technical Terms* gives the following definition of "protoplasm":

protoplasm [CYTOL] The colloidal complex of protein that composes the living material of a cell.

("CYTOL" is the dictionary's abbreviation for cytology.) This precise definition gives you the general class of objects (its genus) that protoplasm belongs to (a colloidal complex of protein) and then gives you its specific differentiation from that class (the complex is that specific one in the living material of a cell).

The usefulness of such a definition depends on your level of knowledge. If, for instance, you don't know the definition of "colloidal" and "protein," you won't learn much from the definition. So your next step would be to look up "colloid," which is

the phase of a colloidal system made up of particles having dimensions of 10–10,000 angstroms (1–1000 nanometers) and which is dispersed in a different phase.

Next you would need to find that a "colloidal system" is

an intimate mixture of two substances, one of which, called the dispersed phase (or colloid), is uniformly distributed in a finely divided state through the second substance, called the dispersion medium (or dispersing medium); the dispersion medium or dispersed phase may be a gas, liquid, or solid.

By now, you understand that a colloid is a suspension of fine particles in a continuous medium. By checking a standard dictionary, you would also find that the Greek root for "colloid" is *kollodes,* meaning "glutinous"; knowing this would confirm your suspicion that protoplasm is somehow jelly-like.

Then, to finish your understanding of the definition, you might need to look up "protein":

protein [BIOCHEM] Any of a class of high-molecule-weight polymer compounds composed of a variety of amino acids joined by peptide linkages.

Definitions from *McGraw-Hill dictionary of scientific and technical terms.* 1984. 3d ed. Edited by Sybil P. Parker. New York: McGraw-Hill.

Unless you had just finished a course in organic chemistry, you would probably need to check the definitions of "polymer," "amino," and "peptide" to understand this definition of protein.

That is, understanding the simple, straightforward definition of terms means understanding the words that compose the definition. Learning a new vocabulary, of course, is often part of the learning process in any discipline.

Defining and Explaining Processes

Entire books can be devoted to defining and explaining processes, such as a textbook on the chemistry of cell biology. But the essence of a process can often be defined succinctly, as in the definition for "ablation":

ablation [AERO ENG] The carrying away of heat, generated by aerodynamic heating, from a vital part by arranging for its absorption in a nonvital part, which may melt or vaporize and then pass away, taking the heat with it. Also known as ablative cooling. [GEOL] The wearing away of rocks, as by erosion or weathering. [HYD] The reduction in volume of a glacier due to melting and evaporation. [MED] The removal of tissue or a part of the body by surgery, such as by excision or amputation.

This definition shows you that

1. Definitions often define processes or reactions, not just things in themselves.
2. One word can often have different, though related, meanings in different fields.
3. Being aware of the roots of words can often explain how they can be used in different fields, with a related meaning. "Ablation" comes the Latin "ab" + "latus," meaning "to be carried away from," which shows the concept that all the above definitions have in common.

The following excerpts offer a variety of definitions in science and technology.

Scientists and engineers often find it useful to try to define an entire discipline, thereby describing its appeal and importance. In the following article, Ralph A. Raimi, professor of mathematics at the University of Rochester, does just that.

What Is Mathematics?

Ralph A. Raimi

It seems to me no bad thing that a mathematician should be called upon once in a while to defend himself and his science before a popular audience that knows little or nothing about the subject. Indeed, I have more than once suggested, to some nearby dean or provost, that *every* professor of every subject should occasionally give a public lecture of this nature. Too often, public lectures are given by medical or social scientists, explaining the latest useful (or potentially useful) results in genetics or demography—things that can more or less be understod by a general audience. Mathematics cannot do this; and while medicine or economics in some sense can, the popularization even in these subjects tends, by concentrating on useful applications, to falsify the spirit that truly drives a scholar, in no matter what field of study. It is this spirit that is hard to convey, I think, while the rest is either impossible or in the newspapers.

Raimi, R. A. Winter 1985–86. What is Mathematics? *Rochester Review,* 13–15. Reprinted by permission of Rochester Review, University of Rochester.

What follows here is more or less the text of a lecture I have given to several Rochester audiences, including the group of parents of our mathematics graduates on Commencement Day, when our department, together with Statistics and Computer Science, distributes the actual diplomas that represent four years' study. I try to answer the question: What have your children been doing here for the past four years? Why mathematics? Who needs it?

Well, mathematics (I explain) has enjoyed a good reputation for the past three thousand years and more, but it is surprising to most people to discover how recently this good reputation has been due to the practical uses that can be made of it. I don't even have to say this of statistics and computer science; they didn't exist a century ago, at least not by name. But of mathematics it surely needs saying.

Now of course there is practical value in ordinary arithmetic. There is no doubt that commercial slaves in ancient Egypt, three thousand years ago, were able to add and multiply and do whatever is needed to keep the accounts when ordering food for an army, or loading a ship with enough oil to last out the voyage. But this is hardly the beginning of mathematics. Today we have learned all that by the fourth year in primary school, and a sensible person can go through life in most occupations with no more formal mathematical training than that.

The famous Rhind Papyrus of 1600 B.C. (now on display at the British Museum in London) is an Egyptian document that contains, however, an enormous number of calculations of considerable subtlety, whose application to navigation or war or astronomy or religion is not apparent. For example, what we construe as a simple fraction $2/7$ was construed in the Rhind Papyrus as a problem: How do you divide two by seven? The "answer" is given as $1/4 + 1/28$, which adds to $2/7$ to be sure, but why? What is there about this "answer" that is any better than the original notion of "two parts in seven"?

Certainly it was not particularly practical to be able to express fractions with non-unit numerators as sums of fractions that do have unit numerators—but it does offer a challenge, if one is also constrained to have all the denominators in the "answer" of different sizes. Practical or not, this challenge is taken up at length in the Rhind Papyrus, showing from even the earliest of civilizations that puzzle-solving was sufficiently honored to be taught to students and systematically recorded on expensive materials (papyrus in this case).

This is far from the only example, nor will I offer very many from history. It is sufficient to consider the following list of mathematical terms: "isosceles triangle" (sixth century B.C.), "parabola" (fifth century B.C.), "sine" (second century B.C., in some sense), "definite integral" (250 B.C. or 1680 A.D.—Archimedes or Leibniz—according to how you look at it), "complex exponential" (1750 A.D.), "transfinite number" (1880), "locally convex topological linear space" (1940).

Here is a list that goes straight from Pythagoras to Bourbaki, to name two more or less mythical figures in the history of our science, and not one item in the list was seen as having the slightest practical value at the time of its invention. It is true that even in ancient times some of these things had application of a sort. Babylonian astronomy was legendary, and did make use of numerical tables of some sophistication, but to say it was "practical" is going too far: It was used in religion, and while religion "used" astronomy which "used" mathematics, religious holiday calculation is not what we today understand as an example of the usefulness of mathematics.

Again, the Platonic philosophy "used" geometry in much the same way. Plato's (religious) doctrine of ideals held geometry as its model, a way for the human mind to understand how a non-existent thing like a perfect circle can be more real and more important than a "genuine" wheel, or mark in the sand. Newton's calculus was essential to his physics; did that make it "useful"? Nobody did anything with Newton's physics either, for that matter. It grew no wheat, signed no treaties, clothed no army. The Industrial Revolution of the eighteenth and nineteenth centuries—the work of the great builders of bridges and steam engines, right up to the time of Henry Ford, George Eastman, and Thomas Edison—was conducted by geniuses, perhaps, but not by people who knew or cared anything at all for mathematics, whether the mathematics of Newton or of the Rhind Papyrus.

Suddenly, mainly in the twentieth century, mathematics of a nonelementary kind has become the backbone of science, and science of technology, and technology of production and the wealth of nations. Suddenly we mathematicians no longer have to defend ourselves; we can point at space satellites or superconductors and announce that we stand behind them. But actually this is nothing but an accident of our times; it is not the reason we study mathematics.

There was a mathematician of the thirteenth century named Fibonacci, whose most famous book carefully explained the convenient Arabic numeral system (decimal system) and its advantages in arithmetic over the cumbersome Roman numerals. But this truly practical advice to Europe, though influential and important, was not Fibonacci's main interest. The same book is also full of problems of the sort that schoolchildren puzzle over to this day: Two men are digging a trench. One could do it in five hours, the other in three. How long will it take them if they work together—and do not get in each other's way?

Practical? Not a bit. I have never in my life met a man who needed to know how to solve that problem. Yet every mathematician will instantly recognize this problem as having greater interest than the substitution of the symbol *12* for the symbol *XII*.

I have one last example, much more recent. A few years ago I was visiting my brother in Detroit. He is a business man, five years older than me, and not versed in mathematics or science. We had eaten dinner together, and were walking around the neighborhood he lives in, as one does after such a dinner.

"Ralph," he said (he calls me Ralph), "Ralph, you are a mathematician." I said, "Yup," meaning "Yes, of course, go on." Casual, automatic agreement, as to something obvious. Sometimes I pronounce it "Yep."

"Well then, you know the square numbers? Like 4, which is the square of 2, or 9, which squares 3?" "Yup," I said. "And then there is 16—the square of 4?" "Yup," I said. "And 25 is the next one; that's 5 squared, and 6—that gives 36. . . ." "Yup." (It is sometimes hard to stop my brother, so I let him go on.) "And then 49 is 7 squared?" "Yup."

"Well now," he said, "look what happens when you subtract these numbers from each other in order: Start with 4. If you subtract 1 (1 is the square of 1, after all)—if you subtract 1 from 4 you get 3; if you subtract 4 from 9 you get 5; if you subtract 9 from 16 you get 7. . . ."

"Yup," I said. "And if you—if you subtract 16 from 25 you get 9, and 25 from 36 is 11, and 36 from 49 is 13." "Yup."

"Well Ralph, don't you see what we're getting? Those differences are the odd numbers! We get 3, 5, 7, 9, 11 *in that order* when we subtract the successive squares from each other." I said, "Yup." "Well," he said, "does it go *on* like that?" (My brother was almost breathless in his anxiety to make clear what the question was.)

"Yup." I answered this one as quickly and casually as the others.

There was a short silence while my brother thought over the answer. "Gives you the creeps," he said.

Now the phenomenon my brother discovered is not very deep. It was not deep at the time of Fibonacci either, and in the time of Pythagoras every philosopher, at least, was acquainted with this fact. My brother did not ask me to prove it; he accepted my word. But in fact any twelve-year-old should be able to prove it, using the simple algebraic formula $(N+1)^2 - N^2 = 2N+1$. And even a younger child should be able to deduce it pictorially from the diagram [Figure 5–10], which I will not explain further since I consider its message self-evident.

The task of the mathematician is, in part, to provide such proofs, since nobody wants to live by guesswork and uncertainty, but the emotion that accompanies the *discovery* of such a pattern is what drives a mathematician to his work.

The formula on the difference of successive squares is only slightly deeper as mathematics than the calculations of daily life, but my brother's reaction to that extra bit of depth illustrates the enormous difference between performing a learned ritual (addition or multiplication) and discovering a pattern in a place where before appeared to be randomness or disorder.

Much more than the recent practical value that mathematics has been exhibiting, this love of symmetry, of hidden relationships, of unexpected logical implications, is what mathemati-

FIGURE 5-10

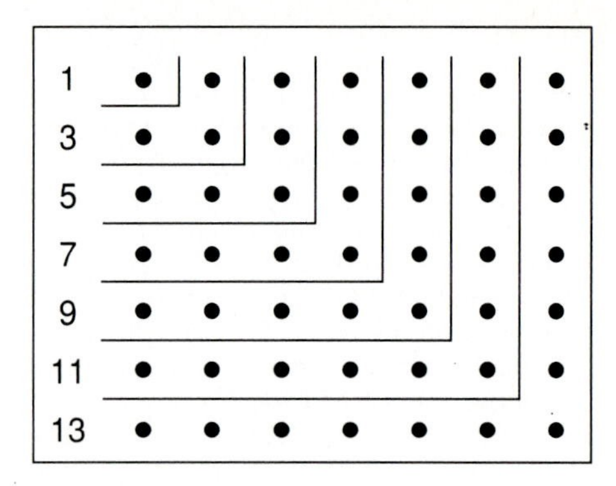

cians live by. I believe much the same is true of statisticians and computer scientists, who work in more obviously applicable branches of mathematical science. They seek to find order—or *create* order—in confused domains.

If what we mathematical scientists did had less practical value than it has, there would surely be fewer of us, naturally, since society cannot afford to feed and clothe a very large number of worthless dreamers. But the few of us who remained would behave no differently than we do now. Even in very poor societies, like that of India in 1910, where the genius Ramanujan appeared, or that of Pisa in the thirteenth century, where Fibonacci appeared, there are always some few whose passion for mathematics cannot be stifled. And what is even more curious, there is always a public, perhaps not large, but a public just the same, willing in one way or another to see that these mathematicians can flourish. The social arrangements run from inherited wealth to payment for teaching, or in modern times government fellowships and prizes, but somehow, in every society that values human thought for its own sake, there will be mathematics.

There is an old saying, "Necessity is the mother of invention." Like most old sayings, this one is demonstrably false. The starving man has no time to invent new methods of agriculture, and Isaac Newton was fleeing a plague year in Cambridge when he invented the Binomial Theorem and not a cure for plague. Inventions are composed by people of imagination, who have the leisure and the materials necessary to the task, not by those oppressed by urgent need.

Just the same, like most old sayings, "Necessity is the mother of invention" is also true. One must understand something different, however, by the word "necessity." There is a necessity in the human spirit, once it is free from its most pressing and fundamental wants, a necessity to see order in the world or to put it there, to create structures of order, such as cathedrals and concertos—and spacecraft and microprocessors too, for that matter. This need is as old as mankind, and mathematics is one example, among others to be sure, of how humans go about its satisfaction.

That mankind as a whole has always appreciated this labor and its purpose, even when it understood very little of its content, has been a fortunate thing for modern technology and prosperity, but it is even more heartening in another direction: It permits a certain degree of optimism to those of us who wish to think well of mankind and its spiritual future.

Scientists often need to define the methods they are using in their research. In the medical field of psychiatry, contemporary researchers are concentrating more on treating mental illness with a combination of medicine and interpersonal therapies, than on the traditional method of psychoanalysis. In the following abstract, two such therapies are defined, in relationship to medications that accompany them. Doctors are using these methods to treat depression.

NIMH Treatment of Depression Collaborative Research Program: An Abstract

Irene Elkin, Morris B. Parloff, Suzanne W. Hadley, and Joseph H. Autry

CB Therapy.—Cognitive behavior therapy is based on the theory that underlying the depressive syndrome are faulty or distorted cognitions or patterns of thinking. Specifically, Beck et al posited a cognitive triad: unrealistically negative views of patients about themselves, about the world, and about the future. According to the manual developed by Beck and his colleagues,

> *Cognitive therapy is an active, directive, time-limited, structured approach. . . . It is based on an underlying theoretical rationale that an individual's affect and behavior are largely determined by the way in which he structures the world. His cognitions (verbal or pictorial "events" in his stream of consciousness) are based on attitudes or assumptions (schemas), developed from previous experiences.*
>
> *The therapeutic techniques are designed to identify, reality-test, and correct distorted conceptualizations and the dysfunctional beliefs (schemas) underlying these cognitions. The patient learns to master problems and situations which he previously considered insuperable by reevaluating and correcting his thinking. The cognitive therapist helps the patient to think more realistically and adaptively about his psychological problems and thus reduces symptoms.*

Behavioral methods may also be used in CB therapy to increase the patient's activity level and to shape other targeted behaviors. The therapy is defined and described in great detail in the treatment manual.

IPT.—Interpersonal psychotherapy is based on the premise that depression occurs in an interpersonal context. The techniques are intended to help patients achieve a better understanding of their interpersonal problems and to improve their social functioning. The rationale is that if improvement in interpersonal relations can be effected, improvement in other areas will follow. Although IPT is largely derived from a psychodynamic model, according to the manual developed by the New Haven–Boston group.

> *The nature of IPT is interpersonal—not intrapsychic—with focus on the "here-and-now" rather than on early developmental experiences. The overall goals of treatment are to encourage mastery of current social roles and adaptation to interpersonal situations.*
>
> *. . . Interpretation and personality reconstruction are not attempted but rather, reliance is upon familiar techniques such as reassurance, clarification of internal emotional states, improvement of interpersonal communication, and reality testing of perceptions and performance.*

A detailed description of the theoretical basis of the approach and the techniques used can be found in the treatment manual.

Rationale and Description of Comparison/Control Conditions.

Imipramine-CM.—In order to evaluate the effectiveness of the psychotherapies, they are being compared with a reference treatment condition, i.e., a treatment that has already been found to be efficacious with this patient population. Ideally, such a reference condition would be a standard or established form of psychotherapy. There is, however, no clearly defined form of psychotherapy that research has already established as efficacious with depressed patients. Therefore, an antidepressant drug serves as the reference condition in this study,

---

From Elkin, I., M. B. Parloff, S. W. Hadley, and J. H. Autry, 1985. NIMH treatment of depression collaborative research program. *Archives of General Psychiatry* 42:310–11.

since considerable research evidence is available for the efficacy of this treatment. There is, of course, a problem in using a reference condition that involves different expectations and mechanisms of action but, nevertheless, at present, this seems to provide the best available standard against which to compare the experimental psychotherapy treatment conditions.

The best reference drug to use would be the drug with the longest history of use and for which a large amount of efficacy data exists with this patient population. Imipramine best meets these requirements. Although a great deal of the research literature on imipramine involves inpatient populations, a number of studies have been conducted on outpatients. Most of these demonstrate the efficacy of imipramine when compared with placebo controls.

Details of the imipramine-CM condition are provided in a *Clinical Management-Imipramine-Placebo Administration Manual* and in the addendum to that manual. The manual outlines a fairly flexible dosage schedule, with the general goal of achieving a dose of 200 mg/day by the third week of treatment (or physician's choice of a lower dose if necessitated by the severity of side effects). The pharmacotherapist may increase the dose up to a maximum of 300 mg/day, based on the patient's response to treatment. The manual includes guidelines for determining the patient's response to medication (both therapeutic response and side effects) and for adjusting dosage throughout the study.

Imipramine is administered within the context of a clinical management (CM) session. Guidelines for the conduct of this session are provided in the manual. The CM component is intended not only for the purpose of medication management but also to provide a generally supportive atmosphere and to enable the psychiatrist to assess the patient's clinical status. Ideally, the pharmacotherapist provides sufficient support and concern to maintain patient motivation and to achieve patient compliance with the treatment regimen but not to create a significant overlap with the two psychotherapy conditions. Due to the CM component, this condition may perhaps be more realistically considered a medication plus minimal supportive therapy condition. Blood levels are obtained (at 2, 4, 8, 12, and 16 weeks) to determine, retrospectively, whether a therapeutic blood level of the drug has been achieved and to ascertain compliance with the drug regimen.

P-CM.—Imipramine can serve as a useful "reference" condition, against which to compare the psychotherapies, only if it is indeed an effective treatment in this study. To establish the efficacy of imipramine at these sites, with these patient (and therapist) samples, it is important to include a pill-placebo condition in the present design. As indicated previously, the psychopharmacology literature does, on the whole, support the effectiveness of imipramine in the treatment of depression. That is why it was chosen as a reference condition in this study. There is a fair amount of variability in response rates reported, however, both for drug and placebo conditions. In addition, the majority of the studies involved inpatient, rather than outpatient, populations. We cannot be certain how effective imipramine will be in the present settings and, therefore, how appropriate as a reference condition until it is tested in these settings, using a pill-placebo control condition.

In addition, the pill-placebo may also provide a partial control condition against which to compare the psychotherapies. (That there is no inert drug being administered in the psychotherapy conditions and no concomitant expectation of effects due to a drug does, of course, limit the interpretations that can be made.) The manual and training for this condition include guidelines for providing support and encouragement to the patient and giving direct advice when necessary. This CM component thus approximates a "minimal supportive therapy" condition, and the placebo condition serves as a control both for expectations due to administration of a drug and for contact with a caring, supportive therapist. The condition thus provides a most stringent test of the specific effectiveness of the psychotherapy conditions.

It should be pointed out that NIMH staff and consultants originally considered the inclusion of other control conditions, especially those that might constitute a "psychotherapy placebo" control, and, for varying reasons, rejected all of them. If, on the one hand, the proposed control conditions were far removed from actual treatment, ethical and plausibility questions would be raised; if, on the other hand, a condition were to meet ethical and plausibility concerns, it would constitute a new treatment condition in its own right, rather than a "placebo control."

In reaching the decision to include the pill-placebo condition, ethical as well as scientific issues were faced. Two major questions addressed were (1) would it be ethically defensible to withhold active treatment known to be effective for depressed outpatients? and (2) would placebo patients be at increased risk for suicide or other severe clinical problems? After extensive deliberations, we concluded that (1) as discussed previously, there was not yet sufficient knowledge about the efficacy of the tricyclics for this particular population to warrant dropping the placebo control, and (2) the study could provide sufficient safeguards for all patients against risk of suicide or other severe problems.

These safeguards include screening out imminently suicidal patients, regular contacts with a therapist, periodic independent evaluations, and clearly specified mechanisms for withdrawing patients from the research protocol, if necessary, and providing them with other clinically appropriate treatment. Finally, the CM component of the pharmacotherapy conditions was developed to ensure adequate clinical care, especially because of the inclusion in the study of the placebo condition.

The imipramine-CM and P-CM conditions are administered double-blind, with neither patients nor therapists knowing in which condition a patient has been placed. Therapists are instructed to adjust dosage on the assumption that an active medication is being administered.

Writing Exercises

1. Following the example of Raimi's definition of mathematics, write a definition of your major field of study. Try to define your field by describing why practitioners get excited about it and how it tries to find order and beauty in the world.

2. Take a concept that is common to more than one field—for instance, stress in human psychology and in structural engineering, blood pressure and fluid pressure in a pipe, homeostasis in biology and social units, the fight-or-flight response in animals and humans, the algorithm in mathematics and in problem-solving with computers, etc. Write definitions of the concept for each field, showing relationships.

3. Select a field of study that you don't know much about, research it, and write a definition of five of the key concepts in that field.

Reference List

McGraw-Hill Dictionary of Scientific and Technical Terms 1984. 3d ed. Edited by Sybil P. Parker. New York: McGraw-Hill.

McGraw-Hill Encyclopedia of Science and Technology 1987. 6th ed. New York: McGraw-Hill.

REPORTING TRIALS, EXPERIMENTS, AND EMPIRICAL RESEARCH, INCLUDING THE UNDERGRADUATE LAB REPORT

Professional journals, as well as books, are filled with reports of trials and experiments. In some fields, such as coal gasification or enhanced oil recovery, most of the literature falls into this category.

The next journal article reports on testing in the field of civil and structural engineering: the method of using gauges to monitor the construction and performance of bridges.

Monitoring a Bridge's Pulse

Philip Di Vietro

Five years after the collapse of the Sunshine Skyway Bridge, the Florida Department of Transportation is building a replacement. With a 1,200 ft long main span, the new bridge will be the longest cable-stayed bridge in America and possibly one of the most closely monitored bridges in the world.

The latest field instrumentation and microcomputer technology, set up by Construction Technology Laboratories, Skokie, Ill., is providing background data for quality assurance during fabrication of bridge segments and other construction phases. Most of the results will come in over the long term and thus benefit posterity. The actual instrumentation of the bridge will not be functioning one hundred percent until the bridge is completed. Instruments are installed in the bridge segments before the concrete is cast. About 30% are in place and functioning; the rest are not yet operative.

Initially, CTL aided Florida DOT in the design of the bridge by evaluating long-term properties of the proposed concrete. Compressive strength, coefficient of thermal expansion, creep, and shrinkage were determined at various concrete ages. The measured concrete properties were then compared with design values to determine any discrepancies. These same data will also be used to predict long-term bridge behavior.

Instrumentation

CTL, which developed, designed and installed the instrumentation program, is training FDOT staff how to digest the data and to run the instrumentation network. CTL will turn everything over to FDOT about two years after construction is completed.

The two main piers supporting the cable-stayed spans were instrumented first. Six sections were instrumented in the north main pier. With the exception of the pier section 8 ft below the roadway, all gages in the north pier were installed after the pier was cast. [See Figures 5–11 and 5–12.]

Strain gages for the pier section 8-ft. below the roadway were installed before concrete casting. Five sections in the south pier were also instrumented in a similar layout. However, no strain sensors were installed at the pier section 8 ft below the roadway of the south pier.

Three pylon sections on top of the north pier have also been instrumented at 8 ft, 85 ft, and 145 ft above the roadway. Both temperature and strain sensors were installed before concrete was cast.

Seventeen large roadway segments were to be instrumented, with sensors installed before casting in the precasting plant. 534 gages are being used—228 concrete strain meters and 306 temperature sensors.

Readings will be taken manually until an automatic data acquisition system is installed. Readings will be taken before and after each event that can cause changes of strain or stress in the instrumented sections.

CTL uses a single microcomputer to read and record over 500 remotely located sensors within minutes. By positioning the remote signal processors at eight strategic locations in the piers and roadway segments, monitoring of all sensors can be centralized from a single location.

Traditionally, readings of installed sensors are taken manually by at least two technicians. With sensors widely scattered, near instantaneous manual readings are impossible. However, the remote signal processors eliminate time delays caused by manual techniques. Signals are digitized (similar to the way a telephone system digitizes voices) before data transmission. This permits the transmission of data undisturbed over long cable lengths (over ¾ mile). The scan gages are located more than one half a mile apart.

From Di Vietro, P. March 1986. Monitoring a bridge's pulse. *Civil Engineering* 56, no. 3: 54–55. Reprinted with permission of ASCE.

FIGURE 5-11
Monitoring the Longest Cable-Stayed Bridge

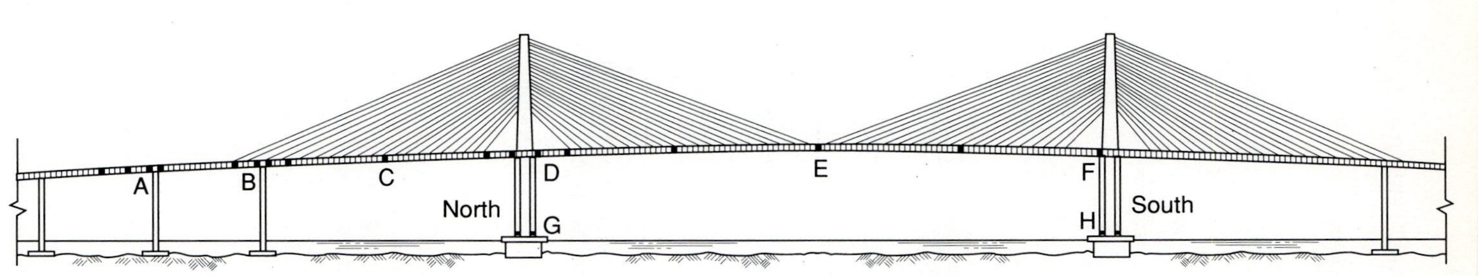

Over 500 gages closely monitor the new Sunshine Skyway bridge during construction and will continue to do so after construction.

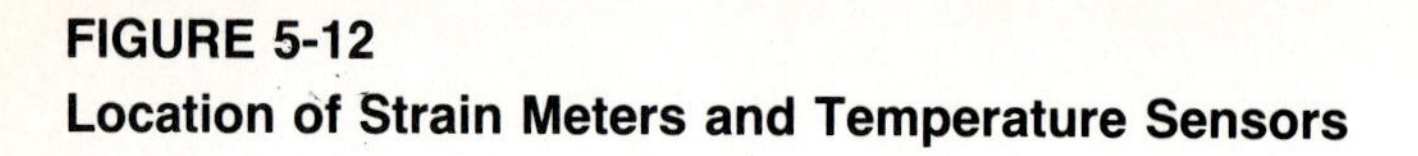

FIGURE 5-12
Location of Strain Meters and Temperature Sensors

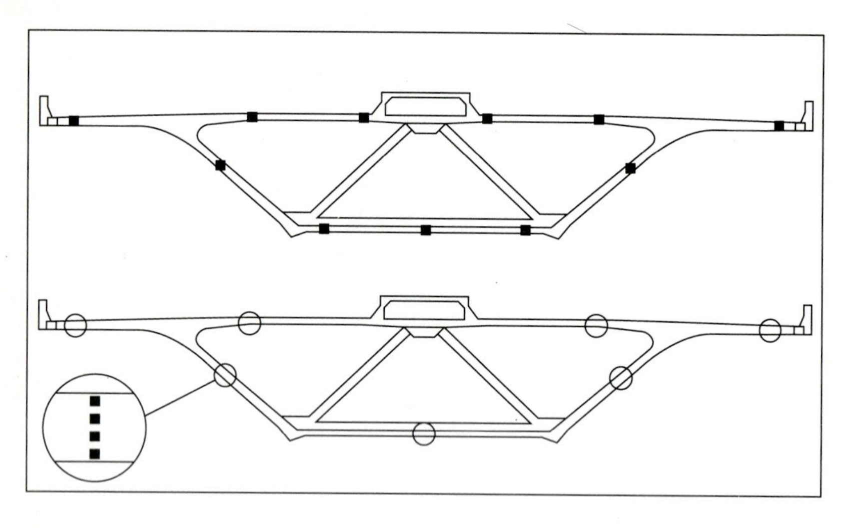

Above: Top segment shows concrete strain meters; bottom segment shows temperature sensors.
Below: Blackened cubes indicate precast segments with combinations of strain meters & thermocouples. Letters indicate signal processors.

Signal processors are connected in a closed loop arrangement. Accidental damage of the data communication cable at any point will not disturb the communication link of the remote signal processors to the microcomputer. This ensures uninterrupted recording.

The microcomputer has a built-in real time clock which allows readings at preprogrammed times and dates. Data can be displayed in commonly accepted engineering units or in unreduced formats. Data are also stored on a magnetic cartridge. Cartridges are sent back to CTL for analysis without interrupting field recording.

Results

The instrumentation cost Florida DOT some $500 thousand, according to District Engineer William Trefz, who suggested that the investment to monitor the $200 million bridge seemed worthwhile in view of the information gained.

Trefz says that engineers wanted to see "how close their design assumptions were to actual conditions, and to be able to adjust the bridge before opening, if that seemed appropriate. They particularly wanted to develop correlations between in-place shrinkage and creep as compared to lab tests for those conditions, and to see that the bridge would function as designed." In addition, Trefz says that they were able to check specs such as those limiting heat differentials due to hydration to 35 F for each concrete pylon placement.

The consequences of accidental or abnormal construction loading can be measured. Early modifications of design or timely adjustments of details can circumvent costly time delays and repairs. Impacts to construction can be quickly determined and the schedule modified accordingly.

Long-term bridge behavior will be predicted. Certain design assumptions related to time-dependent behavior of bridges can be verified. The state-of-the-art design procedure on cable-stayed bridge technology can be improved. Effects of temperature, wind, time-dependent movements, support settlement, accidental impact, overloads, and other factors can be assessed with the in-place instrumentation.

Instrumentation can also be part of a maintenance and inspection program. Bridge integrity can be confirmed by measuring concrete strain variations, changes in vertical deflection of the bridge spans, and rotations.

Florida DOT can record over 500 readings within minutes at any preselected interval. Cost savings in performing regular inspection by using installed instrumentation is substantial, especially for large and complex structures such as the 21,878 ft long Sunshine Skyway Bridge.

Researchers often use statistical analysis to confirm hypotheses. The following article from the *New England Journal of Medicine* convincingly argues that mortality from automobile crashes is highest in areas of low population density.

Geographic Variations in Mortality From Motor Vehicle Crashes

Susan P. Baker, M.P.H., R. A. Whitfield, M.A., and Brian O'Neill, B.Sc.

Abstract

Using a new technique to study the mortality associated with motor vehicle crashes, we calculated population-based death rates of occupants of motor vehicles during the period 1979 through 1981 and mapped them according to county for the 48 contiguous states of the United States. Mortality was highest in counties of low population density ($r = -0.57$; $P<0.0001$) and was also inversely correlated with per capita income ($r = -0.23$; $P<0.0001$). Death rates varied more than 100-fold; for example, Esmeralda County, Nevada, with 0.2 residents per square mile (2.6 km^2), had a death rate of 558 per 100,000 population, as compared with Manhattan, New York, with 64,000 residents per square mile and a death rate of 2.5 per 100,000. Differences in road characteristics, travel speeds, seat-belt use, types of vehicles, and availability of emergency care may have been major contributors to these relations. (N Engl J Med 1987; 316:1384–7.)

Motor vehicle crashes are the leading cause of death in the United States among persons 1 to 34 years of age. Detailed geographic analyses of the mortality associated with such crashes have not been undertaken, despite the fact that county maps such as those showing cancer "hot spots" have contributed to our understanding of other important health problems and have helped to identify high-risk populations.

State maps reveal major regional variations in death rates related to motor vehicle crashes but lack the specificity of analysis according to county. For example, in such states as New York, where much of the population lives in urban areas, the death rate for the state is determined largely by the rates of cities, in which death rates of occupants of motor vehicles are low.

In this study, the population-based mortality of occupants of motor vehicles was calculated and mapped for all counties in the United States, and correlations with population density and income were determined. Knowledge of the relations between death rates and population density or income can contribute to our understanding of motor vehicle–related deaths. In addition, detailed maps call attention to similarities among counties with especially high or low death rates.

From Baker, S. P., R. A. Whitefield, and B. O'Neill. 28 May 1987. Geographic variations in mortality from motor vehicle crashes. *New England Journal of Medicine* 316, no. 22:1384–87. Reprinted by permission. *Note:* The selection footnotes have been omitted.

Methods

County-specific death rates associated with motor vehicle crashes for the period 1979 through 1981 were calculated on the basis of data on deaths recorded in the National Highway Traffic Safety Administration's Fatal Accident Reporting System (FARS) and the 1980 population of the United States. Three-year average rates were calculated and mapped for all deaths of occupants of motor vehicles (127,110 deaths in the three-year period). The FARS data base was used in preference to mortality data from the National Center for Health Statistics (NCHS), because motor vehicle occupants are not always distinguishable in NCHS data and because deaths in FARS are tabulated according to the county in which the crashes occur. For purposes of comparison, death rates of occupants were also calculated from NCHS mortality data according to the county of residence.

Correlations with population density and per capita income were analyzed according to the rank order of each observation. Density and income data for each county were based on the 1980 census. Rates were not standardized for the composition of the county population according to age and sex, but relevant age distributions of the population were examined.

Maps were produced with use of a cartographic file from a 1970 geographic base file produced by the U.S. Bureau of the Census. This file contains the boundaries of counties and county equivalents in the 48 contiguous states plus the District of Columbia. Four unequal categories were used to identify counties with more extreme rates. The shading pattern on the maps represents percentile distributions as follows: 90th percentile and above, solid black; 75th through 89th percentile, dark gray; 10th through 74th percentile, light grey; and below the 10th percentile, white.

Results

The death rate of occupants of motor vehicles was 18.7 per 100,000 population for the entire United States, but it varied dramatically from one county to another: 10 percent of all counties in the 48 contiguous states had death rates of less than 13.5 per 100,000, whereas another 10 percent had rates of 57.3 per 100,000 or higher. Comparison of Figures 5–13 and 5–14 shows that the mortality is inversely correlated with the population-density pattern, with the highest death rates seen predominantly in counties with fewer than five people per square mile ($r = -0.57$; $P < 0.0001$). In most of Nevada, for example, death rates were in the highest two categories (75th through 89th and $\geq$90th percentile) in all the counties in which population density was in the lowest category ($<$10th percentile). The reverse was true near Las Vegas and Reno, where the population density is highest and death rates were low; Figure 5–15 illustrates the correlation.

The more than 100-fold variation in death rates and the remarkable association with population density are illustrated by a comparison of specific counties. For example, Manhattan (New York County) has more than 64,000 residents per square mile (2.6 km^2) and had a death rate of 2.5 per 100,000. Philadelphia, Pennsylvania, and Hudson, New Jersey, each with 12,000 residents per square mile, had death rates of 4.1 and 4.3 per 100,000, respectively. In contrast, the 777 residents of Esmeralda County, Nevada, spread over 3587 square miles (9290.33 km^2) (0.2 resident per square mile), had a rate of 558 per 100,000. The very highest death rate—1456 per 100,000—occurred in Loving County, Texas, which lost 4 of its 91 residents in motor vehicle crashes during the period 1979 through 1981. Although sparsely settled counties obviously have small populations, which are subject to wide fluctuations in death rates, the results were remarkably consistent; none of the 15 counties with the highest death rates had a population density of more than two persons per square mile.

Similarly, death rates of motor vehicle occupants calculated from NCHS data according to the county of residence of the deceased were highest in rural areas: the rank order of these

FIGURE 5-13
Death Rates of Occupants of Motor Vehicles per 100,000 Population According to County, 1979 through 1981.

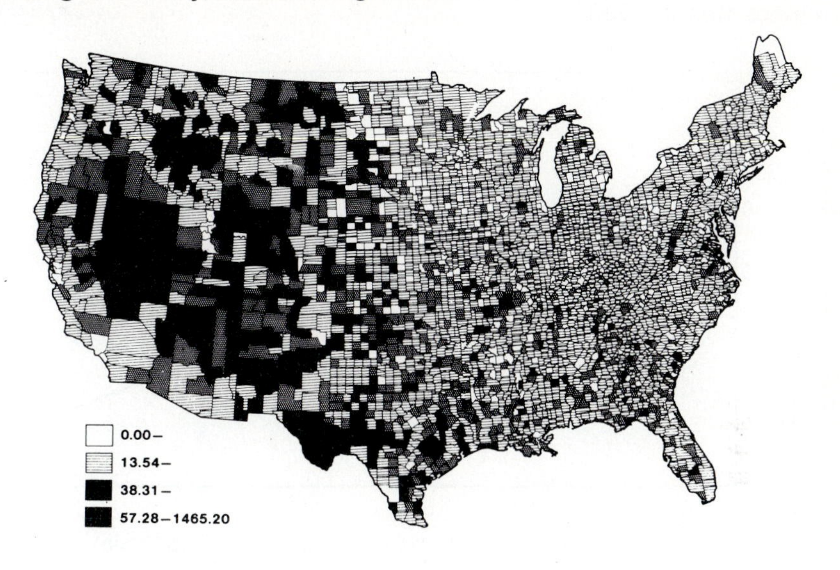

FIGURE 5-14
Population Density per Square Mile According to County, 1980.

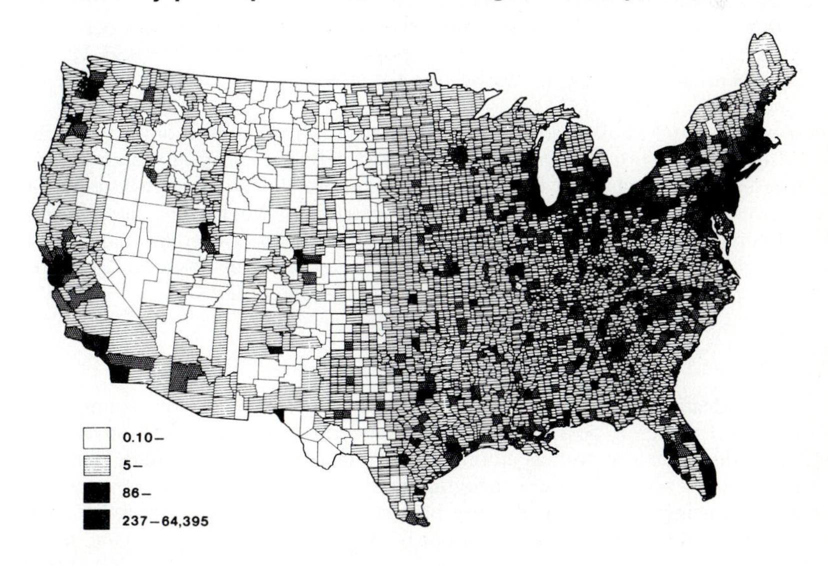

FIGURE 5-15

Death Rates of Occupants of Motor Vehicles per 100,000 Population in Nevada Counties, 1979 through 1981. According to the Population Density per Square Mile in 1980.

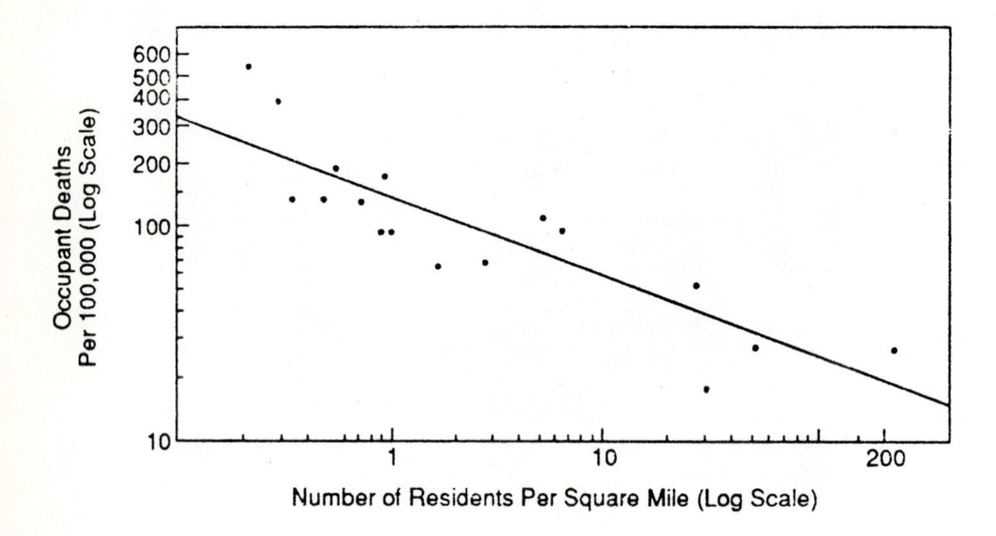

rates correlated directly with the FARS rates, calculated according to the county in which the crash occurred ($r = 0.67$; $P<0.0001$); thus, the high death rates in rural counties are not explained by travel in rural areas by residents of more densely settled areas. Neither are high rural rates explained by the large volume of travel on major routes: for example, the counties in California with the highest death rates do not include the counties traversed by California Route 99 and Interstate 5, which run in a northwesterly direction west of the Sierra Nevada mountains. In fact, none of the federal interstate highways are distinguishable on the map as connecting a string of counties with high death rates.

Death rates of motor vehicle occupants also correlated inversely with per capita income, although the correlation was weaker than for population density ($r = -0.23$; $P<0.0001$). The correlation with income may be partly due to an interaction with density, which was not controlled for.

Since 40 percent of all deaths in motor vehicle occupants occur among persons 15 to 24 years of age, we explored the possibility that this high-risk age group might be substantially overrepresented in rural areas. On the contrary, in both New England, where motor vehicle–related death rates were lowest, and the mountain states, where such rates were extremely high, 19 percent of the population was 15 to 24 years old. County-specific analyses yielded similar results: In Nevada, about 18.5 percent of the population was 15 to 24 in the counties with low population density as well as in the more urban counties.

Discussion

Mapping death rates of motor vehicle occupants according to the county in which the crash occurred makes it possible to demonstrate the substantial variation among counties within individual states, as well as among the various regions of the United States.

The method chosen for these geographic analyses has advantages over such potential alternatives as using NCHS data. The use of data from FARS rather than NCHS makes it possible to identify the county in which a crash occurred, which is important in determining potential contributing factors and developing preventive measures. Because of the close correlation between the county in which the crash occurred and the county of residence, county populations can be used as denominators. County-specific data on age distribution make it possible to determine that the results were not due to an interaction between age and population density or income.

High motor vehicle–related death rates in western states have at times been attributed to the greater distances driven in those states. Even when the data were adjusted for the amount of travel, however, rural areas still had high rates; death rates of occupants of motor vehicles per million vehicle miles of travel were lowest in the Northeast and highest in rural western states. The greater distances between emergency facilities and the reduced access to major trauma centers, however, undoubtedly had an adverse effect. Inadequate care of injured patients has been documented in the rural mountain West and probably contributes to deaths from trauma in many other rural areas.

Poor roads may play a major part in raising death rates in areas of low population density. The road's gradient, curvature, lane width, lighting, striping, signs, and signals; ditches and fixed objects near the roadway; and the presence or absence of adequate shoulders and guardrails all contribute to the likelihood and severity of crashes. The highway design standards governing such factors vary with the amount of traffic and are minimal in rural areas where traffic volume is very low. (About one fourth of all public roads are not even paved.) Illustrating this effect are the death rates of Native Americans living on reservations that are especially rural and have few improved roads: They are 4 to 6 times the average rate on all Indian reservations combined and 8 to 12 times the average for the United States. The low death rates on interstate highways, for which the death rate per million vehicle miles is less than half the national average, attest to the success of improved engineering of major roads.

Speed of travel is also an important determinant of death rates. At high speeds, drivers have less time to avert crashes, and resulting injuries are more severe: The ratio of deaths to injuries among occupants of motor vehicles increases more than sixfold as the posted speed limit increases from 30 to 55 mph. Travel at extremely high speeds (65 mph or higher) is especially common in rural areas. In 1981, an estimated 8.8 percent of the travel on rural interstate highways was at speeds exceeding 65 mph, as compared with 4.0 percent on urban interstate highways. Travel at high speeds on other arteries was also twice as common in rural areas. Ironically, the current impetus for raising the 55-mph speed limit comes primarily from states in the West, where fatality rates on rural interstate highways are more than twice the national rate.

Utility vehicles (jeep-like vehicles) and pickup trucks are associated with high death rates. Their popularity in rural states may contribute substantially to the high fatality rates in those states. In particular, small utility vehicles are associated with a death rate per 10,000 vehicles that is more than twice the rate in small cars and four times the rate in large cars; more than 60 percent of the deaths involve "roll-overs," and ejection is a major problem. Roll-over crashes and high-risk vehicles are about 5 to 10 times as frequent in fatal crashes in sparsely settled states as in the states with the highest population density [Table 5–1].

The use of seat belts is less common in rural areas and varies dramatically with income: In metropolitan areas of Maryland, there was a fourfold difference in the use of belts, and seat-belt use was found to correlate inversely with median property values. Thus, seat-belt use is probably a factor in the inverse relation between the mortality rate of occupants of motor vehicles and income. In addition, very old vehicles may be more commonly used in low-income areas, contributing both to the likelihood of crashes if brakes are not well maintained and, in the case of cars that predate the 1968 occupant-protection standards, to reduced safety in crashes.

In part, the inverse relation with income is due to the fact that per capita income is higher in metropolitan areas, where traffic generally moves more slowly and public transportation is more likely to be available. Population density and income, however, have been shown to have independent effects on county-specific rates of death from unintentional injury, including motor vehicle-related injury. The effect of income itself is clearly illustrated above by the difference in seat-belt use within densely populated Maryland counties.

Table 5-1 Population Density in Relation to Vehicle Type and "Rollovers" in Fatal Crashes

STATE	POPULATION PER SQUARE MILE*		% OF OCCUPANT DEATHS IN "SPECIAL" VEHICLES OR LIGHT TRUCKS†‡		% OF FATAL CRASHES INVOLVING ROLL-OVER†	
	NO.	RANK	%	RANK	%	RANK
New Jersey	986	1	8	49	3	48
Rhode Island	898	2	7	50	4	45
Massachusetts	733	3	9	48	2	50
United States	64	—	18	—	9	—
Montana	5	48	35	3	29	2
Wyoming	5	49	40	2	41	1
Alaska	1	50	43	1	24	6

*From the 1980 census.
†From 1983 FARS data.
‡Light trucks include pickups and vans. "Special" vehicles include utility vehicles (jeep-like vehicles), which are involved in 15 percent of the deaths of motor vehicle occupants in Alaska and 10 percent in Wyoming, as compared with 3 percent in the United States as a whole.

In summary, a variety of factors may underlie the strong inverse correlation between death rates of occupants of motor vehicles and population density or income. They probably include higher travel speeds in rural areas, worse roads, less seat-belt use, more use of high-risk utility vehicles and travel in open pickup trucks, and poorer access to trauma care. These factors are all subject to modification and should receive special attention wherever the population is at excessive risk of death or injury on the highway.

Jane Goodall has pioneered research on primates during the last two decades. In this excerpt, she reports on her work in Gombe National Park, Tanganyika, and describes how her data is relative to the observational conditions under which she recorded it.

Research at Gombe

Jane Goodall

Field Methods

During my early months at Gombe, as the chimpanzees gradually became accustomed to my presence, I was able to piece together, bit by bit, the overall pattern of chimpanzee life, watching through binoculars and writing everything in notebooks. Then in 1962 a mature male chimpanzee, David Greybeard, visited my base camp to feed on the fruit of an oil-nut

Goodall, J. 1986. *The chimpanzees of Gombe: Patterns of behavior.* Cambridge: Harvard University Press, Belknap Press, 51–59. Excerpted by permission of the publishers. From *The chimpanzees of Gombe* by Jane Goodall.

palm that grew there. He returned each day, occasionally accompanied by another male, Goliath, until the fruit was exhausted. In the course of one such visit he took a bunch of bananas from my tent. After this I asked my African cook to leave bananas lying out; if David came, he always took them. Three months later, when the fruits of another palm in the camp ripened, David returned, sometimes accompanied not only by Goliath but by William as well. When this second palm had finished fruiting, the three males still came at times for the bananas we continued to leave out for them. Eventually other chimpanzees followed, including an adult female, Flo, and her family. When in July 1963 Flo became sexually receptive, eight mature males followed her into camp.

That was when I decided to set up an artificial feeding area. This was primarily in order to facilitate the filming at close quarters of chimpanzee behavior, for by then I had been joined by Hugo van Lawick, working for the National Geographic Society. But I was also able for the first time to make fairly regular observations on the various individuals. The number of chimpanzees visiting the feeding area, or *camp,* gradually increased. In 1964 the first infants were born to our regular camp visitors, and it became possible to study details of infant development on an almost-daily basis.

It was in 1964 that the first research assistant joined me at Gombe: the beginning of what was to become the Gombe Stream Research Center. By the end of that year a second assistant had arrived, and slowly the number grew. In 1967 the first independent project was initiated by a doctoral candidate. By 1975 there were twenty-two American, European, and Tanzanian students (undergraduate and graduate) and one postdoctoral researcher, working at Gombe—some on the behavior of baboons.

In 1968 the first Tanzanian field assistant was employed, and in the years that followed he was joined by others. Initially the responsibility of these individuals was simply to accompany students, unused to the African bush. Quite soon it became apparent that the field assistants could provide our research with an extremely important component: a core of individuals, with long-term commitment to the work (the students seldom stayed longer than eighteen months), who were totally familiar with the chimpanzees, the terrain, and the food plants. By 1970 the field assistants were receiving basic training in simple data collection. . . .

The advantages and drawbacks of the banana feeding have been discussed in detail elsewhere. . . . At the beginning, a chimpanzee was in principle given bananas each time he or she visited camp [Figure 5-16]. By 1967 fifty-eight chimpanzees of both sexes and all age

FIGURE 5-16

William, Goliath, and David Greybeard share a cardboard box in my first camp.

groups were regular visitors, and some came almost daily. It was during the four-year period from 1964 to 1967 that the disadvantages of the feeding system became more and more apparent. Ranging and grouping patterns, feeding, and aggression were increasingly influenced. Moreover, because camp was at the periphery of the ranges of some individuals, chimpanzees who met frequently in order to feed on bananas would otherwise probably have associated only occasionally. The quality of the relationships of these and other chimpanzees was undoubtedly affected by this more-frequent association.

In 1968 the feeding system was extensively revised. Each chimpanzee was fed only a small number of bananas (five or six) every seven to ten days, and only when alone or when in a small group of compatible individuals. This resulted almost at once in a marked drop in the frequency of aggression in camp (Wrangham, 1974) and in a return to grouping and ranging patterns similar to those I had observed prior to the establishment of the feeding regime. Individuals began to visit camp much less often; some stopped coming altogether. Today chimpanzees visit camp only three or four times in a year despite the fact that they almost always receive bananas when they do come.

From 1964 on, data were recorded on the interactions of the various chimpanzees who visited camp. The reproductive state of females and the health of all individuals who visited were carefully noted and, starting in 1970, they were weighed regularly.

From 1964 on, data were recorded on the interactions of the various chimpanzees who visited camp. The reproductive state of females and the health of all individuals who visited were carefully noted and, starting in 1970, they were weighed regularly [Figure 5-17].

As the chimpanzees became increasingly tolerant of their human observers, it was possible to follow a selected individual from the time he (or she) left his nest in the morning until the time he retired for the night. Over the years there have been a number of consecutive-day "long follows." The first of these was in 1968, when the pregnant Flo was followed for sixteen days in the hope of observing parturition (which, in fact, took place at night). In 1974 the top-ranking male Figan was followed for fifty consecutive days (Riss and Busse, 1977). In 1976 another chimpanzee, Fifi, was followed daily for forty-five days. In 1977 there were two marathons: sixty-nine days of watching the female Passion (to monitor her interactions with new mothers, in view of her known cannibalistic tendencies) and fifty-five days of following Melissa after she gave birth to twins. There have been a number of other follows of two to three weeks (two of them follow-ups on Figan). Most follows, however, last only one or two days, and many are less than a day.

Throughout twenty-five years of research I have paid special attention to the building up of long-term records. Individuals collecting data for dissertations or special projects were required to leave copies on file at Gombe. Students using check sheets were asked to write qualitative summaries emphasizing major points of interest. Files were maintained on various aspects of behavior such as predation, tool use, interactions with baboons, and so on. From all the available information a "character file" was compiled for each chimpanzee, with extensive cross-referencing. These records were made available to all students who worked at Gombe, in order that they might benefit from the longitudinal nature of the research.

It was in the early 1970s that the value of collaborative, cooperative field research became obvious. The pooling of data collected by a number of observers threw new light on puzzling or seldom-seen aspects of chimpanzee behavior and added significantly to our understanding of social structure. Access to pooled current data and also to back records was especially valuable for students working on ranging and feeding behavior . . . , on adolescence . . . , and on weaning. . . .

Relations between Chimpanzees and Observers

Reactions of the Subjects To what extent does the presence of humans, and all that it entails, affect the behavior of our chimpanzee subjects? This is not an easy question to answer, for if we removed human observers we should not be in a position to record any changes in chimpanzee behavior that might result. However, a few comments are possible.

We have seen that the providing of bananas on a daily basis caused some changes in chimpanzee behavior. To what extent the long-term residue of those changes is still operating today is difficult to assess. Camp still represents an unnatural food source, with its supply available in one place, month after month, year after year. It is certain that if camp were not there, the chimpanzees would not visit that particular clearing with anything like the regularity with which they now pass through. Camp provides, in addition to bananas, a meeting place—an area where the probability is high that a chimpanzee will encounter his companions. A juvenile who has lost his mother, for instance, is likely to visit camp when searching for her and may wait for quite considerable periods, looking anxiously in all directions. Or he may leave, only to return after a short while several times in succession. Without doubt some individuals meet one another much more frequently than they would if camp did not exist. On the other hand, the five or six bananas to which an individual is "entitled" once every week or so have very little significance in the overall diet. (In the days of bountiful feeding I watched an adult male consume between fifty and sixty of the fruits at one sitting, after he had raided the store.) And I have already pointed out that some individuals visit camp very rarely, despite the fact that they are rewarded with bananas when they do. Some chimpanzees, even after becoming accustomed to regular bananas, have moved away and never returned, as we shall see in subsequent chapters.

We should also ask how disturbing it is for a chimpanzee to be followed through the forest by one or two humans, sometimes for days on end. Some chimpanzees show what appears to be a total lack of concern, of interest even, in the close proximity of one or more humans. Others are far more anxious. In part it depends on the behavior of the human observers: those who are insensitive to the behavior of the animal they are following, who move noisily when he is trying to listen, who make sudden movements when he is resting, who approach too closely when he is traveling, and so on, are likely to affect the behavior of their subject. The surprising degree to which almost all the chimpanzees tolerate the observers is something that developed slowly; through the years I have emphasized the importance of preserving and encouraging this remarkable chimpanzee-human relationship. If a particular subject appears nervous or irritable during a follow, the observers are instructed to fall back and watch from farther away, even if this means losing their target animal. And if a chimpanzee really wants to escape, he can do so—all too easily. Given the rugged nature of the terrain at Gombe, it is virtually impossible for a mere human to keep up with a chimpanzee who is determined to get away. Indeed, even when the target is *amenable* to being followed, observations frequently end when the subject is lost (for example, he slips easily through a patch of dense, thorny undergrowth while the humans force their way painfully after him, often on their bellies, or search for an easier route).

Individual chimpanzees differ considerably in their attitude toward humans. There is one female, Nope, who first visited camp in 1965 and has been a regular visitor ever since. Yet even today, unless she is with a number of other chimpanzees, she is too nervous to be followed. Except by me. I find her as calm and tolerant of my presence as any other habituated female. One adult male, Humphrey, often expressed dissatisfaction with his observers by throwing large rocks at them. It seemed that Humphrey was much more tolerant of male observers, and most of his threats were directed toward women. In general, chimpanzees are more likely to take liberties with humans of the female sex. Some adolescent males, for instance, clearly attempt to dominate female researchers at the time when they are struggling to dominate females of their own kind.

In the early years of the research I actively encouraged social contact—play or grooming—with six different chimpanzees (Goodall, 1971). For me personally, those contacts were a major breakthrough: they meant that I had won the trust of creatures who initially had fled when they saw me in the distance. However, once it became evident that the research would continue into the future, it was necessary to discourage contacts of this sort. Not only could such interaction distort the behavior of the chimpanzees, but it could be dangerous for the humans, for chimpanzees are much stronger than we are. Accordingly, researchers were asked not to approach their subjects closer than about 5 meters and to try to ignore, or move away from, any friendly—or unfriendly—advances that might be made. Despite our efforts to maintain a respectable distance between observer and observed, many young chimpanzees make things very difficult with their determined attempts to touch, poke, slap, lick, threaten, or try to play with the persons making observations. Usually such attentions peter out during adolescence, apart from the occasional attempts to impress observers of the female sex.

The above behaviors, extremely attention provoking when they occur, are the exceptions rather than the rule. For the most part the chimpanzees show very little interest in humans or their behavior. We have become as much a normal part of their lives as the baboons and other animals with whom they share the forest. Those of us working day after day with the chimpanzees feel, intuitively, that by and large our presence has surprisingly little effect on their behavior. . . .

Certainly the presence of humans, and the ongoing research, and the way the research is conducted mean that the Gombe chimpanzees are not living in an undisturbed environment. Nevertheless, each of the individuals being studied is subjected to more or less the same degree of disturbance. From early on my own special research interest has been in individual differences. Why is male A so much more aggressive than male B? Why is mother C less tolerant of her infant than mother D? To what extent are differences between adults attributable to differences in their upbringing, family life, and childhood experience?

Answers to such questions are, in a sense, equally meaningful no matter what the social and physical environment of the group under study. The important this is to take such factors into account when making generalizations that apply to the species as a whole—especially as they relate to the ultimate causation of behaviors. This book is in essence a description of one community of chimpanzees; but in order to keep the broader picture in mind I have, whenever possible, compared the behavior of the Gombe chimpanzees with behavior reported from other study sites. It is my hope that the information will be sufficiently comprehensive to allow the reader to judge in what ways (if any) Gombe chimpanzees show behavior that is atypical of the species as a whole.

By inventing the dynamo and the electric motor, Michael Faraday ushered in our electric age. Faraday describes here how he used electricity for one of his most famous experiments: the discovery of the two gases that compose water.

The Decomposition of Water

Michael Faraday

We have the power of arranging the zinc which you have seen acting upon the water by the assistance of an acid, in such a manner as to cause all the power to be evolved in the place where we require it. I have behind me a voltaic pile, and I am about to show you its character and power. I hold here the extremities of the wires which transport this power from behind me, and which I shall cause to act on the water.

A great power of combustion is possessed by potassium, or zinc, or iron-filings; but none of them show such energy as this. I will make contact between the two terminal wires of the battery: what a brilliant flash of light is produced! This light is, in fact, produced by a forty-zinc power of burning: it is a power that I can carry about in my hands, through these wires, at pleasure—although, if I applied it wrongly to myself, it would destroy me in an instant, for it is a most intense thing, and the power you see here put forth, if I allow the spark to last while you count five, is equivalent to the power of several thunderstorms, so great is its force.

I am now going to apply this force to water to pull it to pieces, to see what else there is in the water besides hydrogen; because if we pass steam through an iron tube, we by no means get the weight of water back which we put in, in the form of steam, though we have a very large quantity of gas evolved. We have now to see what is the other substance present.

From Faraday, M. 1839–1855. The decomposition of water. In *Experimental researches in electricity.* 3 vols. London: R. & J. E. Taylor.

What effect has an electric current on water? Here are two little platinum plates which I intend to make the ends of the battery, and this is a little vessel so shaped as to enable me to take it to pieces and show you its construction. In those two cups I pour mercury, which touches the ends of the wires connected with the platinum plates. In the vessel I pour some water containing a little acid (but which is put only for the purpose of facilitating the action; it undergoes no change in the process), and connected with the top of the vessel is a bent glass tube which now passes under the jar.

I have now adjusted this apparatus, and we will proceed to affect the water some way or other. In the other case, I sent the water through a tube which was made red-hot; I am now going to pass the electricity through the contents of this vessel. Perhaps I may boil the water; if I do boil the water, I shall get steam; and you know that steam condenses when it gets cold, and you will therefore see by that whether I do boil the water or not. Perhaps, however, I shall not boil the water, but produce some other effect. You shall have the experiment and see. There is one wire which I will put to this side, and here is the other wire which I will put to the other side, and you will soon see whether any disturbance takes place. Here it is seeming to boil up famously; but does it boil? Let us see whether that which goes out is steam or not. I think you will soon see the jar will be filled with vapour, if that which rises from the water is steam. But can it be steam? Why, certainly not; because there it remains, you see, unchanged. There it is standing over the water, and it cannot therefore be steam, but must be a permanent gas of some sort. What is it? Is it hydrogen? Is it anything else? Well, we will examine it. If it is hydrogen, it will burn. I will now apply a light to it. You see it is certainly combustible, but not combustible in the way that hydrogen is. Hydrogen would not have given you that noise; but the colour of that light, when the thing did burn, was like that of hydrogen: it will, however, burn without contact with the air. That is why I have chosen this form of apparatus, for the purpose of pointing out to you what are the particular circumstances of this experiment.

In place of an open vessel I have taken one that is closed; and I am going to show you that that gas, whatever it may be, can burn without air, and in that respect differs from a candle, which cannot burn without the air. And our manner of doing this is as follows: I have here a glass vessel which is fitted with two platinum wires through which I can apply electricity; and we can put the vessel on the air-pump and exhaust the air, and when we have taken the air out we can fasten it on to this jar, and let into the vessel that gas which was formed by the action of the voltaic battery upon the water, and which we have produced by changing the water into it—for I may go as far as this and say we have really, by that experiment, changed the water into that gas. We have not only altered its condition, but we have changed it really and truly into that gaseous substance, and all the water is there which was decomposed by the experiment. As I screw this vessel on here and make the tubes well connected, and when I open the stopcocks, if you watch the level of the water you will see that the gas will rise. I will now close the stopcocks, as I have drawn up as much as the vessel can hold, and I will pass an electric spark, from an induction coil, through the gas. The vessel was quite clear and bright at first, but it has now become dim with a deposit of water. I will again connect it to our gas reservoir, for that is what the jar really is: and as I open the stopcocks you see that the water rises: this indicates that the glass vessel must be filling. "But why is the jar empty after each explosion?" you may ask. Because the vapour or gas into which that water has been resolved by the battery explodes under the influence of the spark, and changes into water; and by and by you will see in this upper vessel some drops of water trickling down the sides and collecting at the bottom.

We are here dealing with water entirely, without reference to the atmosphere. The water of the candle had the atmosphere helping to produce it; but in this way it can be produced independently of the air. Water, therefore, ought to contain that other substance which the candle takes from the air, and which, combining with the hydrogen, produces water.

I will now dip the poles—the metallic ends of this battery—into water, and see what will happen when they are kept far apart. I place one here and the other there, and I have little shelves with holes which I can put upon each pole, and so arrange them that whatever escapes from the two ends of the battery will appear as separate gases; for you saw that the water did not become vaporous, but gaseous. The wires are now in perfect and proper connexion with the vessel containing the water; and you see the bubbles rising; let us collect these bubbles and see what they are. Here is a glass cylinder; I fill it with water and put it over one end of the pile; and I will take another and put it over the other end of the pile. And so now we have a double apparatus, with both places delivering gas. Both these jars will fill with gas. There they go, that to the right filling very rapidly; the one to the left filling not so rapidly. I should have twice as much in this as I have in that. Both these gases are colourless; they stand over the water without condensing; they are alike in all things—I mean in all apparent things; and we have an opportunity of examining these bodies and ascertaining what they are. Their bulk is large, and we can easily apply experiments to them. I will take this jar first, and will ask you to be prepared to recognize hydrogen.

Think of all its qualities—the light gas which stood well in inverted vessels, burning with a pale flame at the mouth of the jar—and see whether this gas does not satisfy all these conditions. If it be hydrogen, it will remain here while I hold this jar inverted. It burns when a light is applied to the mouth of the jar; it is evidently hydrogen.

What is there now in the other jar? You know that the two together made an explosive mixture. But what can this be which we find as the other constituent in water, and which must therefore be that substance which made the hydrogen burn? We know that the water we put into the vessel consisted of the two things together. We find one of these in hydrogen: what must that other be which was in the water before the experiment, and which we now have by itself? I am about to put this lighted splinter of wood into the gas. The gas itself will not burn, but it will rekindle the glowing splinter. See how it invigorates the combustion of the wood, and how it makes it burn far better than the air would make it burn; and now you see by itself that very other substance which is contained in the water, and which, when the water was formed by the burning of the candle, must have been taken from the atmosphere. What shall we call it, A, B, or C? Let us call it O—call it "oxygen": it is a very good, distinct-sounding name. This, then, is the oxygen which was present in the water, forming so large a part of it.

We shall now begin to understand more clearly our experiments and researches; because, when we have examined these things once or twice, we shall soon see why a candle burns in the air. When we have in this way analysed the water—that is to say, separated, or electrolysed its parts out of it—we get two volumes of hydrogen, and one of the body that burns it. And these two are represented to us on the following diagram, with their weights also stated; and we shall find that the oxygen is a very heavy body by comparison with the hydrogen. It is the other element in water.

Oxygen	88.9
Hydrogen	11.1
Water	100.0

As an undergraduate, you'll be writing the trial of an experiment when you write a standard lab report, a required form in most undergraduate courses in science and engineering. Lab reports traditionally have at least four parts: introduction, theory, procedure, and conclusions.

The introduction states the topic of the experiment and some of its constraints. The theory section describes relevant theoretical considerations and the hypothesis to be tested. The procedure portion details how the experiment was conducted. Sometimes this section is divided into "Materials" and "Methods". And the conclusions section states the results of the experiment and discusses any difficulties or anomalous data. It is often divided into "Results" and "Conclusions," where "Results" summarizes the data collected. An abstract and a list of equipment often accompany a lab report.

The following undergraduate lab report in electrical engineering describes flow in resistive circuits.

Signal Measurement 1: Experiment No. 1

John Mathews

Abstract

This first experiment was a very straight-forward exercise in meter design and error measurement. Two series resistive circuits were connected, each used as a voltage divider circuit. The voltage across a resistor was determined by three methods; theoretically--no meter considered, ideally--a perfectly accurate meter, and finally, the actual measured voltage as read on the meter.

The two circuits used different resistive values (one being ten times the other), but they maintained the same ratio between resistors so that the theoretical output was the same for both circuits. This allowed a direct comparison between high meter loading and low meter loading of the circuit under study.

It was found that the circuit using higher value resistors was more seriously affected by the loading of the meter which lead to very high error in measurement. The concept of error and precision was considered carefully in this experiment and it was concluded that such terms must be defined carefully in advance in order to give meaning to claims of "such and such percent error occurred in reading x."

Introduction

This experiment will deal with two similar resistive circuits, each used as a voltage divider circuit. The values of resistance changed by a factor of ten from one circuit to the other, but the ratio between resistors in each voltage divider circuit remains the same.

A rather poor quality volt meter--20,000 ohms per volt--is used to measure the output voltage from the dividers. This meter will present additional circuit resistance across the output, thus changing the normal voltage level. The amount of change varies from one circuit to the other, with the higher value resistors being more affected than the lower value circuit.

Theory

The circuit shown in [Figure 5-17] is called a voltage divider. The output, taken from $\underline{r_2}$ in this circuit is determined by the formula:

From Mathews, J. 1987. Signal measurement 1: Experiment no. 1. Student paper, West Virginia University. Reprinted by permission of the author.

FIGURE 5-17
Voltage Divider

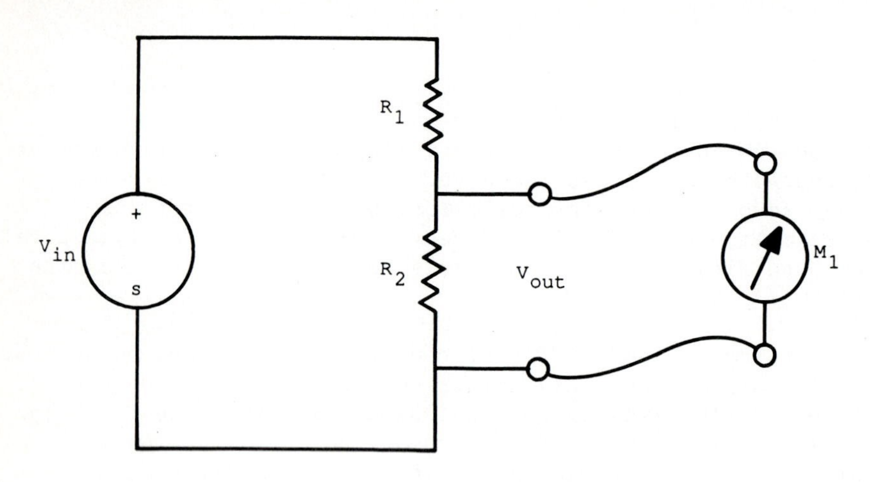

$$V_{in} * (R_2)/(R_2 + R_1) = V_{out} \quad (1)$$

It can be seen that when any external current path is added across R_2, the output must be recalculated, taking into consideration the new resistance value which will be present in parallel with R_2. This is shown in the figure above by the meter M_1 connected by dotted lines.

The introduction of the meter modifies the original current path and this is accounted for by the formula for parallel resistors:

$$R_t = 1/\{1/r_a + 1/r_b\}. \quad (2)$$

Combining (1) and (2), the calculation for the output voltage in the divider becomes:

$$V_{in} * 1/[1/R_2 + 1/R_m] / [R_1 + 1/(1/R_2 + 1/R_m)] = V_{out} \quad (3)$$

R_m in the above equation is the resistance added to the output by the meter circuit. It can be seen that if the meter resistance is varied, the output will correspondingly vary. This effect is called meter loading and it is the root cause of errors during measurements on experimental circuits. Note, for instance, if the meter resistance were nearly zero, very little voltage potential could appear across R_2 without extreme current levels being present.

A great deal of thought was given to errors in measurement, and how they should be determined. Three basic calculations can be made with each circuit. The original voltage divider will develop a voltage ideally of V_r^2 across R_2. However, we know that the addition of the meter will change this voltage and we can calculate this change exactly. This new value--the expected output voltage R_m, will differ from the ideal circuit and this may be considered an introduced error.

In addition to the above, the actual voltage measured on the meter may differ somewhat from the calculated reading (the one that assumes the meter loading). This reading error can not be calculated in advance, and must be determined experimentally. This is because, while the meter may claim a certain ohms/volt rating, this may not, in fact, be the correct value. Operator error can also effect these results--such as parallax caused by reading the meter slightly off from one side of dead center.

There are therefore three separate output voltages across R_2 to be considered. The ideal voltage without a meter to effect the circuit, the expected voltage which takes into account the resistance of the meter and thus its effect, and the measured value which is a function of the particular meter used, and the operator taking the readings.

Procedure

Initially, the meter to be used in the experiment was examined for the ohms/volt rating. This data is used along with the full scale voltage of each range to determine the meter's expected internal resistance. This also leads to the meter's full scale deflection current. Table 5-1 on the following page shows the results of this calculation for the three voltage ranges used; 1.6v, 8v, & 40v. The meter current is found simply from V/r = I and is the same for all scales.

The two voltage divider circuits were then setup as shown in figures 2 & 3. The output voltage was taken as across the lower

Table 5-1 Meter Resistance

VOLTAGE SCALE:	INTERNAL RESISTANCE	METER CURRENT
1.6 volts	1.6 * 20,000 = 32k ohms	1.6 / 32k = 50_u amp
8 volts	8 * 20,000 = 160k ohms	8 / 160k = 50_u amp
40 volts	40 * 20,000 = 800k ohms	40 / 800k = 50_u amp

FIGURE 5-18a
Voltage Divider #1

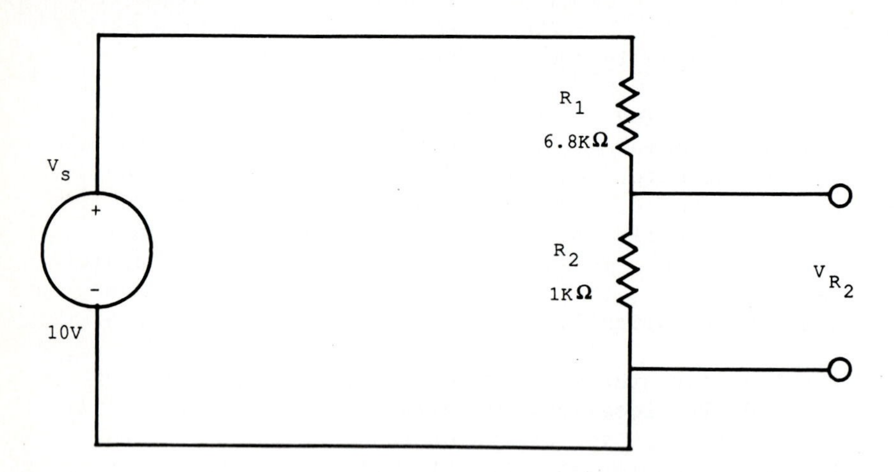

resistor in each case and labeled V_r^2. A 10 volt dc power source was used to provide the drive for the circuit.

Using equation (1), the ideal output voltage was determined for the two circuits. Then, the meter resistance was considered and using equation (3) the expected output voltage was found.

Calculations of percent error were performed in three ways. First the expected error from expected readings was found by using the ideal voltage as the reference and the expected reading as the error figure thus:

FIGURE 5-18b
Voltage Divider #2

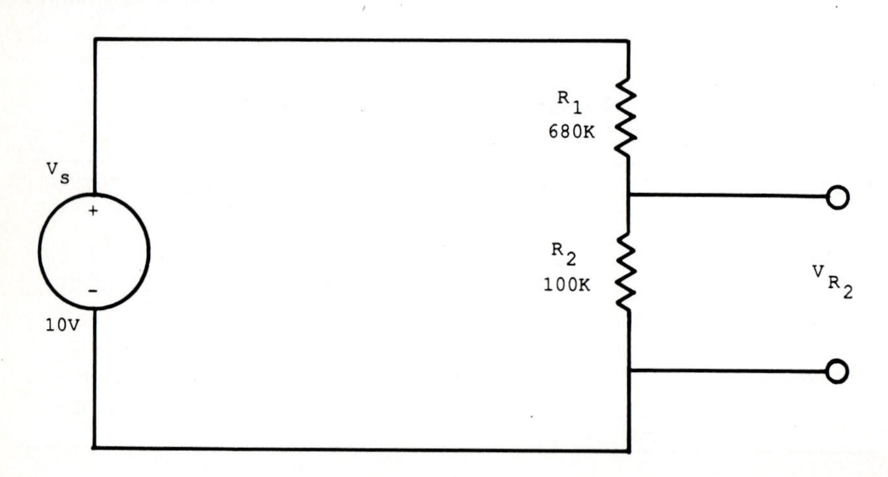

$$(V_r^2 - V_m)/V_r^2 * 100 = \%error \quad (4)$$

where V_r^2 is the ideal voltage and V_m is the expected meter reading.

Next, the meter is used to take the readings on the circuits with each of the 3 range scales. The ideal voltage is again used as the reference and the observed voltage is the error thus:

$$(V_r^2 - V_o)/V_r^2 * 100 = \%error \quad (5)$$

where V_r^2 is the ideal voltage and V_o is the actual observed voltage.

Finally, the expected meter reading is taken as the reference and the observed reading as the error thus:

$$(V_m - V_o)/V_m * 100 = \%error \quad (6)$$

As a last effort, the experiment calls for a probe to be designed to double the scale of the 1.6 volt range on the meter. This was simply done by using the ohms / volt rating of the meter, and the desired scale of 3.2 volts. A simple calculation of 20k * 3.2 = 64k ohms shows that the required internal resistance of 64k ohms must be made up from the known internal resistance on the 1.6volt scale of 32k ohms and the probe resistance which must also be 32k ohms for a total of 64k ohms as required. This was implemented and the meter showed a full scale deflection at 3.2 volts from the power supply.

Table 5-2 . . . shows the ideal voltage for the two circuits calculated with equation (1). Table 5-3 shows the expected voltage with the meter in-circuit, calculated with equation (3). Table 5-4 shows the observed voltage as measured by the meter.

Next on Table 5-5, the error of the meter is shown as found by equation (4). Table 5-6 shows the error between ideal voltage and observed voltage using equation (5). Last, Table 5-7 shows the

Table 5-2 Ideal Output Across V_{r2} (no meter load)

DIVIDER CIRCUIT	IDEAL VOLTAGE R^2
CKT #1	1.282 volts
CKT #2	1.282 volts

Table 5-3 Expected Readings

VOLTAGE SCALE:	CIRCUIT #1	CKT #2
1.6 volt	1.248 volts	0.344 volts
8 volt	1.275 volts	0.830 volts
40 volt	1.281 volts	1.156 volts

Table 5-4 Observed Readings

VOLTAGE SCALE:	CIRCUIT #1	CKT #2
1.6 volt	1.31 volts	0.34 volts
8 volt	1.38 volts	0.90 volts
40 volt	1.25 volts	1.20 volts

Table 5-5 Percent Error of Ideal vs Expected $(V_{r^2} - V_m)/V_{r^2}$

VOLTAGE SCALE:	CIRCUIT #1	CKT #2
1.6 volts	2.65%	73.15%
8 volts	.546%	35.27%
40 volts	.078%	9.82%

Table 5-6 Ideal vs Observed $(V_{r^2} - V_o)/V_{r^2}$

VOLTAGE SCALE:	CIRCUIT #1	CKT #2
1.6 volts	2.18%	73.48%
8 volt	7.64%	29.79%
40 volts	2.50%	6.40%

Table 5-7 Expected vs Observed $(V_m - V_o)/V_m$

VOLTAGE SCALE:	CIRCUIT #1	CKT #2
1.6 volts	4.97%	1.22%
8 volts	8.24%	8.45%
40 volts	2.48%	3.94%

error between the expected reading and the observed reading from equation (6).

Conclusions

It was immediately apparent in this experiment that the term %error was ambiguous unless strictly defined. A glance through the data sheets will show that the three types of error vary greatly and their meaning must be defined.

If a circuit under study must be measured, the device used for the measurement will introduce error simply by using some of the circuit's power for the measurement. This is called meter loading in the voltmeter case and is seen in this experiment as an error in the 70 percentile range between the circuit value without the meter and the new value with the meter. This loading is smaller if the meter resistance is high compared to the circuit element--as for instance when using the oscilloscope to measure a DC voltage on circuits similar to those used in this lab. The scope's input impedance of 10 megohms or more would be negligible to the operation of the voltage dividers. Consider the situation in the opposite extreme however; if the meter resistance were to be at or near zero--perhaps a homemade voltage indicator of some type--the loading would be extreme and the error introduced would be astronomical.

Clearly, if the effect of the meter is understood and allowed for in the computations then the difference between ideal voltage and expected voltage becomes zero since we are now assuming the load of the meter in our calculations and thus the ideal voltage is the loaded down output the meter should read. This becomes the expected voltage and error should now be noted between the observed voltage and the expected voltage.

If for instance, we calculate that the output voltage should be .3442 volts after correcting for the loading of the meter, and we then observe the reading to be .34 volts--as was the case for circuit #2 on 1.6 volt scale--then the error is quite small, only

2% in fact. This same circuit and measurement represents a 73% error if we consider that the ideal voltage across the resistor should be 1.282 volts without the load of the meter connected to it.

One must conclude from this that while meter loading may be a problem and should be avoided if possible, it does not in itself cause errors in measurement. When differences occur between the expected value and the observed value, such factors as faulty connections, visual parallax error, and faulty equipment should be considered. The severe loading of the circuit is, after all, just one more parameter which can be mathematically considered and corrected for.

The final proof of reliability--or precision if you want--is seen in the last table (Table 7) which shows the percent error between the expected and observed data to be very small indeed, even with the more severely loaded down circuit #2. Indeed, circuit #1 and #2 now seem about equal in error after the loading has been accounted for.

Equipment List:

Simpson 262 voltmeter SN #0A28337
(2) Heathkit decade resistance box model ln-3117 sn #17 & #42
Viz power supply model WP706
Fluke DVM model 801A

Writing Exercises

These four exercises are exploratory kinds of writing, so just imagine you are writing to someone interested in the topic. Adjust the length to the demands of the material.

1. Reproduce one of Michael Faraday's experiments and write up the results in a form appropriate to your discipline.

2. Describe one of the pieces of testing equipment used in your discipline.

3. Design an experiment to analyze the interrelationship between selected factors having to do with your friends' normal behavior. Select the factors according to some hypothesis you have about this behavior. For instance, describe the relationship among these factors: grade point average, total hours spent in dating practices (preparation, travel, working for the money to go out, the event itself, after-events, etc.), and average amount of spending money available each week and its source. Write up the results.

4. Interview one of your professors who is currently doing empirical or statistical research and write a description of that research (goal, method, results, etc.).

6

Traditional Science and Technology, II: Explanation, Analysis, and Persuasion

In Chapter 5 we introduced the traditional form of writing about science and technology and provided several examples of descriptive works. We continue the discussion here, illustrating the forms of explanation, analysis, and persuasion.

WHY OR HOW A PROCESS OR MECHANISM WORKS

Explaining the workings of a process or mechanism is an important function in the literature of science and technology. But what does it mean to explain something? In the next section, we will talk about explaining in terms of abstract concepts. In this section, we will limit our discussion to mechanisms and processes.

We explain something by describing the interrelationship between or among its parts. In the following excerpt, Sir Charles Galton Darwin (1887–1962), a British physicist and mathematician and the grandson of Charles Darwin, explains how an electron microscope works. Darwin wrote this description in the very early days of electron microscopy, and many of his promises about the potential of this kind of microscope have been realized.

Revealing the Invisible

Sir Charles Galton Darwin

I am going to tell something about a new instrument that promises to be very useful in all sorts of ways. This is the electron microscope, and its great virtue is that it can magnify things something like fifty times as much as an ordinary one. This makes it reveal a lot of things that were quite invisible before. For example, there were many diseases which were known to be due to microbes, but the microbes were too small to see in the best existing microscope, whereas now we can photograph them.

From Darwin, Sir. C. G. [1960] 1971. Revealing the invisible. Reprinted in *Classics in science,* edited by E. N. Da C. Andrade, 289–94. Washington, N.Y.: Kennikat Press.

When you think of a microscope you probably think of a man sitting at a table in front of the window squinting with one eye through a little brass tube. The electron microscope is nothing like that. It stands about 9 feet from the floor: the working part is in a column about a foot in diameter and there are various parts of its gear alongside so that altogether it covers about three feet square on the ground. The object you want to study is put on a very little thin sheet of celluloid in a small chamber about 6 feet from the ground, and at about knee height there are a number of small windows at the sides of the column that you can look down through to see the image formed on a green luminescent screen.

Now, as to how it is done. The essential things in this microscope, as in an ordinary microscope, are the lenses, but the lenses are quite different here. To make an enlarged image of anything the essential thing is to be able to bend the rays coming from it. If it is possible to bend the rays in any way at all, then by being ingenious in arranging how they are bent it is possible to make an enlarged image of some sort, and by very skilful designing to make a really good enlarged image. When the rays are rays of light the way this is done is by glass lenses, as in a telescope or a camera, and you will appreciate what I said about skilful designing when I tell you that it could easily take a team of highly trained men a year to work out the shapes of the little lenses—often eight of them one after the other—that go to make up a new pattern of high-class microscope. But in the electron microscopic we are not using light, and so the lenses have to be quite different. We are making a beam of electrons, which are the ultimate stuff of electricity, tiny particles which go to make up a great part of matter. They were discovered by J. J. Thomson, and it is only because they are working in the valves of your wireless sets that you can listen to broadcast sound. The electrons are given off by a heated wire. In a radio valve they go comparatively slowly but in the microscope they are speeded up to about a quarter of the speed of light, which means that if you could make a pipe right round the earth and shoot them through it, they would come round and hit you in the back in about half a second. At this high speed the electrons can go right through a very thin sheet of celluloid but will be stopped or scattered by any rather thicker object such as a microbe that you put on the celluloid. Here we have a set of rays which can give a picture of the object if we can bend them. Electrons can be bent either by electric fields or by magnetism, and so we have either electric or magnetic lenses. Both work, but on the whole the magnetic ones have been more used. The lens consists of a circular coil of wire in a specially designed iron frame. It carries electric current, and this curls up the tracks of the electrons in a rather complicated way that has just the same property as a glass lens has for light, of bringing them to a focus.

We have now got the two requisites for a microscope, light (only here it is not light) and lenses, and we must put them together to make the microscope. The electrons start from a heated wire at the top of the microscope column and are speeded up so as to travel downwards. There are three lenses in the microscope. The first is called the condenser, and it concentrates the beam of electrons on the object you are examining. In an ordinary microscope there has to be a condenser lens too—though people often forget about it—to concentrate the light on to the microscope slide; otherwise there would not be enough light to see by. The electrons now come to the object, say a microbe on a thin sheet of celluloid. The ones that hit the microbe are stopped, but the rest go on. Next, just below, there is another magnetic lens which could produce an image about 18 inches below, magnified about 100 times. For the small things you want to look at this is still much too small, so you do not form the image there but instead put another lens below which magnifies it again 100 times. So the final image which is looked at or photographed is about ten thousand times magnified, and as this is often still a bit small, you enlarge the photograph you have taken perhaps five or ten times more by ordinary photography. There is one peculiarity in the image of a magnetic lens that I might mention. In a camera or, for the matter of that, in your own eyes too, the image is exactly upside-down, and so it is in some telescopes, though in others a lens is put in specially to make it right way up, but with a magnetic lens it is twisted round so as to point in some other direction. You have to focus of course, which you do by altering the current in the lens, and as you turn the handle to do this you see the image slowly twisting round as it gets focused.

Now why is this new elaborate gear better than the old microscope? The answer is that in the old microscope, though you can magnify the image indefinitely you gain nothing by it beyond a certain point, as you see only a large blurred image instead of a small blurred image. There is a definite limitation to the size of the things you can see, which is the wave-length of the light you use. The wave-length of light is about a fifty-thousandth of an inch, and if you try to magnify up anything as small as this you cannot help losing all the sharp edges and you get only indefinite blobs. By very hard work, using ultra-violet light, something rather better can be done, but, roughly speaking, it is not worth magnifying more than about four thousand times. There is a similar theoretical limitation for electrons, but it only comes in for sizes many thousands of times smaller, and if it were the only limitation, we could hope to be just about able to see individual atoms. However, there is another limitation which is not so fundamental, but which threatens to be practically more serious, and this is that it seems unlikely that anyone will succeed in making a good enough lens to go beyond about a hundred times smaller than an ordinary microscope can do. At present most of the successful photographs are magnified not much beyond twenty thousand, though I have seen some very good ones at a hundred thousand.

As to what the microscope can be used for, there is the trouble that the object has to be in a vacuum, and so it is much easier to work with dry things. Particles of smoke of various kinds can be seen; some smokes are stringy, some are in little cubes and some in needles. Another trick is to study the roughnesses of an apparently smooth surface of metal or anything else, by coating it with a thin layer of resin which you afterwards peel off and put in the microscope. But probably the chief interest is in microbes and such things. Many of these can be seen with an ordinary microscope, but we now know that some of them, that looked like a blob, really had a swimming tail. There were many diseases known to be caused by microbes which were too small to be seen at all, but now we can see them. I have seen one beautiful photograph made in Germany, which shows the mysterious thing called bacteriophage, a beast which attacks and kills bacteria. You can see a large black object, the bacteria, and a crowd of things like tadpoles round it swimming towards it and attacking it. Without the electron microscope they were simply not seen at all. We have also ourselves made photographs of those other mysterious things called viruses which cause certain kinds of diseases. Altogether I think we can be pretty sure that in quite a few years a great deal more will be known about disease.

I expect you want to know who invented this remarkable instrument, but I have got to disappoint you. It is the thing everyone asks about any new invention, and it is nearly always impossible to answer. Inventions are hardly ever like that. It does not matter whether it is the telegraph or photography or wireless or radiolocation, you cannot say who was the inventor, because it is a gradual process of one man seeing something but not how to use it, another pushing it on a little way, and so on. Of all the really big inventions of the last hundred years there is only one where I should care definitely to name the undisputed discoverer, and that one is Röntgen's discovery of X-rays, and they were found more by luck than by judgment. The electron microscope is not in that class. The theory behind it was quite well known to physicists a long time ago, and the main question was to see that there could be practical results. I remember myself not so many years ago hearing about the business and thinking it was a pretty game but that nothing would come of it. Perhaps the earliest practical work was done by a number of Germans ten years ago or so, but quite a number of other workers were on to it, and laboratory instruments had been made and used in several countries both here and abroad. Now the thing is being manufactured and can be bought in America and we are fortunate in having half a dozen of their instruments in this country. They are being used for a great variety of purposes, and whatever may come of the work there can be no doubt that we shall within a very short time know a great deal more about what goes on in the world at the size of a millionth of an inch.

Harold M. Schmeck, Jr., a science writer for the *New York Times,* shows how parts interrelate by describing how the brain can affect the immune system in the body's fight against disease.

By Training The Brain, Scientists Find Links To Immune Defenses

Harold M. Schmeck Jr.

In the classic experiments of the Russian physiologist Ivan Pavlov, dogs salivated when a bell rang because they had been trained so that their brains associated the sound with the presence of food.

Now, American scientists have evoked a similar conditioned reflex to show that the brain can exercise direct control over cells of the immune defense system, the body's main bulwark against disease. [See Figure 6–1].

Other recent experiments here and abroad have shown that the two hemispheres of the brain influence the immune defenses in different ways and that some brain chemicals have specific effects on immune cells.

It has long been known that the body's two most important windows on the outside world are the brain and immune defense system.

Everything we hear, feel, see or imagine comes through the brain. And almost every virus, microbe or other foreign particle that invades the body triggers some kind of immune response. But the specific effects these two master systems exert on each other have been relatively little explored, partly because the interactions are so complex.

Some ingeniously designed research is now beginning to reveal details of the brain's effects on the immune defenses. The findings have raised hopes that eventually the research may lead to better treatments for disorders in which the immunological defense system is deficient or active in some abnormal way.

One series of experiments has shown that the activity of certain immune defense cells called natural killer cells can be greatly enhanced by the brain's trained response to a totally extraneous stimulus from the outside world—a strong odor. The killer cells are part of the body's surveillance system that protects against invasion and probably against cancer.

The research was designed by Dr. Novera Herbert Spector of the National Institute of Neurological and Communicative Disorders and Stroke, a unit of the National Institutes of Health. The experiments were done mainly at the University of Alabama medical school in Birmingham by Brent Solvason, Dr. Vithal Ghánta and Dr. Raymond Hirahito.

Mice were exposed for three hours at a time to the odor of camphor. The scientists showed that exposure to this odor, by itself, had no detectable effect on the immune system. But in the experiments, some of the mice were also given injections of a synthetic chemical called poly I:C (for polyinosinic-polycytidilic acid), which is known to enhance the activity of natural killer cells. The exposures were repeated nine times in a strategy similar to that of the Pavlovian conditioning in which dogs were given food every time a bell rang. In each session of the immunity experiments, the mice were exposed to the odor and given injections of the chemical.

Then, in the 10th session, the mice were exposed only to the odor of camphor. They received no injections at all. Nevertheless, every mouse showed a large increase in natural killer cell activity.

The effect, Dr. Spector said, was comparable to that of Pavlov's experiments in which animals could be made to salivate simply at the ringing of a bell. In the new case, the animals' brains evidently activated the immune defense without waiting for the poly I:C, just as the dogs had begun to salivate in the Pavlovian experiments without waiting for the food to appear.

In the new experiments, several different control groups of mice were given treatments that were similar but not identical to the injections and exposure to odor that were the crux of the research. This was done to make sure that the effect was really what it seemed to be and was not a result of some unforeseen aspect of the experiments. For example, some

---

From Schmeck, H. M., Jr. 1 January 1985. By training the brain, scientists find links to immune defenses. *New York Times* I, 13.

FIGURE 6-1
Interactions between the nervous system and the immune system

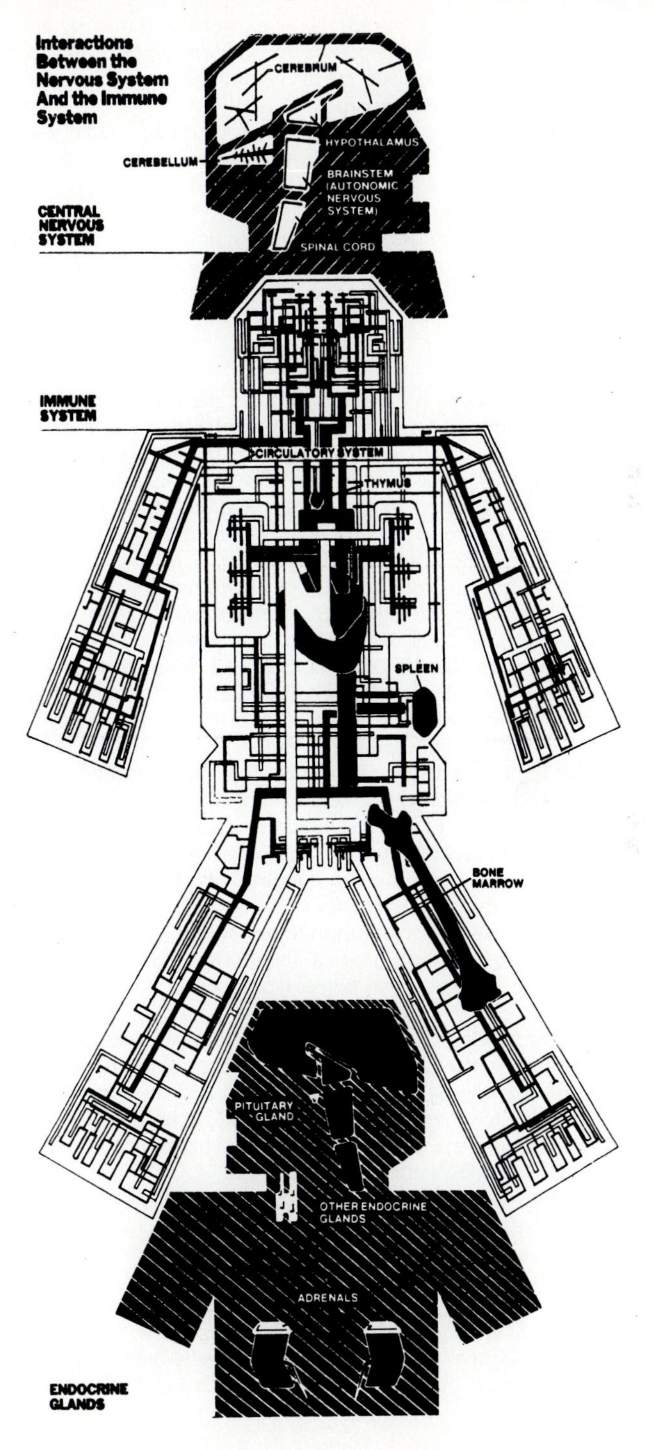

animals were exposed to the odor of camphor and given the injections in each of nine sessions. Then, in the 10th session, they were not exposed to the odor at all, but were given injections of harmless salt water.

When the two groups were compared, the scientists found that the animals in the experimental group—those given only the exposure to odor of camphor in the 10th session—had natural killer cell activity three times as great as the animals in the "control" group that lacked the exposure to the odor in the final session.

This threefold difference was strongly significant, Dr. Spector said. Furthermore, the animals that had been "conditioned" to the odor had 39 times as much activity of natural killer cells as another control group that was exposed to the odor of camphor and given injections of salt solution in each of 10 sessions.

The research was reported in December at the National Institutes of Health in Bethesda, Md., at the First International Workshop on Neuroimmunomodulation, a word coined by Dr. Spector to embrace studies of the links between the brain and the immune defense system.

Effects of Left-Handedness

Another report at the workshop demonstrated that the two hemispheres of the brain have different effects on the immune defenses, a circumstance that had been suspected since scientists such as the late Dr. Norman Geshwind of Harvard University reported that left-handed people were more likely than right-handers to suffer from disorders of the immune system.

The new experiments, reported by Dr. Gerard Renoux of the Medical School of Tours in France, showed that the destruction of parts of the brains of mice had different consequences, depending on which side of the brain was affected.

The experiments were done with a breed of mice known to lack any strong dominance of one brain hemisphere over the other. Nevertheless, when a large portion of the front surface portion of the animals' left hemisphere was removed, the number and activity of certain immune defense cells called T cells was reduced. When comparable surgery was done on the right hemisphere instead, no difference in T cell numbers was found, but the activity of T cells appeared to be increased, Dr. Renoux reported at the meeting.

In laboratory experiments, Dr. Michael R. Ruff and Dr. Candace Pert of the health institutes discovered that important brain chemicals including beta endorphin, enkephalin and dynorphin tend to attract the scavenger cells called macrophages. These are cells that the immune defense system sends to injured tissue to help heal wounds. Why these cells of the immune system should react to chemicals that are thought to be part of the brain's means of dealing with pain is not known. But the circumstance hints at previously unknown links between the brain and the immune defenses.

Earlier research demonstrated that lymphocytes, which are immunologically active white blood cells, have receptors that fit some of the brain chemicals of the neuroactive peptides class. There is even evidence that some lymphocytes actually produce some of these chemicals.

These findings offer specific clues to a generality that has been obvious to scientists for a long time: That the brain is not only the organ that controls behavior, but is ultimately the monitor and governor of every aspect of body function and chemistry. For example, the brain and nervous system influence the digestive system and orchestrate the complex chemistry of the endocrine glands. In turn, the functioning of these vital systems can have profound effects on behavior—necessarily through effects on the brain.

The new discoveries suggest many specific interactions of a similar kind between the brain and immune defenses. The brain's messages to the several hormone-producing glands and the circulatory system affect the immune system; in turn, activity of the immune defenses, directly or indirectly, influences the brain.

And the sheer power of the effects the brain on the body as a whole is amazing. From the dawn of history, medicine and folklore have been replete with illustrative anecdotes, even though the means by which these effects work have usually defied explanation.

Scientists have confirmed that a witch doctor can cause death by convincing the victim that he or she is going to die. Many studies have shown that some patients will stop feeling pain after a doctor has administered an inert drug called a placebo, provided the patients believe they have received powerful painkillers.

Science Versus Shamanism

One of the most dramatic cases of such mind control over health was reported a few years ago in The Journal of the American Medical Association. It was the thoroughly documented case of a 28-year-old woman from the Philippines who had systemic lupus erythematosus, a serious immunological disease. Her disorder was diagnosed at a hospital in Longview, Wash., after two clinics documented a spectrum of troubles that go with the disease, including weakness, skin rash, anemia and disorders of the liver and lymphatic system.

Treatment with the hormone drug prednisone helped for several weeks, but other complications developed, including kidney and thyroid problems. Other drugs were used to combat this situation, but more complications arose that called for still other drugs and for evaluation at other medical centers.

Finally, in the face of yet another recommended drug treatment, "the patient instead elected to return to the remote Philippine village of her birth," her physician reported. Medical tests, done the day she left, confirmed what the doctors already knew about her disease.

"Much to the surprise of distraught family members and skeptical physicians," the report continued, "the patient returned three weeks later." She said the village witch doctor had removed a curse placed on her by a previous suitor.

The woman was in outwardly good health and stayed that way. Two years later she gave birth to a healthy girl. "Even now," said the report, "she insists that her lupus was cured by removal of the 'evil spirit.' "

What internal mechanisms did the witch doctor's suggestions tap, asked the reporting doctor, and how did he do it?

"The answers to these perfectly reasonable questions are not yet available, but the type of research needed to provide them is clear," Dr. Spector said in a report summarizing the state of research in the puzzling but important field that today is exploring the brain's powerful influences over the immune defenses.

Writing Exercise

Following the examples provided by Darwin and Schmeck, explain the workings of a process or a mechanism by describing the interrelations of the parts and how those interrelationships make up the whole of the process or mechanism.

Imagine that your explanation is a section of a longer work that either (1) shows how to manipulate the process for human ends or (2) describes how to fix or maintain the mechanism.

Your instructor will assign an appropriate length.

EXTENDED DEFINITION OF AN ABSTRACT CONCEPT

Scientists and engineers often must define abstract concepts, usually those describing basic laws and principles of mathematics, biology, physics, chemistry, and engineering. Such concepts are often called scientific laws, such as the law of gravity or the laws of thermodynamics. These laws are mental constructs: verbal descriptions of forces or truisms that cannot be observed empirically but do predict or account for the behavior of natural phenomena.

Writing a clear definition of a concept requires intellectual facility and rhetorical and grammatical precision. Have you ever tried, for instance, to explain to an inquisitive seven-year-old how "pressure" (an abstract concept) enables the water in your house to get from the water heater in the basement to the shower on the second floor, without the use of a water pump? If you have, you know that definition often involves explanation: you use the definition of the abstract concept "pressure" to explain the physical process.

The Difficulty of Writing Definitions: "Gravity"

As an example of the difficulty of writing definitions, let's look at the definition of "gravity" in the *McGraw-Hill Dictionary of Scientific and Technical Terms:*

gravity [MECH] The gravitational attraction at the surface of a planet or other celestial body.

This definition is circular: a linguistic relative ("gravitational") of the term-to-be-defined ("gravity") is used to define the term. We learn from this definition that "gravity" is the word to use when talking about gravitation as it affects objects on the surface of a planet. The next step, then, is to find the definition of "gravitation":

gravitation [PHYS] The mutual attraction between all masses in the universe. Also known as gravitational attraction.

The dictionary doesn't give the definition of a physical thing, as it would have done had we looked up "valve." Instead, it defines "gravitation" as a *relationship that exists between two or more objects,* not as a wave or particle or in any other terms of physics. Certainly, we are dealing here with *an idea about an interrelationship.*

Understanding Related Terms

We then take the next step of looking around in the dictionary for supplementary material. Two items are relevant:

gravitational constant [MECH] The constant of proportionality in Newton's law of gravitation, equal to the gravitational force between any two particles times the square of the distance between them, divided by the product of their masses. Also known as constant of gravitation.

and

Newton's Law of Gravitation [MECH] The law that every two particles of matter in the universe attract each other with a force that acts along the line joining them, and has a mag-

---

Definitions from *McGraw-Hill dictionary of scientific and technical terms.* 1984. 3d ed. Edited by Sybil P. Parker. New York: McGraw-Hill.

nitude proportional to the product of their masses and inversely proportional to the square of the distance between them. Also known as the law of gravitation.

If we then extend our search to the *McGraw-Hill Encyclopedia of Science and Technology,* we find a formula for determining the force of gravity:

Put into symbols, the gravitational force F exerted between two particles with masses m_1 and m_2 separated by a distance d is given by Eq. (1), where G is called the constant of gravitation.

$$F = Gm_1m_2/d^2 \tag{1}$$

Definition and Explanation: Making Connections from the Familiar to the Unfamiliar

It is difficult to explain concepts and laws because they are abstractions (and thus cannot be easily described in tactile terms) and because they assert complex relationships. Such assertion of relationships is what we call an explanation. "Explanation" comes from the Latin "explanus," meaning to spread something out flat (and thus to enable it to be seen plainly). It requires you to make a connection between the unfamiliar—the concept or process to be explained—and the familiar—the knowledge that the audience already has. If, for instance, you wanted to explain how water gets to your third-floor shower without the use of a water pump, you would have to connect the unfamiliar idea of water going up to the familiar idea of water going down as it follows the force of gravity (another abstract concept, since "gravity" itself has never been observed, only its supposed effects).

Explaining something means placing a new concept (something that seems random, unrelated, or confusing) into a familiar or older context, thereby expanding that context. This gradual expansion of contexts constitutes learning. When you explain something, you integrate previously unrelated events into a larger pattern. And this pattern, which can then be applied fruitfully to other events, becomes a type of practical definition.

Metaphors: A Common Device for Definition

To build bridges for explanation and definition, scientists and engineers often use metaphors. "Metaphor" comes from two Greek roots ("meta" + "pherein"), meaning to carry across. Metaphors carry meaning across the communication bridge from the familiar to the unfamiliar, thus building a link in the chain of new knowledge.

Jacob Bronowski succinctly describes the importance of metaphor, or analogical thinking.

From *McGraw-Hill encyclopedia of science and technology*. 1987. 6th ed. New York: McGraw-Hill.

Science as Foresight

Jacob Bronowski

Man has only one means to discovery, and that is to find *likenesses* between things. To him, two trees are like two shouts and like two parents, and on this likeness he has built all mathematics. A lizard is like a bat and like a man, and on such likenesses he has built the theory of evolution and all biology. A gas behaves like a jostle of billiard balls, and on this and kindred likenesses rests much of our atomic picture of matter.

In looking for intelligibility in the world, we look for unity; and we find this (in the arts as well as in science) in its unexpected likenesses. This indeed is man's creative gift, to find or make a likeness where none was seen before—a likeness between mass and energy, a link between time and space, an echo of all our fears in the passion of Othello.

So, when we say that we can explain a process, we mean that we have mapped it in the likeness of another process which we know to work. We say that a metal crystal stretches because its layers slide over one another like cards in a pack, and then that some polyester yarns stretch and harden like a metal crystal. That is, we take from the world round us a few models of structure and process (the particle, the wave, and so on), and when we research into nature, we try to fit her with these models.

Yet one powerful procedure in research, we know, is to break down complex events into simpler parts. Are we not looking for the understanding of nature in these? When we probe below the surface of things, are we not trying, step by step, to reach her ultimate and fundamental constituents?

We do indeed find it helpful to work piecemeal. We take a sequence of events or an assembly to pieces: we look for the steps in a chemical reaction, we carve up the study of an animal into organs and cells and smaller units within a cell. This is our atomic approach, which tries always to see in the variety of nature different assemblies from a few basic units. Our search is for simplicity, in that the distinct units shall be few, and all units of one kind identical.

And what distinguishes one assembly of these units from another? the elephant from the giraffe, or the right-handed molecule of sugar from the left-handed? The difference is in the organization of the units into the whole; the difference is in the structure. And the likenesses for which we look are also likenesses of structure.

This is the true purpose of the analytic method in science: to shift our gaze from the thing or event to its structure. We understand a process, we explain it, when we lay bare in it a structure which is like one we have met elsewhere.

The use of metaphors in science and technology is common. Among other things, mathematics is a metaphoric language for expressing relationships or functions between or among complexes of numbers. Mathematical symbols and notation systems express these relationships more concisely and precisely than words could. (Have you ever tried, using only words, to explain a complex function in calculus?)

Geologists use the term "plate tectonics" to describe the movement of surface units on the earth because the idea of a plate is useful for visualizing these sections. Similarly, electrical engineers use two common metaphors when they say that electricity "flows" along a "circuit." First they compare the movement of electricity to the flow of water, and then they assert that this current flows along a circuit, a pathway for a prescribed journey or race, the traversing of which requires returning to the starting point.

Metaphors abound in the disciplines of physics and astronomy. Physicists often use metaphors when the concepts they are trying to explain seem unimaginable in the realm of common sense. For instance, in describing Einstein's concept of the general theory of relativity (which extends his special theory of relativity to include gravity), Fritjof Capra says, "To understand the meaning of curved space-time, we have to use curved two-dimensional surfaces as analogies." Capra's explanation is masterful.

The Tao of Physics

Fritjof Capra

The theory of relativity discussed so far is known as the "special theory of relativity." It provides a common framework for the description of the phenomena associated with moving bodies and with electricity and magnetism, the basic features of this framework being the relativity of space and time and their unification into four-dimensional space-time.

In the "general theory of relativity," the framework of the special theory is extended to include gravity. The effect of gravity, according to general relativity, is to make space-time curved. This, again, is extremely hard to imagine. We can easily imagine a two-dimensional curved surface, such as the surface of an egg, because we can see such curved surfaces lying in three-dimensional space. The meaning of the word curvature for two-dimensional curved surfaces is thus quite clear; but when it comes to three-dimensional space—let alone four-dimensional space-time—our imagination abandons us. Since we cannot look at three-dimensional space "from outside," we cannot imagine how it can be "bent in some direction."

To understand the meaning of curved space-time, we have to use curved two-dimensional surfaces as analogies. Imagine for example, the surface of a sphere. The crucial fact which makes the analogy to space-time possible is that the curvature is an intrinsic property of that surface and can be measured without going into three-dimensional space. A two-dimensional insect confined to the surface of the sphere and unable to experience three-dimensional

FIGURE 6-2
Drawing a "straight line" on a plane and on a sphere

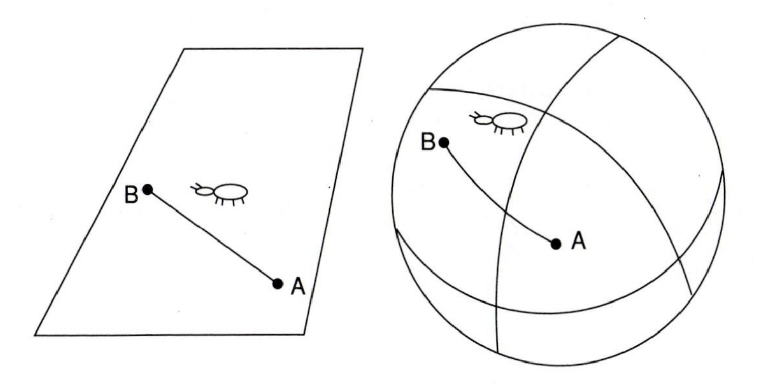

---

space could nevertheless find out that the surface on which he is living is curved, provided that he can make geometrical measurements.

FIGURE 6-3
On a sphere a triangle can have three right angles

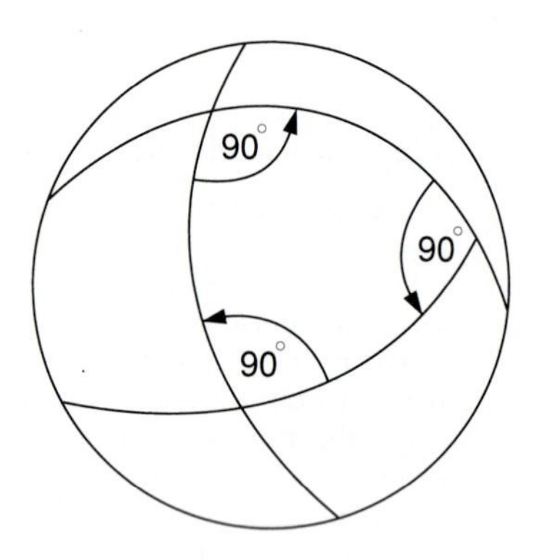

To see how this works, we have to compare the geometry of our bug on the sphere with that of a similar insect living on a flat surface. Suppose the two bugs begin their study of geometry by drawing a straight line, defined as the shortest connection between two points. The result is shown in Figure 6-2. We see that the bug on the flat surface drew a very nice straight line; but what did the bug on the sphere do? For him, the line he drew is the shortest connection between the two points A and B, since any other line he may draw will be longer; but from our point of view we recognize it as a curve (the arc of a great circle, to be precise). Now suppose that the two bugs study triangles. The bug on the plane will find that the three angles of any triangle add up to two right angles, i.e. to 180° [see Figure 6-3]; but the bug on the sphere will discover that the sum of the angles in his triangles is always greater than 180°. For small triangles the excess is small, but it increases as the triangles become larger; and as an extreme case, our bug on the sphere will even be able to draw triangles with three right angles. Finally, let the two bugs draw circles and measure their circumference. The bug on the plane will find that the circumference is always equal to 2π times the radius, independent

FIGURE 6-4
Drawing a circle on a sphere

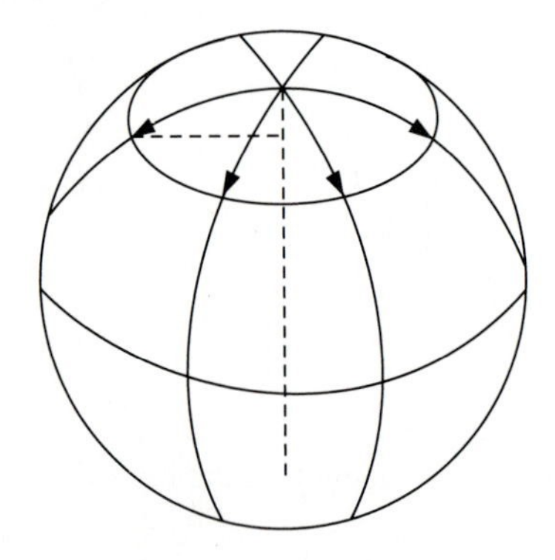

of the size of the circle. The bug on the sphere, on the other hand, will notice that the circumference is always less than 2π times the radius. As can be seen in [Figure 6-4] our three-dimensional point of view allows us to see that what the bug calls the radius of his circle is in fact a curve which is always longer than the true radius of the circle.

As the two insects continue to study geometry, the one on the plane should discover the axioms and laws of Euclidean geometry, but his colleague on the sphere will discover different laws. The difference will be small for small geometrical figures but will increase as the figures become larger. The example of the two bugs shows that we can always determine whether a surface is curved or not, just by making geometrical measurements on the surface, and by comparing the results with those predicted by Euclidean geometry. If there is a discrepancy, the surface is curved; and the larger the discrepancy is—for a given size of figures—the stronger the curvature.

Scientists also use metaphors to describe new concepts in astrophysics and astronomy. Many astrophysicists believe that entire galaxies may reside on the surface of enormous areas of empty space that are shaped like bubbles, and that these galaxies may ride on the surface of these bubbles like islands on the sea. Of course, there aren't really bubbles in space, but thinking as if there were enables us to get a feel for what space and its contents may be like.

Scientists use metaphors to make the unimaginable coherent, to make the unfamiliar familiar. A classic example of metaphors being used to define and explain abstract concepts is the following selection on the concept of relativity, by Bertrand Russell.

Russell (1872–1970) was an English philosopher, logician, Renaissance man-of-letters, and tireless campaigner against war and nuclear armaments. Among his many publications are *Principia Mathematica,* coauthored with A. N. Whitehead (1910–1913), in which he attempted "to derive the principles of mathematics from self-evident logical principles," and *The ABC of Relativity* (1925), a nontechnical introduction to Einsteinean physics, from which the following selection is taken.

To enable a reader to begin to understand the difficult, abstract concept of relativity as defined by Albert Einstein, Russell uses several metaphors: human perceiver as a drugged balloonist, an electron, or a sun; reality as a swarm of bees, an avalanche, and a set of clouds; and two billiard balls hitting each other as being like a comet passing through a solar system. Russell uses these metaphors to explain how differently you have to think in order to begin to understand the idea of relativity.

Touch and Sight: The Earth and the Heavens

Bertrand Russell

Everybody knows that Einstein did something astonishing, but very few people know exactly what it was that he did. It is generally recognized that he revolutionized our conception of the physical world, but the new conceptions are wrapped up in mathematical technicalities. It is true that there are innumerable popular accounts of the theory of relativity, but they generally cease to be intelligible just at the point where they begin to say something important. The authors are hardly to blame for this. Many of the new ideas can be expressed

---

From Russell, B. 1925. *The ABC of relativity.* New York: Harper & Row, 9–15.

in non-mathematical language, but they are none the less difficult on that account. What is demanded is a change in our imaginative picture of the world—a picture which has been handed down from remote, perhaps pre-human, ancestors, and has been learned by each one of us in early childhood. A change in our imagination is always difficult, especially when we are no longer young. The same sort of change was demanded by Copernicus, when he taught that the earth is not stationary and the heavens do not revolve about it once a day. To us now there is no difficulty in this idea, because we learned it before our mental habits had become fixed. Einstein's ideas, similarly, will seem easier to generations which grow up with them; but for us a certain effort of imaginative reconstruction is unavoidable.

In exploring the surface of the earth, we make use of all our senses, more particularly of the senses of touch and sight. In measuring lengths, parts of the human body are employed in pre-scientific ages: a 'foot,' a 'cubit,' a 'span' are defined in this way. For longer distances, we think of the time it takes to walk from one place to another. We gradually learn to judge distance roughly by the eye, but we rely upon touch for accuracy. Moreover it is touch that gives us our sense of "reality." Some things cannot be touched: rainbows, reflections in looking-glasses, and so on. These things puzzle children, whose metaphysical speculations are arrested by the information that what is in the looking-glass is not "real." Macbeth's dagger was unreal because it was not "sensible to feeling as to sight." Not only our geometry and physics, but our whole conception of what exists outside us, is based upon the sense of touch. We carry this even into our metaphors: a good speech is "solid," a bad speech is "gas," because we feel that a gas is not quite "real."

In studying the heavens, we are debarred from all senses except sight. We cannot touch the sun, or travel to it; we cannot yet walk round the moon, or apply a foot-rule to the Pleiades. Nevertheless, astronomers have unhesitatingly applied the geometry and physics which they found serviceable on the surface of the earth, and which they had based upon touch and travel. In doing so, they brought down trouble on their heads, which it was left for Einstein to clear up. It turned out that much of what we learned from the sense of touch was unscientific prejudice, which must be rejected if we are to have a true picture of the world.

An illustration may help us to understand how much is impossible to the astronomer as compared with the man who is interested in things on the surface of the earth. Let us suppose that a drug is administered to you which makes you temporarily unconscious, and that when you wake you have lost your memory but not your reasoning powers. Let us suppose further that while you were unconscious you were carried into a balloon, which, when you come to, is sailing with the wind on a dark night—the night of the fifth of November if you are in England, or of the fourth of July if you are in America. You can see fireworks which are being sent off from the ground, from trains, and from aeroplanes travelling in all directions, but you cannot see the ground or the trains or the aeroplanes because of the darkness. What sort of picture of the world will you form? You will think that nothing is permanent: there are only brief flashes of light, which, during their short existence, travel through the void in the most various and bizarre curves. You cannot touch these flashes of light, you can only see them. Obviously your geometry and your physics and your metaphysics will be quite different from those of ordinary mortals. If an ordinary mortal were with you in the balloon, you would find his speech unintelligible. But if Einstein were with you, you would understand him more easily than the ordinary mortal would, because you would be free from a host of preconceptions which prevent most people from understanding him.

The theory of relativity depends, to a considerable extent, upon getting rid of notions which are useful in ordinary life but not to our drugged balloonist. Circumstances on the surface of the earth, for various more or less accidental reasons, suggest conceptions which turn out to be inaccurate, although they have come to seem like necessities of thought. The most important of these circumstances is that most objects on the earth's surface are fairly persistent and nearly stationary from a terrestrial point of view. If this were not the case, the idea of going on a journey would not seem so definite as it does. If you want to travel from King's Cross to Edinburgh, you know that you will find King's Cross where it has always been, that the railway line will take the course that it did when you last made the journey, and that

Waverley Station in Edinburgh will not have walked up to the Castle. You therefore say and think that you have travelled to Edinburgh, not that Edinburgh has travelled to you, though the latter statement would be just as accurate. The success of this common-sense point of view depends upon a number of things which are really of the nature of luck. Suppose all the houses in London were perpetually moving about, like a swarm of bees; suppose railways moved and changed their shapes like avalanches; and finally suppose that material objects were perpetually being formed and dissolved like clouds. There is nothing impossible in these suppositions. But obviously what we call a journey to Edinburgh would have no meaning in such a world. You would begin, no doubt, by asking the taxi-driver: "Where is King's Cross this morning?" At the station you would have to ask a similar question about Edinburgh, but the booking-office clerk would reply: "What part of Edinburgh do you mean, sir? Prince's Street has gone to Glasgow, the Castle has moved up into the Highlands, and Waverley Station is under water in the middle of the Firth of Forth." And on the journey the stations would not be staying quiet, but some would be travelling north, some south, some east or west, perhaps much faster than the train. Under these conditions you could not say where you were at any moment. Indeed the whole notion that one is always in some definite "place" is due to the fortunate immobility of most of the large objects on the earth's surface. The idea of "place" is only a rough practical approximation: there is nothing logically necessary about it, and it cannot be made precise.

If we were not much larger than an electron, we should not have this impression of stability, which is only due to the grossness of our senses. King's Cross, which to us looks solid, would be too vast to be conceived except by a few eccentric mathematicians. The bits of it that we could see would consist of little tiny points of matter, never coming into contact with each other, but perpetually whizzing round each other in an inconceivably rapid ballet-dance. The world of our experience would be quite as mad as the one in which the different parts of Edinburgh go for walks in different directions. If—to take the opposite extreme—you were as large as the sun and lived as long, with a corresponding slowness of perception, you would again find a higgledy-piggledy universe without permanence—stars and planets would come and go like morning mists, and nothing would remain in a fixed position relatively to anything else. The notion of comparative stability which forms part of our ordinary outlook is thus due to the fact that we are about the size we are, and live on a planet of which the surface is not very hot. If this were not the case, we should not find pre-relativity physics intellectually satisfying. Indeed we should never have invented such theories. We should have had to arrive at relativity physics at one bound, or remain ignorant of scientific laws. It is fortunate for use that we were not faced with this alternative, since it is almost inconceivable that one man could have done the work of Euclid, Galileo, Newton and Einstein. Yet without such an incredible genius physics could hardly have been discovered in a world where the universal flux was obvious to non-scientific observation.

In astronomy, although the sun, moon, and stars continue to exist year after year, yet in other respects the world we have to deal with is very different from that of everyday life. As already observed, we depend exclusively on sight: the heavenly bodies cannot be touched, heard, smelt or tasted. Everything in the heavens is moving relatively to everything else. The earth is going round the sun, the sun is moving, very much faster than an express train, towards a point in the constellation Hercules, the "fixed" stars are scurrying hither and thither like a lot of frightened hens. There are no well-marked places in the sky, like King's Cross and Edinburgh. When you travel from place to place on the earth, you say the train moves and not the stations, because the stations preserve their topographical relations to each other and the surrounding country. But in astronomy it is arbitrary which you call the train and which the station: the question is to be decided purely by convenience and as a matter of convention.

In this respect, it is interesting to contrast Einstein and Copernicus. Before Copernicus, people thought that the earth stood still and the heavens revolved about it once a day. Copernicus taught that "really" the earth rotates once a day, and the daily revolution of sun and stars is only "apparent." Galileo and Newton endorsed this view, and many things were thought to prove it—for example, the flattening of the earth at the poles, and the fact that bodies are heavier there than at the equator. But in the modern theory the question between Copernicus and his predecessors is merely one of convenience; all motion is relative, and there is no difference between the two statements: "the earth rotates once a day" and "the heavens revolve about the earth once a day." The two mean exactly the same thing, just as it means the same thing if I say that a certain length is six feet or two yards. Astronomy is easier if we take the sun as fixed than if we take the earth, just as accounts are easier in decimal coinage. But to say more for Copernicus is to assume absolute motion, which is a fiction. All motion is relative, and it is a mere convention to take one body as at rest. All such conventions are equally legitimate, though not all are equally convenient.

There is another matter of great importance, in which astronomy differs from terrestrial physics because of its exclusive dependence upon sight. Both popular thought and old-fashioned physics used the notion of "force," which seemed intelligible because it was associated with familiar sensations. When we are walking, we have sensations connected with our muscles which we do not have when we are sitting still. In the days before mechanical traction, although people could travel by sitting in their carriages, they could see the horses exerting themselves, and evidently putting out "force" in the same way as human beings do. Everybody knew from experience what it is to push or pull, or to be pushed or pulled. These very familiar facts made "force" seem a natural basis for dynamics. But Newton's law of gravitation introduced a difficulty. The force between two billiard balls appeared intelligible because we know what it feels like to bump into another person; but the force between the earth and the sun, which are ninety-three million miles apart, was mysterious. Newton himself regarded this "action at a distance" as impossible, and believed that there was some hitherto undiscovered mechanism by which the sun's influence was transmitted to the planets. However, no such mechanism was discovered, and gravitation remained a puzzle. The fact is that the whole conception of "force" is a mistake. The sun does not exert any force on the planets; in Einstein's law of gravitation, the planet only pays attention to what it finds in its own neighbourhood. The way in which this works will be explained in a later chapter; for the present we are only concerned with the necessity of abandoning the notion of "force," which was due to misleading conceptions derived from the sense of touch.

As physics has advanced, it has appeared more and more that sight is less misleading than touch as a source of fundamental notions about matter. The apparent simplicity in the collision of billiard balls is quite illusory. As a matter of fact the two billiard balls never touch at all; what really happens is inconceivably complicated, but is more analogous to what happens when a comet penetrates the solar system and goes away again than to what common sense supposes to happen.

Most of what we have said hitherto was already recognized by physicists before Einstein invented the theory of relativity. "Force" was known to be merely a mathematical fiction, and it was generally held that motion is a merely relative phenomenon—that is to say, when two bodies are changing their relative position, we cannot say that one is moving while the other is at rest, since the occurrence is merely a change in their relation to each other. But a great labour was required in order to bring the actual procedure of physics into harmony with these new convictions. Newton believed in force and in absolute space and time; he embodied these beliefs in his technical methods, and his methods remained those of later physicists. Einstein invented a new technique, free from Newton's assumptions. But in order to do so he had to change fundamentally the old ideas of space and time, which had been unchallenged from time immemorial. This is what makes both the difficulty and the interest of his theory. . . .

Richard Dawkins teaches evolutionary biology at Oxford University. He is the author of *The Selfish Gene* and *The Blind Watchmaker,* from which the following selection is taken. Dawkins explains the arrangement of genetic structure and the genetic storage of information by comparing them to storing information on a computer disk.

The Blind Watchmaker

Richard Dawkins

. . . This kind of argument [how "new" genes acquire favor] is not limited to biochemistry. We could make the same kind of case for clusters of compatible genes building the different parts of eyes, ears, noses, walking limbs, all the cooperating parts of an animal's body. Genes for making teeth suitable for chewing meat tend to be favoured in a "climate" dominated by genes making guts suitable for digesting meat. Conversely, genes for making plant-grinding teeth tend to be favoured in a climate dominated by genes that make guts suitable for digesting plants. And vice versa in both cases. Teams of "meat-eating genes" tend to evolve together, and teams of "plant-eating genes" tend to evolve together. Indeed, there is a sense in which most of the working genes in a body can be said to cooperate with each other as a team, because over evolutionary time they (i.e. ancestral copies of themselves) have each been part of the environment in which natural selection has worked on the others. If we ask why the ancestors of lions took to meat-eating, while the ancestors of antelopes took to grass-eating, the answer could be that originally it was an accident. An accident, in the sense that it could have been the ancestors of lions that took up grass-eating, and the ancestors of antelopes that took up meat-eating. But once one lineage had *begun* to build up a team of genes for dealing with meat rather than grass, the process was self-reinforcing. And once the other lineage had begun to build up a team of genes for dealing with grass rather than meat, *that* process was self-reinforcing in the other direction.

One of the main things that must have happened in the early evolution of living organisms was an increase in the numbers of genes participating in such cooperatives. Bacteria have far fewer genes than animals and plants. The increase may have come about through various kinds of gene duplication. Remember that a gene is just a length of coded symbols, like a file on a computer disc; and genes can be copied to different parts of the chromosomes, just as files can be copied to different parts of the disc. On my disc that holds this chapter there are officially just three files. By "officially" I mean that the computer's operating system tells me that there are just three files. I can ask it to read one of these three files, and it presents me with a one-dimensional array of alphabetical characters, including the characters that you are now reading. All very neat and orderly, it seems. But in fact, on the disc itself, the arrangement of the text is anything but neat and orderly. You can see this if you break away from the discipline of the computer's own official operating system, and write your own private programs to decipher what is actually written on every sector of the disc. It turns out that fragments of each of my three files are dotted around, interleaved with each other and with fragments of old, dead files that I erased long ago and had forgotten. Any given fragment may turn up, word for word the same, or with minor differences, in half a dozen different places all around the disc.

The reason for this is interesting, and worth a digression because it provides a good genetic analogy. When you tell a computer to delete a file, it appears to obey you. But it doesn't actually wipe out the text of that file. It simply wipes out all *pointers* to that file. It is as though a librarian, ordered to destroy *Lady Chatterly's Lover,* simply tore up the card from the card index, leaving the book itself on the shelf. For the computer, this is a perfectly economical way to do things, because the space formerly occupied by the "deleted" file is automatically available for new files, as soon as the pointers to the old file have been removed. It would be a waste of time actually to go to the trouble of filling the space itself with blanks. The old file won't itself be finally lost until all its space happens to be used for storing new files.

But this re-using of space occurs piecemeal. New files aren't exactly the same size as old ones. When the computer is trying to save a new file to a disc, it looks for the first available

---

fragment of space, writes as much of the new file as will fit, then looks for another available fragment of space, writes a bit more, and so on until all the file is written *somewhere* on the disc. The human has the illusion that the file is a single, orderly array, only because the computer is careful to keep records "pointing" to the addresses of all the fragments dotted around. These "pointers" are like the "continued on page 94" pointers used by the *New York Times.* The reason many copies of any one fragment of text are found on a disc is that if, like all my chapters, the text has been edited and re-edited many dozens of times, each edit will result in a new saving to the disc of (almost) the same text. The saving may ostensibly be a saving of the same file. But as we have seen, the text will in fact be repeatedly scattered around the available "gaps" on the disc. Hence multiple copies of a given fragment of text can be found all around the surface of the disc, the more so if the disc is old and much used.

Now the DNA operating system of a species is very very old indeed, and there is evidence that it, seen in the long term, does something a bit like the computer with its disc files. Part of the evidence comes from the fascinating phenomenon of "introns" and "exons." Within the last decade, it has been discovered that any "single" gene, in the sense of a single continuously read passage of DNA text, is not all stored in one place. If you actually read the code letters as they occur along the chromosome (i.e. if you do the equivalent of breaking out of the discipline of the "operating system") you find fragments of "sense," called exons, separated by portions of "nonsense" called introns. Any one "gene" in the functional sense, is in fact split up into a sequence of fragments (exons) separated by meaningless introns. It is as if each exon ended with a pointer saying "continued on page 94." A complete gene is then made up of a whole series of exons, which are actually strung together only when they are eventually read by the "official" operating system that translates them into proteins.

Further evidence comes from the fact that the chromosomes are littered with old genetic text that is no longer used, but which still makes recognizable sense. To a computer programmer, the pattern of distribution of these "genetic fossil" fragments is uncannily reminiscent of the pattern of text on the surface of an old disc that has been much used for editing text. In some animals, a high proportion of the total number of genes is in fact never read. These genes are either complete nonsense, or they are outdated "fossil genes."

Just occasionally, textual fossils come into their own again, as I experienced when writing this book. A computer error (or, to be fair, it may have been human error) caused me accidentally to "erase" the disc containing Chapter 3. Of course the text itself hadn't literally all been erased. All that had been definitely erased were the *pointers* to where each "exon" began and ended. The "official" operating system could read nothing, but 'unofficially' I could play genetic engineer and examine all the text on the disc. What I saw was a bewildering jigsaw puzzle of textual fragments, some of them recent, others ancient "fossils." By piecing together the jigsaw fragments, I was able to recreate the chapter. But I mostly didn't know which fragments were recent and which were fossil. It didn't matter for, apart from minor details that necessitated some new editing, they were the same. At least some of the "fossils," or outdated "introns," had come into their own again. They rescued me from my predicament, and saved me the trouble of rewriting the entire chapter.

Writing Assignment

One of the cornerstones of scientific and technical writing is the ability to define an abstract concept and then to use that definition to explain the operation of an everyday object or the occurrence of an operation or process that is so common that its principles are either overlooked or taken for granted.

For instance, if you think about the principle of relativity discussed by Bertrand Russell, you can understand better the Doppler effect: the "sound" of a train whistle does *seem* to change in pitch as the train approaches the station and passes you, the stationary observer, though we all know that the absolute pitch of the whistle remains constant and that it only seems to change as a function of its distance and speed from you.

1. Think of a basic principle (an abstract concept) in engineering or the sciences. Define the principle and then use your definition to explain the workings of a process or natural phenomenon so that your reader can understand how that process or phenomenon can take place.

You might, for instance, define electricity and use your definition to explain how a telephone transmits signals. You might discuss the theory of gravity and the phenomenon of planetary spin in order to explain the occurrence of the tides. Or you might discuss, from the field of medicine and psychiatry, the theory of the Oedipus and Electra complexes and how that theory explains and predicts the behavior of children toward parents (and parent substitutes) and the behavior of those in and out of power.

Use a metaphor to develop your definition and explanation, as Bertrand Russell did. Keep in mind that the reason you use metaphors is to make what you're explaining more accessible to a nonspecialist audience of college-educated readers.

2. Using writing exercise No. 1, place your theory and explanation into a problem-solution format. Write a letter to someone who has just written to you asking about a problem that he or she is having with a piece of equipment or a natural phenomenon. Imagine that you are a specialist writing to a nonspecialist (someone with less technical and scientific expertise than you have). Explain to that person why he or she needs to understand the theory behind the problem before being able to understand the solution you present.

Here is the body of a letter that a senior majoring in chemistry wrote in response to writing exercise No. 2.

In your letter, you ask if it would be possible to use our 20-amp fuse instead of our 10-amp fuse in your ticker tape master control unit. You say that the old fuse should be about ready to burn out, and that the only fuse you have left in stock that fits the machine is a 20-amp fuse. You add further that the 10-amp fuse you require is no longer available on the market except through this company, so you are waiting for the proper replacement, which does, in fact, accompany this letter.

In order to understand the proper use of fuses and their reasons for being in electrical circuitry, it is necessary to understand the meaning of electric current. Electric current is the flow of negatively charged electric particles, called electrons, in an electric circuit. This flow of electrons is caused by a difference of electrical potential (called "voltage") between two points. A point with an excess of electrons carries a negative charge, and a point with a depletion of electrons carries a positive charge. When a suitable pathway for electron transport, such as a wire, is connected from one of the points to another, the electrons are drawn spontaneously from the negative to the positive point. (The electrons are trying to re-establish a state of equilibrium by flowing to the positive point.) The number of electrons passing through a point in this pathway per second is the quantity called "amps." When the differences in voltage between the two points increases or decreases, the number of electrons flowing per second increases or decreases respectively.

Since more electrons are reaching their destination in the same amount of time when the voltage is raised, there is a greater force being exerted by the motion and friction of

Letter by Greg Barba, West Virginia University, October 22, 1985.

these charged masses (the electrons). Two types of forces exist here: (1) the electro-magnetic force, which is caused by the motion of the negative electric charges of the electrons; (2) friction, which is caused by the rubbing and bumping of these particles into each other and into other particles in their pathway. These forces cause electricity to work, but if they are too great, the electricity can cause damage as well.

The fuse in an electrical circuit can be compared to a river dam. The dam, which lets through the proper flow of water, is called "the 10-amp fuse." When this dam is used, boats can sail on the river, and people can swim with ease. The streams that branch off from the river irrigate the farmlands, and the crops grow.

With these ideas in mind, it is possible to approach your question rationally. The major purpose of a fuse is to protect the circuitry from the possibility of an excessive current surge that could subject the electron pathway to more of an electron flow than it could handle. If the circuit did not have this fuse as a weak link, a power surge might cause a sudden onslaught of electrons on the electrical circuitry, exerting more force than the circuitry could withstand.

The fuse is designed to break open when too many electrons try to crowd through at one time. In this manner, the pathway to the rest of the system is blocked by the hole in the pathway, and the electrons call off the journey to the point of positive voltage. The fuse opening is like the dam closing: in both cases, the pathway is blocked.

Fuses are labeled according to the amount of flow they will allow before cutting off the pathway. If we used a 5-amp fuse, instead of the proper 10-amp fuse, for our metaphorical dam, the river below would not get enough water to serve its intended purpose. The river would be so shallow that the boats would get stuck on the bottom and founder. The streams would not get enough water flow, and the crops would not be able to grow. Conversely, we could use the 20-amp fuse for our dam and be in trouble as well. Too much water would be allowed to flow, and our boats would be swept away, our crops washed out, and the surrounding countryside destroyed by floods.

With these insights into electricity and the workings of fuses in mind, you can see why I would strongly advise against using a 20-amp fuse in place of the 10-amp fuse in your master control unit. I expect that any surge in current with the 20-amp fuse installed would damage some of your equipment. Please install the fuse I'm sending as soon as possible.

Reference List

McGraw-Hill dictionary of scientific and technical terms. 1984. 3d ed. New York: McGraw-Hill.

McGraw-Hill encyclopedia of science and technology. 1987. 6th ed., vol. 6, 371. New York: McGraw-Hill.

Whitehead, A. N., and Bertrand Russell. 1910–13. *Principia mathematica.* 3 vols. Cambridge: At the University Press.

PERSUASION: AUDIENCE-BASED DOCUMENTS

The Literature Review

A review of the most recent literature in the field is essential for the practicing scientist or engineer. It informs the researcher what work has been done and, therefore, what is needed for the future.

Literature reviews are often found in the introductory sections of many articles and reports. They give an intellectual framework to the research or engineering technology being reported and show that the problem being addressed is significant and worthwhile to analyze.

Literature reviews are usually difficult to write, simply because they require the author to possess a complete understanding of the field, both in terms of how it has developed historically and what constitutes its major achievements. Additionally, the authors may organize such reviews by using their own points of view on the matter at hand, especially in regard to what kind of research or technological development has been most useful.

In the following literature review, Sarah Blaffer Hrdy, a primatologist, reviews the literature on theories of sexual selection among primates (how and for what reasons females and males choose their mating partners). Hrdy has also written *The Woman That Never Evolved.*

Empathy, Polyandry, and the Myth of the Coy Female

Sarah Blaffer Hrdy

Anisogamy and the Bateman Paradigm

In one of the more curious inconsistencies in modern evolutionary biology, a theoretical formulation about the basic nature of males and females has persisted for over three decades, from 1948 until recently, despite the accumulation of abundant openly available evidence contradicting it. This is the presumption basic to many contemporary versions of sexual selection theory that males are ardent and sexually undiscriminating while females are sexually restrained and reluctant to mate. My aims in this paper will be to examine this stereotype of "the coy female," to trace its route of entry into modern evolutionary thinking and to examine some of the processes that are only now, in the last decade, causing us to rethink this erroneous corollary to a body of theory (Darwin, 1871) that has otherwise been widely substantiated. In the course of this examination, I will speculate about the role that empathy and identification by researchers with same-sex individuals may have played in this strange saga.

Obviously, the initial dichotomy between actively courting, promiscuous males and passively choosing, monandrous females dates back to Victorian times. "The males are almost always the wooers," Darwin wrote in 1871, and he was very clear in his own writings that the main activity of females was to choose the single best suitor from among these wooers. As he wrote in *The Descent of Man and Selection in Relation to Sex* (1871), "It is shown by various facts, given hereafter, and by the results fairly attributable to sexual selection, that the female, though comparatively passive, generally exerts some choice and accepts one male in preference to the others." However the particular form in which these ideas were incorporated into modern and ostensibly more "empirical" versions of post-Darwinian evolutionary thought derived from a 1948 paper about animals by a distinguished plant geneticist, Angus John Bateman.

Like so much in genetics, Bateman's ideas about the workings of nature were based primarily on experiments with *Drosophila,* the minuscule flies that materialize in the vicinity of rotting fruit. Among the merits of fruitflies rarely appreciated by housekeepers are the myriad of small genetic differences that determine a fruitfly's looks. Bred over generations in a laboratory, distinctive strains of *Drosophila* sporting odd-colored eyes, various bristles, peculiar crenulations here and there, grotesquely shaped eyes, and so forth can be produced by scientists, and these markers are put to use in tracing genealogies.

Bateman obtained various lots of differently decorated *Drosophila* all belonging to the one species, *Drosophila melanogaster.* He housed three to five flies of each sex in glass containers and allowed them to breed. On the basis of 64 such experiments, he found (by

---

From Hrdy, S. B. 1986. Empathy, polyandry, and the myth of the coy female. In *Feminist approaches to science,* edited by Ruth Bleier, 120–29, 142–46. New York: Pergamon Press.

counting the offspring bearing their parents' peculiar genetic trademarks) that while 21% of his males failed to fertilize any female, only 4% of his females failed to produce offspring.

A highly successful male, he found, could produce nearly three times as many offspring as the most successful female. Furthermore, the difference between the most successful and the least successful male, what is called the *variance* in male reproductive success, was always far greater than the variance among females. Building upon these findings, Bateman constructed the centerpiece to his paradigm: whereas a male could always gain by mating just one more time, and hence benefit from a nature that made him undiscriminatingly eager to mate, a female, already breeding near capacity after just one copulation, could gain little from multiple mating and should be quite uninterested in mating more than once or twice.

From these 64 experiments with *Drosophila,* Bateman extrapolated to nature at large: selection pressures brought about by competition among same-sexed individuals for representation in the gene pools of succeeding generations would almost always operate more strongly upon the male than upon the female. This asymmetry in breeding potential would lead to a nearly universal dichotomy in the sexual nature of the male and female:

> *One would therefore expect to find in all but a few very primitive organisms . . . that males would show greater intra-sexual selection than females. This would explain why . . . there is nearly always a combination of an undiscriminating eagerness in the males and a discriminating passivity in the females. Even in a derived monogamous species (e.g. man) this sex difference might be expected to persist as a rule. (Bateman, 1948, p. 365)*

This dichotomy was uncritically incorporated into modern thinking about sexual selection. In his classic 1972 essay on "Parental Investment and Sexual Selection," Harvard biologist Robert Trivers acknowledged Bateman's paper as "the key reference" (provided him, as it happens by one of the major evolutionary biologists of our time, and Trivers' main mentor at Harvard, Ernst Mayr). Trivers' essay on parental investment, carrying with it Bateman's model, was to become the second most widely cited paper in all of sociobiology, after Hamilton's 1964 paper on kin selection.

Expanding on Bateman's original formulation, Trivers argued that whichever sex invests least in offspring will compete to mate with the sex investing most. At the root of this generalization concerning the sexually discriminating female (apart from Victorian ideology at large) is the fact of anisogamy (gametes unequal in size) and the perceived need for a female to protect her already substantial investment in each maternal gamete; she is under selective pressure to select the best available male to fertilize it. The male, by contrast, produces myriad gametes (sperm), which are assumed to be physiologically cheap to produce (note, however, that costs to males of competing for females are rarely factored in), and he disseminates them indiscriminately.

Two central themes in contemporary sociobiology then derive directly from Bateman. The first theme is the dichotomy between the "nurturing female," who invests very much more per offspring than males, and "the competitive male," who invests little or nothing beyond sperm but who actively competes for access to any additional female (see for example Daly & Margo Wilson, 1983, pp. 78–79; Trivers, 1985, p. 207). As Trivers noted in his summary of Bateman's experiments with *Drosophila,* "A female's reproductive success did not increase much, if any, after the first copulation and not at all after the second; most females were uninterested in copulating more than once or twice" (1972, p. 138). And so it was that "coyness" came to be the single most commonly mentioned attribute of females in the literature on sociobiology. Unlike the male, who, if he makes a mistake can move on to another female, the female's investment was initially considered to be so great that she was constrained from aborting a bad bet and attempting to conceive again. (Criticisms and recent revisions of the notion are discussed later in the section, "The Females Who Forgot to be Coy".) In this respect, contemporary theory remains fairly faithful to Darwin's original (1871) two-part definition of sexual selection. The first part of the theory predicts competition between males for mates; the second, female choice of the best competitor.

The second main sociobiological theme to derive from Bateman is not explicitly discussed in Darwin but is certainly implicit in much that Darwin wrote (or more precisely, did not write) about females. This is the notion that female investment is already so large that it can not be increased and the idea that most females are already breeding close to capacity. If this were so, the variance in female reproductive success would be small, making one female virtually interchangeable with another. A logical corollary of this notion is the incorrect conclusion that selection operates primarily on males.

The conviction that intrasexual selection will weigh heavily upon males while scarcely affecting females was explicitly stated by Bateman, but also appears in implicit form in the writings of contemporary sociobiologists (Daly & Margo Wilson, 1983, Chapter 5; Wilson, 1978, p. 125). It is undeniable that males have the capacity to inseminate multiple females while females (except in species such as those squirrels, fish, insects, and cats, where several fathers can sire a single brood) are inseminated—at most—once each breeding period. But a difficulty arises when the occasionally true assumption that females are not competing among themselves to get fertilized is then interpreted to mean that there will be reduced within-sex competition among females generally (e.g., Freedman, 1979, p. 33).

Until about 1980—and even occasionally after that—some theoreticians were writing about females as though each one was relatively identical in both her reproductive potential and in her realization of that potential. This erroneous generalization lead some workers (perhaps especially those whose training was not in evolutionary biology per se) to the erroneous and patently non-Darwinian conclusion that females are not subject to selection pressure at all and the idea that competition among males is somehow more critical because "leaving offspring is at stake" (Carol Cronin, 1980, p. 302; see also Virginia Abernethy, 1978, p. 132). To make an unfortunate situation worse, the close conformity between these notions and post-Victorian popular prejudice meant that ideas about competitive, promiscuous men and choosey women were selectively picked up in popular writing about sociobiology. An article in *Playboy Magazine* celebrating "Darwin and the Double Standard" (Morris, 1979) comes most vividly to mind, but there were many others.

The Females Who Forgot to be Coy

Field studies of a number of animal groups provide abundant examples of females who, unlike Bateman's *Drosophila,* ardently seek to mate more than once or twice. Furthermore, fertilization by the best male can scarcely be viewed as their universal goal since in many of these cases females were not ovulating or else were actually pregnant at the time they solicit males.

It has been known for years (among some circles) that female birds were less than chaste, especially since 1975 when Bray, Kennelly, and Guarino demonstrated that when the "master" of the blackbird harem was vasectomized, his females nevertheless conceived (see also Lumpkin, 1983). Evelyn Shaw and Joan Darling (1985) review some of this literature on "promiscuous" females, particularly for marine organisms. Among shiner perch, for example, a female who is not currently producing eggs will nevertheless court and mate with numbers of males, collecting from each male sperm that are then stored in the female's ovaries till seasonal conditions promote ovulation. Female cats, including leopards, lions, and pumas are notorious for their frequency of matings. A lioness may mate 100 times a day with multiple partners over a 6–7-day period each time she is in estrus (Eaton, 1976). Best known of all, perhaps, are such primate examples as savanna baboons, where females initiate multiple brief consortships, or chimpanzees, where females alternate between prolonged consortships with one male and communal mating with all males in the vicinity (DeVore, 1965; Hausfater, 1975; Caroline Tutin, 1975). However, only since 1979 or so has female promiscuity been a subject of much theoretical interest (see for example Alatalo, Lundberg, & Stahlbrandt, 1982; Sandy Andelman, in press; Gladstone, 1979; Sarah Blaffer Hrdy, 1979; Susan Lumpkin, 1983; Meredith Small, forthcoming; R. Smith, 1984; Wirtz, 1983), largely I believe because theoretically the phenomenon should not have existed and therefore there was little theoretical infrastructure for studying it, certainly not the sort of study that could lead to a PhD (or a job).

In terms of the order Primates, evidence has been building since the 1960s that females in a variety of prosimian, monkey, and ape species were managing their own reproductive careers so as actively to solicit and mate with a number of different males, both males within their (supposed) breeding unit and those outside it. As theoretical interest increased, so has the quality of the data.

But before turning to such evidence, it is first critical to put sex in perspective. To correct the stereotype of "coyness," I emphasize female sexual activity but, as always in such debates, reality exists in a plane distinct from that predefined by the debate. In this case, reality is hours and hours, sometimes months and months, of existence where sexual behavior is not even an issue, hours where animals are walking, feeding, resting, grooming. Among baboons (as in some human societies) months pass when a pregnant or lactating mother engages in no sexual behavior at all. The same is generally true for langurs, except that females under particular conditions possess a *capacity* to solicit and copulate with males even if pregnant or lactating, and they sometimes do so. At such times, the patterning of sexual receptivity among langurs could not be easily distinguished from that of a modern woman. The same could be said for the relatively noncyclical, semicontinuous, situation-dependent receptivity of a marmoset or tamarin.

With this qualification in mind—that is the low frequencies of sexual behavior in the lives of *all* mammals, who for the most part are doing other things—let's consider the tamarins.

Tamarins are tiny South American monkeys, long thought to be monogamous. Indeed, in captivity, tamarins do breed best when a single female is paired with one mate. Add a second female and the presence of the dominant female suppresses ovulation in the subordinate. (The consequences of adding a second male to the cage are unknown, since such an addition was thought to violate good management practices.) Nevertheless, in the recent (and first) long-term study of individually marked tamarins in the wild, Anne Wilson Goldizen discovered that given the option, supposedly monogamous saddle-backed tamarins *(Saguinus fusicollis)* will mate with several adult males, each of whom subsequently help to care for her twin offspring in an arrangement more nearly "polyandrous" than monogamous (Goldizen & Terborgh, forthcoming). Furthermore the presence of additional males, and their assistance in rearing young may be critical for offspring survival. One of the ironies here, pointed out in another context by Janet Sayers (1982), is that females are thus presumed to commit what is known in sociobiology as a *Concorde fallacy;* that is, pouring good money after bad. Although in other contexts (e.g., Dawkins, 1976) it has been argued that creatures are selected to cut bait rather than commit Concorde fallacies, mothers were somehow excluded from this reasoning (however, see Trivers, 1985, p. 268, for a specific acknowledgement and correction of the error). I happen to believe that the resolution to this contradiction lies in recognizing that gamete producers and mothers do indeed "cut bait" far more often than is generally realized, and that skipped ovulations, spontaneous abortion, and abandonment of young by mothers are fairly routine events in nature. That is, the reasoning about the Concorde fallacy is right enough, but our thinking about the commitment of mothers to nurture no matter what has been faulty.

Indeed, on the basis of what I believe today (cf. Hrdy, 1981, p. 59), I would argue that a polyandrous component is at the core of the breeding systems of most troop-dwelling primates: females mate with many males, each of whom may contribute a little bit toward the survival of offspring. Barbary macaques provide the most extreme example (Taub, 1980), but the very well-studied savanna baboons also yield a similar, if more moderate, pattern. David Stein (1981) and Jeanne Altmann (1980) studied the complex interactions between adult males and infants. They found that (as suggested years ago by Tim Ransom and Bonnie Ransom, 1972) former, or sometimes future, consorts of the mother develop special relationships with that female's infant, carrying it in times of danger and protecting it from conspecifics, possibly creating enhanced feeding opportunities for the infant. These relationships are made possible by the mother's frequent proximity to males with whom she has special relationships and by the fact that the infant itself comes to trust these males and seek them out; more is at issue than simply male predilections. Altmann aptly refers to such males as *god-fathers*. Infants, then, are often the focal-point of elaborate male-female-infant relationships, relationships that are often initiated by the females themselves (Barbara Smuts, 1985).

Even species such as Hanuman langurs, blue monkeys, or redtail monkeys, all primates traditionally thought to have "monandrous" or "uni-male" breeding systems, are far more promiscuous than that designation implies. Indeed, mating with outsiders is so common under certain circumstances as to throw the whole notion of one-male breeding units into question (Cords, 1984; Tsingalia & Thelma Rowell, 1984). My own first glimpse of a langur, the species I was to spend nearly 10 years studying intermittently, was of a female near the Great Indian Desert in Rajasthan moving rapidly through a steep granite canyon, moving away from her natal group to approach and solicit males in an all-male band. At the time, I had no context for interpreting behavior that merely seemed strange and incomprehensible to my Harvard-trained eyes. Only in time, did I come to realize that such wandering and such seemingly "wanton" behavior were recurring events in the lives of langurs.

In at least three different sets of circumstances female langurs solicit males other than their so-called *harem-leaders:* first, when males from nomadic all-male bands temporarily join a breeding troop; second, when *females* leave their natal troops to travel temporarily with all-male bands and mate with males there; and third, when a female for reasons unknown to any one, simply takes a shine to the resident male of a neighboring troop (Hrdy 1977; Moore 1985; filmed in Hrdy, Hrdy, & Bishop, 1977). It may be to abet langurs in such projects that nature has provided them attributes characteristic of relatively few mammals. A female langur exhibits no visible sign when she is in estrus other than to present to a male and to shudder her head. When she encounters strange males, she has the capacity to shift from cyclical receptivity (that is, a bout of heat every 28 days) into a state of semicontinuous receptivity that can last for weeks. Monkeys with similar capacities include vervets, several of the guenons, and gelada baboons, to mention only a few (reviewed by Hrdy & Whitten, 1986).

A number of questions are raised by these examples. First, just exactly why might females bother to be other than coy, that is why should they actively seek out partners including males outside of their apparent breeding units (mate "promiscuously", seek "excess" copulations, beyond what are necessary for fertilization)? Second, why should this vast category of behaviors be, until recently, so generally ignored by evolutionary theorists? As John Maynard Smith noted, in the context of mobbing behavior by birds, "behavior so widespread, so constant, and so apparently dangerous calls for a functional explanation" (1984, p. 294).

To be fair, it should be acknowledged that mobbing behavior in birds is more stereotyped than sexual behavior in wild cats or monkeys, and it can be more systematically studied. Nevertheless, at issue here are behaviors exhibited by the majority of species in the order primates, the best studied order of animals in the world, and the order specifically included by Bateman in his extrapolation from coyness in arthropods to coyness in anthropoids. Furthermore, females engaged in such "promiscuous matings" entail obvious risks ranging from retaliatory attacks by males, venereal disease, the energetic costs of multiple solicitations, predation risks from leaving the troop, all the way to the risk of lost investment by a male consort who has been selected to avoid investing in other males' offspring (Trivers, 1972). In retrospect, one really does have to wonder why it was nearly 1980 before promiscuity among females attracted more than cursory theoretical interest.

Once the initial conceptual block was overcome (and I will argue in the last section that the contributions of women researchers was critical to this phase, at least in primatology), once it was recognized that oh yes, females mate promiscuously and this is a most curious and fascinating phenomenon, the question began to be vigorously pursued. (Note though that the focus of this paper is on male-centered theoretical formulations, readers should be aware that there are other issues here, such as the gap between theoreticians and fieldworkers, which I do not discuss.)

In my opinion, no conscious effort was ever made to leave out female sides to stories. The Bateman paradigm was very useful, indeed theoretically quite powerful, in explaining such phenomena as male promiscuity. But, although the theory was useful in explaining male behavior, by definition (i.e., *sexual selection* refers to competition between one sex for *access* to the other sex) it excluded much within-sex reproductive competition among females, which was not over fertilizations per se but which also did not fall neatly into the realm of the survival-related phenomena normally considered as due to natural selection. (The evolution of sexual swellings might be an example of a phenomenon that fell between definitional cracks and hence went unexplained until recently [Clutton-Brock & Harvey, 1976; Hrdy, 1981].)

To understand female promiscuity, for example, we first needed to recognize the limitations of sexual selection theory and then needed to construct a new theoretical base for explaining selection pressures on females.

The realization that male-male competition and female choice explains only a small part of the evolution of breeding systems has led to much new work (e.g., Wasser, 1983, and work reviewed therein). We now have, for example, no fewer than six different models to explain how females might benefit from mating with different males (see Smith, R., 1984, for a recent review).

These hypotheses, most of them published in 1979 or later, can be divided into two categories, first those postulating genetic benefits for the offspring of sexually assertive mothers, and second, those postulating nongenetic benefits for either the female herself or her progeny. All but one of these (the oldest, "prostitution hypothesis") was arrived at by considering the world from a female's point of view.

Whereas all the hypotheses specifying genetic benefits predict that the female should be fertile when she solicits various male partners (except in those species where females have the capacity to store sperm), this condition is not required for the nongenetic hypotheses. It should be noted, too, that only functional explanations for multiple matings are listed. The idea that females simply "enjoy" sex begs the question of why females in a genus such as *Drosophila* do not appear highly motivated to mate repeatedly, while females in other species apparently are so motivated and have evolved specific physiological apparatus making promiscuity more likely (e.g., a clitoris, a capacity for orgasm brought about by prolonged or multiple sources of stimulation, a capacity to expand receptivity beyond the period of ovulation, and so forth; see Hrdy, 1981, Chapter 7 for discussion). Nevertheless, the possibility persists that promiscuous behaviors arise as endocrinological accidents or perhaps that females have orgasms simply because males do (Symons, 1979), and it is worth remembering that an act of faith is involved in assuming that there is any function at all. (I mention this qualifier because I am not interested in arguing a point that can not currently be resolved.)

Assuming that promiscuous behaviors and the physiological paraphernalia leading to them have evolved, four hypotheses are predicated on genetic benefits for the offspring of sexually assertive mothers: (a) the "fertility backup hypothesis," which assumes that females will need sperm from a number of males to assure conception (Meredith Small, forthcoming; Smith, R., 1984); (b) "the inferior cuckold hypothesis," in which a female paired with an inferior mate surreptitiously solicits genetically superior males when conception is likely (e.g., Benshoof & Thornhill, 1979); (c) "the diverse paternity" hypothesis, whereby females confronted with unpredictable fluctuations in the environment produce clutches sired by multiple partners to diversify paternity of offspring produced over a lifetime (Parker, 1970; Williams, 1975); or (d) in a somewhat obscure twist of the preceding, females in species where litters can have more than one father alter the degree of relatedness between sibs and maternal half-sibs by collecting sperm from several fathers (Davies & Boersma, 1984).

The remaining explanations are predicated on nongenetic benefits for females and do not assume the existence of either genetic differences between males or the existence of female capacities to detect them: (e) the "prostitution" hypothesis, whereby females are thought to exchange sexual access for resources, enhanced status, etc.—the oldest of all the explanations (first proposed by Sir Solly Zuckerman 1932, recently restated by Symons, 1979; see also, Nancy Burley & Symanski, 1981, for discussion); (f) the "therapeutic hypothesis" that multiple matings and resulting orgasm are physiologically beneficial to females or make conception more likely (Mary Jane Sherfey, 1973); (g) the "keep 'em around" hypothesis whereby females (with the connivance of dominant males in the group) solicit subordinate males to discourage these disadvantaged animals from leaving the group (Stacey, 1982); and (h) the "manipulation hypothesis," suggesting that females mate with a number of males in order to confuse information available to males about paternity and thereby extract investment in, or tolerance for, their infants from different males (Hrdy, D. B., 1979; Stacey, 1982).

It is the last hypothesis that I now want to focus on, not because that hypothesis is inherently any better than others, but because I know the most about it and about the assumptions that needed to be changed before it could be dreamed up.

The "manipulation hypothesis," first conceived in relation to monkeys, grew out of a dawning awareness that, first of all, individual females could do a great deal that would affect the survival of their offspring, and second, that males, far from mere dispensers of sperm, were critical features on the landscape where infants died or survived. That is, females were more political, males more nurturing (or at least not neutral), than some earlier versions of sexual selection theory would lead us to suppose.

Writing Exercise

Hrdy's review is a tour de force, written by a knowledgeable professional immersed in the field. Choose a field you know something about, and select a particular research problem in that field, either from your own knowledge or in conference with your professor. Research the field and write a four-page literature review of the current state of knowledge. Include your own point of view about the direction of future research if your professor thinks such an emphasis would be appropriate.

Reference List

Hrdy, S. B. 1981. *The woman that never evolved.* Cambridge: Harvard University Press.

The Feasibility Report

Engineers and scientists in business, industry, and academic institutions commonly use two kinds of reports to describe their work: the feasibility report and the progress report. Each has many variations, which are dictated by the needs of the particular industry or by the editorial boards of academic journals and the national societies that support and validate them. In most cases, managers or contracting agencies establish the specific variation.

The progress report, which is industry's equivalent of the academic formal article, will be covered in the next section.

The feasibility report essentially says whether a proposed line of research or technical innovation is worthwhile to undertake. Such a report presents an argument designed to persuade the reader that the proposal is reasonable. It is usually submitted for review by higher management and, as such, tends to have a standard format, as follows:

Cover letter: A letter of transmittal, sending the report from the research group to other sections of the company. Usually states the basic conclusion of the study.

Executive summary: The equivalent of the abstract in academic journal articles, the executive summary does what its name implies: it summarizes the report for executives, who often do not have time to read the entire report but need to know and understand its contents.

Problem statement: A clear, short statement of the problem under discussion: how the problem arose and what its significance is. It should concentrate on a well-defined problem—that is, a problem that will allow you to recognize a satisfactory solution when you see it. An ill-defined problem is one that is so vague or imprecise that no solution can satisfactorily resolve it.

Background and literature review: An overview of state-of-the-art information about the project. Usually includes a persuasive point of view on past research: what it's accomplished and, by implication, what is the most reasonable next step.

Possibilities and review of options: A description of the various approaches to solving the problem.

Recommendation: A description of the best option and the reasons for its excellence.

References: A list, in correct bibliographical form, of the references used to write the report.

These seven parts are not etched in stone. You will find them with different names and subdivided into other parts. Sometimes the literature review will be placed in an appendix. But, despite the possible variations, the underlying form remains: the problem and its background; reasonable options; and the best option.

Here is a feasibility report on wastewater discharge, written by a student.

A REPORT ON THE POWDERED ACTIVATED CARBON TREATMENT (PACT)
WASTE TREATMENT ALTERNATIVE FOR WHOLESOME DAIRY PRODUCTS, INC.

By
John P. Young III
Project Engineer

Prepared for
Ms. Helen Burkeheart
Director of Research
Wholesome Dairy Products, Inc.

20 November 1986

TABLE OF CONTENTS

From Young, J. P., III. 20 November 1986. A report on the powdered activated carbon treatment (PACT) waste treatment alternative for Wholesome Dairy Products, Inc. Student paper, West Virginia University.

Memo of Transmittal
WHOLESOME DAIRY PRODUCTS, INC.
Intracompany Correspondence

Date: November 20, 1986
From: John P. Young III, Staff Engineer
To: Ms. Helen Burkeheart, Director of Research
Subject: Disposition of our Wastewater Discharge (Effluent)

For some time, our company has piped its waste effluent to the Morgantown Waste Water Treatment Plant for cleanup and sludge disposal. Although the charges for this service, and our yearly pipeline maintenance costs have been rather high, the cost of constructing and operating a private on-site waste treatment or pre-treatment plant has prevented our considering these alternative solutions.

We have just received a new schedule of special service charges from the Morgantown plant, and effective June 1st of next year, our basic cost per 1,000 gallons of effluent will increase by 12%. Furthermore, we are faced with increasing the capacity (through replacement) of our two-mile pipeline, and labor and materials costs for this type of work have increased significantly during the past ten years.

Recently, the successful field testing of a new kind of on-site treatment plant, combining the activated-sludge process with carbon filtration was announced by DuPont engineers. Until now, we could not consider an activated-sludge plant because the waste removal rate was not high enough to permit our dumping the treated wastewater directly into the Monongahela River. The new process, called PACT (Powdered Activated Carbon Treatment), is reported to have a high enough efficiency to allow direct disposal to a public watercourse.

This report contains engineering data on the new PACT process, and compares the overall cost of this alternative to our present situation. Since my primary field is process analysis, the enclosed cost information doesn't include tax considerations, and will have to be studied by the Finance Committee. However, the initial figures indicate that this new waste treatment process will offer our company both savings and flexibility, over our present system.

EXECUTIVE SUMMARY

The DuPont corporation recently announced the success of a new activated-sludge, carbon-filtration process that it calls PACT (Powdered Activated Carbon Treatment). This process differs from previous carbon-treatment schemes in that the carbon is actually in the aeration tank of the activated-sludge process, instead of in a separate absorption column. Furthermore:

1. BOD and Suspended Solids removal is improved to a level that would allow our plant to dispose of treated waste directly into the Monongahela River.
2. The carbon, in contact with the sludge organisms, serves as a buffer, drastically reducing the number of upsets of the biological systems.
3. The needs for carbon regeneration or replacement is greatly reduced, and the sludge is easier to settle and remove.

The Pact process is capable of handling potent chemical pollutants of up to 40 million gallons per day, making it an ideal candidate for an on-site treatment system at our facility. Our 600,000 gallon per day effluent, which consists mostly of Biochemical Oxygen Demand (BOD) and Suspended Solids (SS), could be treated and discharged directly to the Monongahela River. This would eliminate our pipeline construction and maintenance costs, and the charges from the Morgantown Wastewater Treatment Plant. PACT system total startup costs would not exceed $450,000, and total annual expenses (all sources) would be less than $125,000. We currently pay the Morgantown plant $24,363.00/month, or $292,356.00/year. Add to this approximately $36,000/year in pipeline costs, and our current annual expenditures increase to $328,356.00. This means that we could realize an operational cost savings of about $216,954 in our second year of service with the Pact process. ($316,954 if we replace the pipeline).

Until now, on-site treatment was simply not feasible for our company, due to the high startup and annual operating costs of a standard activated-sludge plant (about twice that of a Pact system), and the fact that our output would still be unacceptable for direct discharge after treatment.

Pre-treatment of our effluent, with subsequent discharge to the Morgantown plant, was not considered, since almost all of our current monthly charges (91%) are basic treatment charges that would not change with the pre-treatment option.

Both the Pact plant, and our pipeline require extensive maintenance work every ten years. This cost will be about $100,000 for both options, so I did not include it in the annual cost analysis. It is, however, included in the initial cost savings analysis, as we will have to increase the pipe's capacity if we decide to keep our present disposal method.

BACKGROUND

In late 1968, the Monongahela River Pollution Control Commission passed a mandate that all wastes entering the river contain no more than 70 parts per million (ppm) Biochemical Oxygen Demand (BOD), and 50 ppm Suspended Solids (SS). This action forced our company to examine the prevailing technology at that time, and consider the available options. In 1968, our production was slightly less than half our current level, so we considered several alternatives:

1. Build a small waste treatment facility at our plant site, and discharge the effluent directly to the river.
2. Store our waste in a holding tank, and truck it to the Morgantown plant.
3. Pipe our waste discharge to the Morgantown sewage treatment plant for disposal.

Although our effluent volume was only 250,000 gallons per day, small on-site waste treatment technology was inadequate for handling our discharge. The bulk of our polluted water is associated with the pasteurization of milk and cheese products, combined with washwater from our storage facility, and the high BOD content required the use of rotating biological contactors and digestion tanks. The startup and maintenance cost of one of these facilities was so great, that we did not give this option serious study at that time.

Many small companies store and truck their waste to their local facility, and we gave this option some thought. Because of ground seepage and the proximity of the river, we could only consider an above ground tank instead of a natural reservoir pond. Moreover, we were in a growth period, and our waste discharge could only increase along with our production. These factors, combined with the purchase and maintenance costs of tank trucks, convinced us to discard this alternative in favor of the pipeline option.

The pipeline option was the best one, despite the fact that we had to purchase a strip of land to run our line along the Monongahela riverbank. Total expenditures for the property and pipeline construction totalled only $60,000 1968 dollars, and total monthly charges for waste treatment at the Morgantown Wastewater Plant were less than $9,000.

By 1978, our production had increased 67%, driving our daily waste output up to nearly 420,000 gallons. Waste flow was still within the maximum recommended level for our pipe, but our factor of safety had diminished from 2:1 to 1.2:1. Once again, we looked at the on-site waste plant option, as improvements in the activated-sludge process made a small treatment plant a possibility. A high-rate, diffused-aeration, activated-sludge plant could have reduced our BOD and SS to 50 ppm, without the need for expensive primary clarifiers.

By that time, however, the Environmental Protection Agency had passed the Clean Water Act of 1977, which called for BOD and SS levels of 30 ppm or less. (EPA 1980) This action effectively removed the small on-site option, and we chose to continue with our pipeline. By 1978, monthly charges from the Morgantown Wastewater Plant had risen to $13,000, and, for the first time, a penalty charge of $1000/month was added on, for our excessively high BOD level. This charge, which has now increased to $2252/month, is for BOD levels above 250 ppm, and is based on a charge of $100 per 1000 pounds of BOD/month. The conversion formula for ppm BOD to pounds BOD can be found on the Breakdown of Current Monthly Wastewater Charges calculations page.

PROBLEM

The problem that now faces us is twofold. First, we are now exceeding the maximum recommended capacity of our pipeline by 20%. We have no option but to replace the present pipeline, increasing its diameter from 12 to 18 inches. The estimated total cost for this operation is $50,000/mile, or $100,000.

The second part of the problem involves the steadily increasing basic treatment service charge from the Morgantown sewage treatment plant. We currently pay $1.30/1000 gallons for the first 60,000 gallons of effluent, and $1.22/1000 gallons for amounts over 60,000 gal. (Kelley and others 1981) In June, 1987, these charges will increase by 12%, to $1.46, and $1.37, respectively. The new charges apply to all industrial users, and will surely increase in the future, whether we increase our production or not.

These problems, combined with new advances in on-site waste treatment technology, have prompted us once again to investigate new alternatives for the disposal of our waste effluent.

LITERATURE REVIEW

Since the late 1960s, public concern has been increasing about water pollution, Both state and federal governments have passed legislation to reduce the strength of waste effluents that are discharged into our waterways. Environmental engineers were faced with the problem of reducing biological and chemical wastes to meet the new standards.

Two major companies, DuPont, and Zimpro, began working separately on a process that combined the activated-sludge and carbon-absorption processes. Activated sludge is an effective process for removing high levels of BOD, and is inexpensive to set up and operate, but the process is easily upset by chemical discharges that are unfamiliar to the sludge organisms. The carbon-absorption process doesn't suffer from such upsets, but requires costly filtration columns, and tons of carbon.

DuPont and Zimpro engineers decided to combine the two processes into one, by placing the carbon in the aeration tank of the activated sludge process. The result was a dramatic increase in the removal of BOD, and dissolved organic carbon (DOC), without the need for separate absorption columns. In addition, the PACT plant doesn't need a primary clarifier to settle out the suspended solids. This reduces the cost of a PACT facility to about 10% more than a basic activated-sludge plant.

In addition to providing more efficient removal of waste pollutants, the new powdered activated carbon treatment process (PACT) is not as susceptible to biological upsets as the regular activated sludge process. Although the mechanism that causes this stability is not fully understood, Zimpro engineers have offered two explanations:

1. The carbon, placed in the aeration tank, serves as a buffer to absorb some of the incoming chemicals until they can be digested by the sludge bacteria.
2. The carbon "inhibits the dilution of the sludge bacteria's secreted enzymes, enabling the organisms to assimilate soluble organics more effectively." (Heath 1986, 81)

BOD removal is further improved because the average sludge age is higher with a PACT plant, compared to a municipal treatment plant. PACT sludge averages 5 to 45 days, depending upon the amount of BOD removal needed, as opposed to 3 to 8 days in a municipal plant. As a result of this increased digestion, the PACT plant produces a significantly lower volume of excess sludge than a standard activated-sludge plant. Furthermore, the PACT sludge settles more easily than regular sludge, due to its lower filamentous organism content. This makes it easier to remove, and reduces the PACT plant's operating cost.

DuPont engineers installed a PACT plant at their Chambers Water Works on the Delaware river about ten years ago, and have been testing it with a variety of chemical and organic wastes. Some of the chemical wastes handled are so powerful, that the excess sludge must be kept from the environment in a special secure landfill. In spite of the strength of these effluents (which include acidic dyes), the 40 million gallon Chambers plant has not had a problem with biological upsets, and is capable of treating a broader range of chemical wastewaters than almost any plant in the U.S. If the PACT process is used to treat wastewaters consisting only of BOD and suspended solids, the secure landfill is not necessary.

Although DuPont patented the PACT process first, they have since sold the patent to Zimpro, who is now marketing small "package" plants for private industrial use. Package plants are currently in use in Ohio, Illinois, Connecticut, North Carolina, and New Jersey.

POSSIBILITIES

Since the activated-sludge process by itself would not remove enough BOD to meet current EPA standards, and carbon absorption columns are too expensive, our only option is to investigate the cost of a small PACT package plant. These plants are probably available in pre-sized form, such as a half-million gallon, one million gallon, etc. For our facility, a one-million gallon plant would be an ideal size. Our current production level of 600,000 gallons/day is as high as we can go without building new facilities, and the 400,000 gallon/day excess capacity will help prevent biological upsets. Based on current EPA construction cost figures for a standard activated-sludge plant (EPA 1980), there is little difference in price between a half-million, and one-million gallon plant.

The same difference in elevation that drives fluid through our pipeline will also operate a PACT plant. Therefore, the only electrical requirements with the plant will be for lighting, instruments, and small pumps.

REVIEW OF OPTIONS

At present, there are just two options available to us:

1. Continue to pay the Morgantown Wastewater Treatment Plant, and increase the size of our pipeline next year, or
2. Install our own PACT on-site treatment plant.

Since I have already discussed the above options in text form, I chose to present an economic breakdown of each one, which can be found on the following two pages. The calculations for the pipeline option are based on our current charges, and do not reflect the 12% rate increase next year. Figures for a PACT package plant were not available at the time of this report. However, the price for a high-rate, diffused-aeration, activated-sludge plant, multiplied by 110%, will provide a fairly accurate cost basis for a PACT plant. (Heath 1986) The main reason for this small price difference is that the PACT system does not require a primary clarifier, as does the activated-sludge plant. This helps to offset the price of the PACT's filter carbon.

When studying these figures, please remember these important points:

1. The most recent prices for the activated-sludge plant from the EPA are in 1976 dollars. To adjust for 1986 dollars, I multiplied the 1976 prices by an 8% inflation factor over a ten-year period. This factor is 2.189. (EPA 1980)
2. None of the figures reflect tax treatment, or other economic factors. I know that the cost of treatment at the Morgantown facility, and construction of a new pipeline is income tax-deductible, and most of the construction and operating costs of the PACT plant will be deductible. (The PACT plant may be eligible for special energy credits). However, this area of study is best left to the Finance Committee.

BREAKDOWN OF CURRENT MONTHLY WASTEWATER CHARGES

Basic Charges:

1st 60,000 gallons at 1.30/1000 gallons	$2,347.20
540,000 gallons at 1.22/1000 gallons	$19,764.00
Total basic charges	$22,111.20

BOD Surcharge:

Based on pounds of BOD over 250 ppm (Kelley and others 1981)

$$SC = \frac{(.6\ MGD)(30\ D)(400-250\ ppm)(8.34\ lbs/gal)}{1000\ gallons}\ (\$100/1000\ lbs)$$

Where: MGD = million gallons/day
D = days
Sc = service charge

	Sc = $ 2,251.80
Total wastewater treatment charges	$24,363.00
Current pipeline maintenance expenses	$ 3,000.00
Total monthly expenses	$27,363.00

Total Annual Expenses = ($27,363.00)(12) = $328,356.00

Pipeline maintenance costs are based on annual expenses divided by 12 months, and includes annual sludge cleanout charge, and winter maintenance.

BREAKDOWN OF EXPECTED PACT ALTERNATIVE COSTS (1 MGD)

Initial construction: (EPA 1980)

Base, 1976	$125,000.00
1986 dollars factor	× 2.189
Expected 1986 construction cost	$273,625.00
10% for powdered carbon	27,363.00
Sub-total	$300,988.00
Site preparation (5%)	$ 15,049.40
Sludge holding tank w/installation	$ 25,000.00
Sludge tank truck and access ramp	$ 75,000.00
Total startup cost	$416,037.40

Annual operating costs: (1980)		
Electricity (100,000 KWH at $.0694/KWH		$ 6,939.00
Maintenance	$10,000	
1986 cost factor	×2.189	$ 21,890.00
Supervisory (1 engineer & 2 assistants) (15% of startup cost)		$ 62,405.61
Sludge disposal ($100/1000 lb)(12,000 lbs/yr.)		$ 1,200.00
Tanker truck maintenance		$ 7,200.00
Total annual operating costs		$ 99,634.61
Unexpected contingencies (15%)		$ 14,945.19
Total annual expenses		$114,579.80

$$\text{Total monthly expenses} = \frac{\$114,579.80}{12} = \$\ 9,548.32$$

RECOMMENDATIONS

Based on the preceeding analysis, I recommend that we switch from our present waste treatment system to an on-site PACT package plant. The PACT system will eliminate our pipeline, along with its associated expenses, and allow us to discharge the treated effluent directly to the river, while meeting EPA requirements. Although our initial expenditures will be higher with the PACT option, we should realize the following advantages:

1. Future operating costs will be lower with the PACT system than with our current method.
2. The Finance Committee will be able to forecast future annual costs more accurately, as we gain experience with the new system. At present, we have only a year's notice of rate increases from the Morgantown Wastewater Treatment Plant, which makes budgeting difficult.
3. A one-million gallon PACT plant will have plenty of reserve capacity, in case we expand our production facilities.

A comparison of projected PACT costs and savings vs. our present situation can be found on the next page.

COMPARISON OF PACT COSTS AND SAVINGS VS. OUR PRESENT SYSTEM
(assumes no increase in operating or construction costs)

	PACT system	Present system	Annual Savings
1987 Costs:			
New Const.	$416,037.40		
Operating	$111,402.48	$328,356.00	
Total	$527,439.88	$328,356.00	−$199,083.88

1988 Costs:			
New Const.		$100,000.00	
Operating	$111,402.48	$328,356.00	
Total	$111,402.48	$428,356.00	+$316,953.52
1989 Costs:			
New Const.			
Operating	$111,402.48	$328,356.00	+$216,953.52
		Three-year savings	$334,823.16

REFERENCES

Environmental Protection Agency. February 1980. Innovative and Alternative Technology Assessment Manual. CD-53

Heath, H. W. 1986. Bugs and Carbon Make a PACT. Civil Engineering 56 (4):81–83

Kelley, Gidley, Blair, & Wolfe, Inc. Sanitary Board: City of Morgantown, West Virginia. Consulting Engineers Financial Feasibility Report for 1981 Bond Issue.

McPhee, Smith, & Rosenstein Engineers. April 1978. The Buffalo Industrial Waste Program Planning and Implementation. Buffalo, N.Y. 14202.

Writing Exercise

As a major class project, write a feasibility report. Choose an interesting problem in your field of study, invent a professional position for yourself, and do the appropriate research. You will want to consult with your professor several times during the project to be sure that you stay on the right track.

You will find it easier to write such a report if you construct a dramatic scenario for it. The scenario should imitate a real-world environment. Once you determine the problem to be solved, you might choose some other students to work with; you might, that is, form a research and writing group, imitating the group effort often used in industry. Some of you could be in charge of research; some, of writing the report itself; and others, of editing, proofreading, and producing the final copy.

The Progress Report

The progress report describes work in progress or experimental work that has reached a publishable conclusion or a point where its results are significant enough to merit being included in a quarterly or monthly journal. This kind of report can actually be given a number of different names: research report, research update, technical status report, interim lab report, etc. Common to American industry, the progress report can have a compact or expanded format. The compact version usually has two parts: work completed and work projected. The longer, expanded version shares some features with the feasibility report:

Abstract: A summary of the article or report. Often written in the third person ("The research team . . ."), but sometimes in the first person plural ("We . . ."). Abstracts are of two kinds: descriptive and informative. Descriptive abstracts describe the problem and the methodology used, leaving out the final conclusions. Informative abstracts add a statement of the conclusions to the descriptive abstract.

Introduction: Usually gives an overview of the problem confronted, the theory behind the methodology used, and a statement about the significance or importance of the current research.

Methodology: Describes how the experiment or research was conducted. Usually written in the passive voice to emphasize the work done and not the researchers *per se.* This section can have an extremely wide range of other titles and subsections. Among these subcategories are materials, methods, data accumulation, experimental facility, instrumentation, run-sample programs, experimental observations, assumptions, theories, nomenclatures, computer programs, statistical data bases, correlation techniques, project planning (description, budget, milestone schedule), project definition, design options and final design, economics and manufacturing criteria, experimental procedure, etc. The purpose of this section is to enable other researchers to duplicate your experiment and thus to verify its data.

Results: A description of the data accumulated, either research or experimental.

Discussion: Sometimes combined with the results section. Interprets the data in terms of the hypothesis of the experiment.

Conclusion: Sums up the experiment. Sometimes called Recommendations. What was learned? Where do we go from here? What are the limits of interpretation?

References: A list of the books and articles referred to in the text. Opposed to a *bibliography,* which is much longer and lists most relevant sources on the topic.

Appendix(es): Optional. Often a place to put original data, plans for equipment, calendar of activities, summaries of research efforts that were later updated, special drawings and plans, etc. If you use an appendix, mention of it should appear in the table of contents, with the individual sections listed.

In some industrial reports, as opposed to journal articles, a table of contents is added, complete with a list of figures and a list of tables, with each providing page numbers of the items listed. Equations are usually numbered throughout the text, the numbers appearing in parentheses at the right margin of the same line of the equation.

The following example is an academic article from a professional journal, which reports on the progress of a research project. Notice that the section "Methodology" is broken down into smaller subsections with headings appropriate for the subject matter.

Erosion of Metallic Plate by Solid Particles Entrained in a Liquid Jet

M. T. Benchaita, P. Griffith, and E. Rabinowicz

An experimental and theoretical study on erosion of a metallic plate by solid particles entrained in a liquid jet has been performed. The test section involved a two-phase flow jet of liquid water and silica-sand impinging on a metallic plate upon which erosion occurred. In combining the fluid mechanics of particle suspension in a liquid and a model of weight removal from the plate (by a single abrasive particle), an analytical approach is developed to determine the distribution and amount of erosion along a metallic plate. This model of erosion distribution along the metallic plate can be used as a tool to predict erosive damage in industrial equipment such as pipe bends, elbows, subsurface safety valves, and pipe contractions.

Introduction

Erosion is a phenomenon which occurs when a stream of solid particles (such as sand grains) strikes a metallic plate and results in material removal from the metallic surface by mechanical action. The damaging effect of erosion substantially reduces the useful life of parts of industrial installations and can become a primary concern from the point of view of cost for conveying systems such as oil-well casings, pipe bends, pumps, and slurry pipelines [1, 2].

The general problem of erosion involves two interdependent phenomena: the fluid mechanics of the solid-fluid two-phase flow (or momentum exchange between the solid and liquid phase) and the wear of materials by collisions of abrasive particles on the metallic surface. For example, when a fluid stream containing solid particles is turned as in pipe bends or tees, the abrasive trajectories are different from those of the fluid motion, and for a large ratio of inertia force to drag force acting on the solid particles, almost all abrasives impact on the wall. In the case of suspension particles, turbulent fluctuations in the flow velocity propel the abrasives to the wall and thus increase the rate of erosion, especially for fine particles, for which case the frequency of collisions with the metallic surface is much greater than in the case of large particles.

In this paper, we are mainly interested in developing a model for the erosion of a metallic plate by the impingement of an abrasive jet (sand particles plus water) at very low concentration (i.e., 0.3 percent by weight). In this approach, one could combine theories of fluid mechanics of solid particles in suspension in a conveying fluid stream, and the wear of metals by abrasives. The method consists of two steps: First we write the equations of motion of the solid particles in the liquid jet, and "track" each individual abrasive in its motion toward the metallic wall. Secondly, once the velocity components of the abrasive at impact on the metallic surface are known, we will derive a simple model of weight loss from the plate by the impact of a single abrasive particle. The procedure from determining the distribution of erosion depth along the metallic plate when acted upon by an abrasive jet looks as follows:

(*a*) evaluation of the liquid velocity distribution in the plane jet (two-dimensional jet) for the case of normal impingement

(*b*) determination of the abrasive trajectories in the jet in terms of the liquid velocity distribution, the abrasive size, the stand-off distance of the metallic plate from the nozzle exit

(*c*) application of the model of erosion to evaluate the rate of mass removal by a single individual abrasive at the point of impact, and then summing up the contribution of all abrasives to determine the erosion distribution along the metallic plate

---

From M. T. Benchaita, P. Griffith, and E. Rabinowicz. August 1983. Erosion of metallic plate by solid particles entrained in a liquid jet. *Journal of Engineering for Industry* 105:215–22. Reprinted by permission of ASME.

The completion of this analysis would allow one to design the flow passages in a bend, a pipe fitting, an elbow, or a stormchoke for a given life or allow one to do preventive maintenance on a pipe that was close to failure. In addition, quantitative estimates of the effect of loading, system size, particle size, liquid velocity, and abrasive size can all be made.

Experimental Program

The Test Loop. The general arrangement of the experimental apparatus is shown in [Figure 6-5]. Sand abrasives are fed into a screw conveyor, where they pass through two suitable screens in order to select a particle size (that corresponds approximately to the average of the hole sizes of the two screens). The selected sand particles are then fed into a mixing tank, into which a known volume of water has been supplied. The abrasives and water are then stirred in order to get a uniform concentration in the mixing tank. A surge pump (or plunger pump) supplies the well stirred mixture of abrasives and water to the region of observation (impingement of the abrasive jet on the copper plate). The use of a surge pump for solid mixture (up to 20 percent solid concentration, by volume) is helpful because of the erosion problems that occur in the valves of ordinary centrifugal pumps. The valves of a surge pump are made of rubber. As the flow is unsteady, a smoother flow is required. In these experiments, the surges in flow velocity are damped out by using a tank which has an enclosed air space above the two-phase flow surface. This is shown in [Figure 6-5]. The flow circuit is closed.

The Test Section. The test section is a flat copper plate onto which a two-phase jet of abrasives and water impinges. The dimensions of the metallic plate are 10 cm × 7.5 cm × 0.6 cm. The abrasive jet is issuing from a nozzle whose stand-off distance from the plate is 5.0 cm. The jet impact is normal to the plate, and the nozzle cross section is made square (2.0 cm × 2.0 cm). In these experiments, a square cross section for the nozzle, which is placed between two vertical plates, gives a T-shaped jet geometry. The choice of this nozzle shape provides a two-dimensional jet which simplifies the theoretical analysis.

FIGURE 6-5
General setup of the experiment

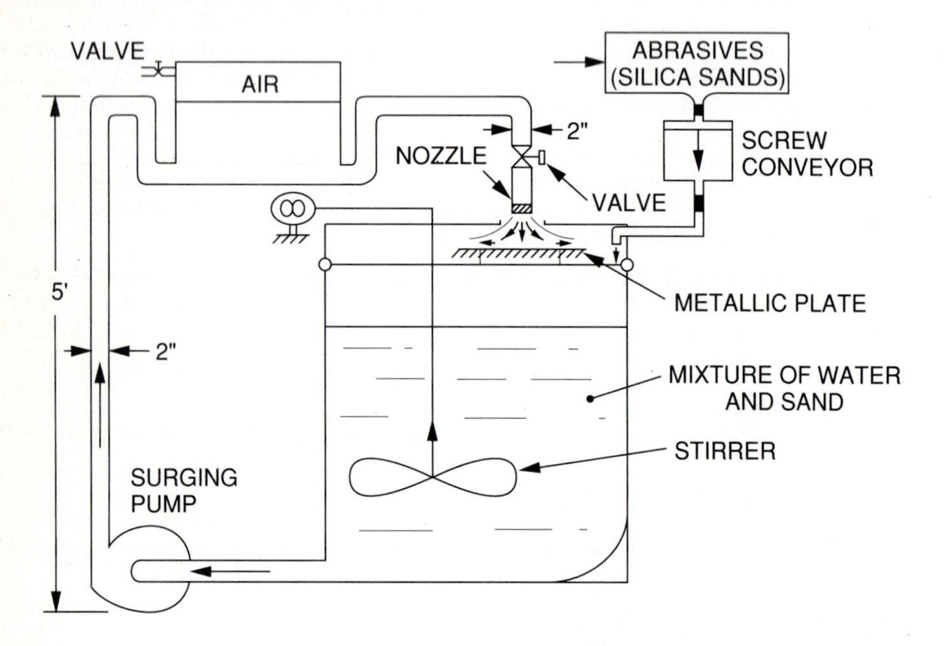

FIGURE 6-6
Measuring device of erosion-depth on copper plate

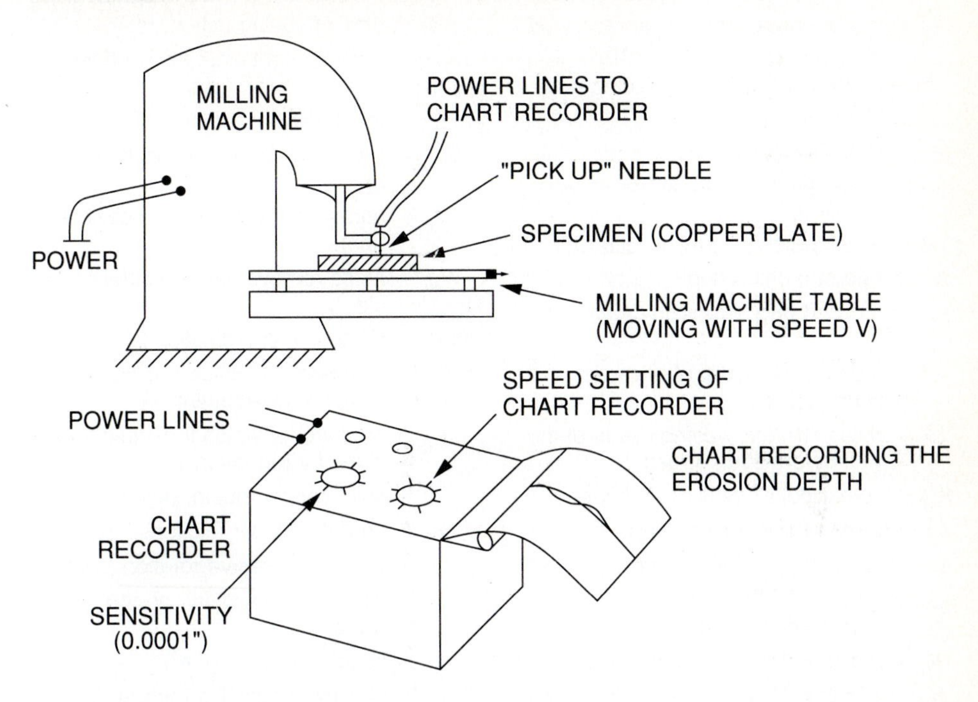

Instrumentation and Erosion Measurement. In this experiment, we are interested in the distribution of erosion depth along the metallic plate. To determine erosion distribution, we used a measuring device for surface roughness adapted to a milling machine [Figure 6-6]. Before the experiment, the metallic surface is made flat. This is achieved by the usual methods of micro-machining and polishing to get a flat surface whose roughness is less than 1 μm. The milling machine table (which is moveable) is used to move the specimen under the "pick-up needle." The pick-up needle is connected to a Twin-Viso Recorder equipped with a two-channel chart recorder (such as a Talysurph). The speed of the chart recorder is adjusted in order to get a recording length equal to the real length of the plate to be tested. These tests were carried out for a relatively long period of time (~1 hr) in order to minimize the effect of very small amounts of debris of erosion and sand that can accumulate on the plate. It is possible that mass can be added to the plate during erosion, especially in the transition (start) of erosion process if the test times are too short.

Experimental Results

Erosion Distribution Along the Metallic Plate (at Normal Impact of the Jet). The experimental results are given for a specified set of system parameters (size of silica-sand, solid concentrations, and jet velocity). The variation of erosion distribution with distance from stagnation point (or radial distance) [Figure 6-7] is very significant: Starting from the stagnation point (minimum erosion), erosion depth increases until it reaches a maximum and then decreases to an asymptotic value at larger distances from the stagnation point. This is probably due to the increase in the horizontal component of velocity of the abrasive and decrease in the normal component of its velocity (at impingement) for increasing radial distance from stagnation point. The information about the maximum erosion depth (along with the asymp-

Nomenclature

A_p = frontal area of the abrasive in streamwise direction, (μm^2)
b = width of the abrasive leading edge, (10 μm)
C_D = drag coefficient
d_m = mean diameter of abrasive particle, (μm)
g = acceleration of gravity, (9.8 m/s^2)
h_m = rate of erosion depth, (μm/hr)
$h_{m,r}$ = reduced rate of erosion depth, (μm/hr)
I = moment of inertia of abrasive particle, ($\mu g.\mu m^2$)
k = ratio of x- and y-components of the wall deformation stress
$K_{l,e}$ = coefficient of local erosion depth
L = width of the nozzle, (cm)
L_m = distance along the plate where all abrasives have impinged, (cm)
M = parameter of the liquid flow, (s^{-1})
m_p = mass of single abrasive, (μg)
N_p = number of abrasive particles impinging on the plate
p = Vickers hardness of the metal plate, (kgf/mm^2)
r_m = mean radius of abrasive particle
t_c = time of cutting action, (s)
(u,v) = x-component and y-component of liquid velocity, (m/s)
U_l = liquid velocity, (m/s)
V_{exit} = liquid velocity at the nozzle exit, (m/s)
V_p = abrasive velocity at the nozzle exit, (m/s)
W_{sp} = mass removal by one single particle, (μg)
(x,y) = Cartesian coordinates
x_{fs} = abscissa for which the liquid flow is given by (equation 1)
y_{fs} = ordinate for which the liquid flow is given by (equation 1)
γ = solid concentration, (kg/kg)
δ_e = boundary layer thickness, (μm)
θ = angle of abrasive rotation, (rad)
ν_l = kinematic viscosity of the liquid, (m^2/s)
ρ_l = liquid density, (kg/m^3)
ρ_p = density of abrasive particle, (kg/m^3)
ρ_s = density of the metallic plate, (kg/m^3)
ϕ = geometric factor at interface abrasive-metal plate

FIGURE 6-7
Variation of erosion depth along copper plate (not to scale)

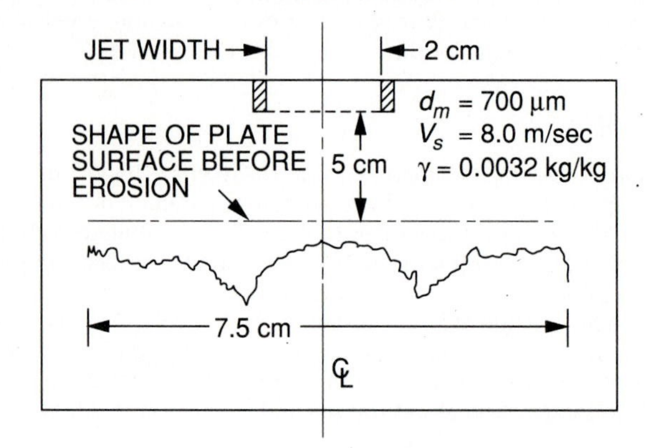

totic value of erosion depth at large radial distance) is useful for designing pipe bends, tees, elbows, and so on.

Other experimental results were obtained for different values of the system parameters (such as jet velocity, particle size) and will be reduced and plotted along with the theoretical results (for comparison purposes) later in this paper.

Theoretical Analysis

In this analysis, we would like to predict the distribution of erosion depth along the copper plate. To carry out this study we will proceed as follows [Figure 6-8].

(*a*) Modeling of the liquid flow and determination of trajectories of solid particles in the abrasive jet until the abrasive impacts on the wall. This is accomplished by resolution of the equations of motion.

(*b*) Derivation of the weight loss from the copper plate by the impingement of a single idealized abrasive particle.

(*c*) Summing up the contributions of all abrasive particles (of a given size) contained in the liquid jet, we will evaluate the distribution of erosion depth along the copper plate.

Model of the Liquid Flow. In order to model the liquid jet, the flow field is assumed to be a two-dimensional, steady potential flow (vorticity-free) in which the turbulent fluctuations are neglected. The detailed study of this type of liquid jet is shown in reference [3]. The flow model of the jet is divided into three regions [Figure 6-9].

FIGURE 6-8
Flow chart for calculation of erosion distribution along the plate

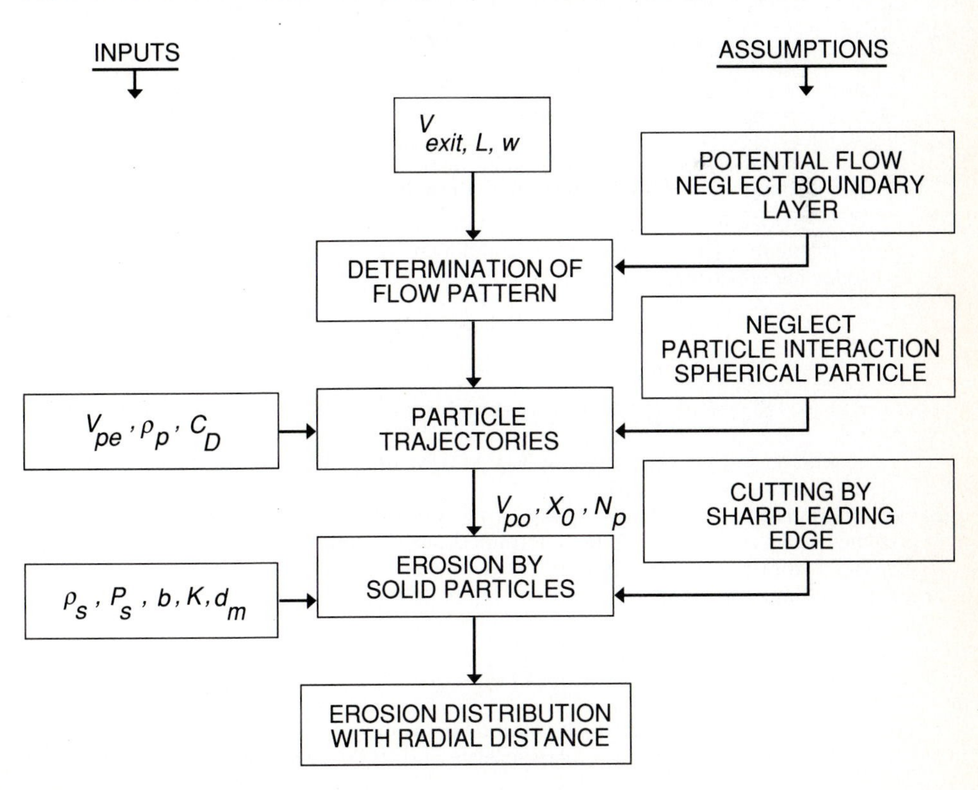

FIGURE 6-9

Two-phase flow (water-solid particles) impinging on a flat plate at normal attack in two-dimensional flow

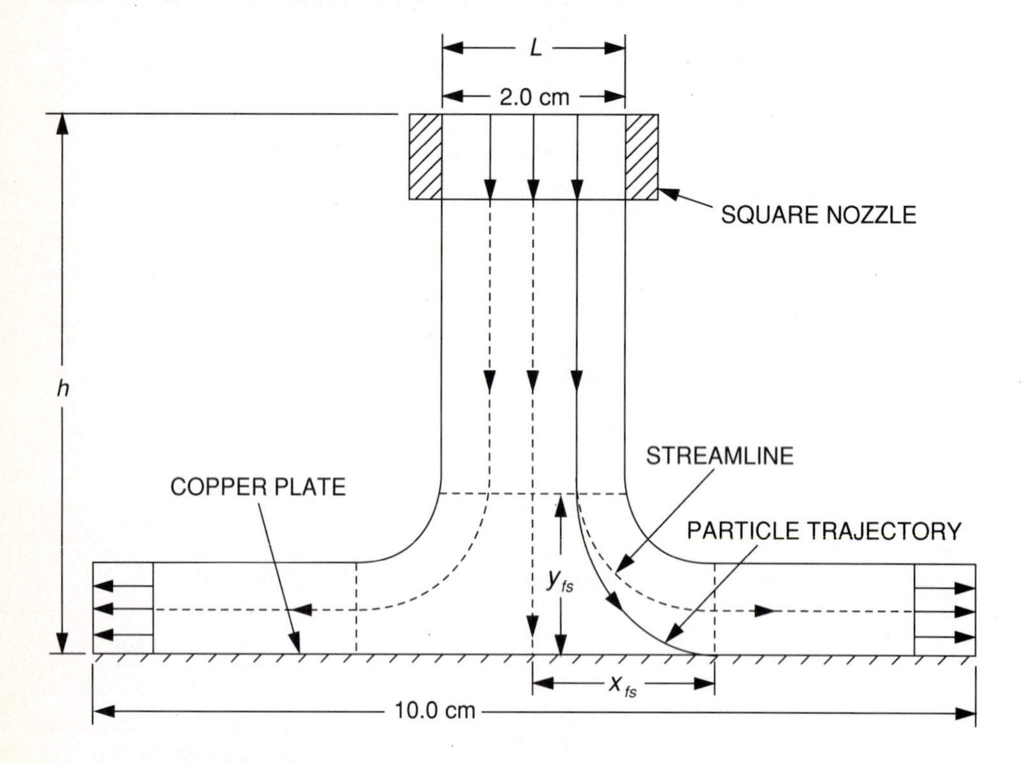

- an incoming parallel and uniform flow from the exit plane of the nozzle down to the position $y = y_{fs}$;
- a streamlined jet flow where the component of the liquid velocity is given by the following equations

$$\left.\begin{aligned} U &= Mx \\ V &= -My \end{aligned}\right\} \tag{1}$$

where (x,y) are the system coordinates measured from the stagnation point, and M is a flow parameter;

- a uniform and parallel outgoing flow beyond the position $x = x_{fs}$ where the velocity normal to the wall is zero.

In fact, a boundary layer develops starting at the stagnation point (at $x = 0$, $y = 0$). The thickness of the boundary layer should be determined, since it could play an important role in the motion of the abrasives at their approach to the metallic plate. The boundary layer is constant in this type of liquid flow [3], and it is given by the following expression

$$\delta_e \simeq 2.4\sqrt{\frac{\nu_l}{M}}$$

where M is the flow parameter which depends upon the jet velocity, the nozzle width, and the stand-off distance between the metallic plate and the nozzle. For the conditions of the experiments the jet velocity was set at 8.0 m/s. The stand-off distance is equal to 5.0 cm, and the nozzle width is 2.0 cm. In this case the boundary layer thickness is equal to 120 μm (viscosity $v_l = 1.14 \times 10^{-6}$ m²/s, flow parameter $M = 444.5$ s^{-1}). Consequently, and since we are interested in abrasive size larger than 300 μm (typical industrial conditions), it should be reasonable to assume that the slowing down effect of the solid abrasives due to the boundary layer can be neglected. . . .

Influence of Abrasive Size on Erosion Profile. In some other studies [4, 5] the total weight loss of the metallic plate is found to be practically independent of the abrasive size (case of $d_m > 200$ μm), when keeping the same loading (i.e. same total mass of impacting particles).

Now, we are interested in evaluating the distribution of erosion depth along the copper plate for different abrasive sizes. The method for determining the local erosion distribution is the same as in the previous paragraph, and the erosion depth along the metallic plate is given by equation (9).

The analysis was carried out for different abrasive sizes, i.e. $d_m = 700$ μm and $d_m = 400$ μm. The jet velocity is kept constant at $V_e = 8.0$ m/s. The results are plotted in [Figure 6-10] (for $d_m = 700$ μm), and [Figure 6-11] (for $d_m = 400$ μm) using an erosion coefficient $K_{l,e} = 0.18$. It can be seen that the experimental erosion profile is adequately correlated by the theoretical model, especially the position of maximum erosion depth. Comparing [Figures 6-10 and 6-11], one could notice that the maximum erosion depth is greater for the larger abrasive size, but larger spread of erosion along the metallic plate is obtained for smaller abrasive size. This important feature of grit size effect on erosion distribution is probably due to the

FIGURE 6-10
Rate of erosion depth along the metallic plate

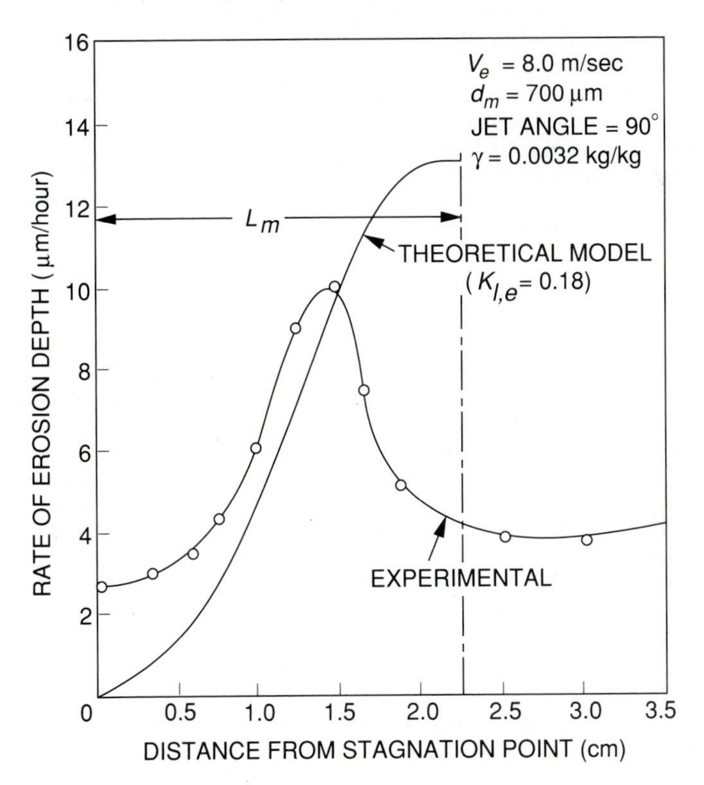

FIGURE 6-11
Erosion distribution along the copper plate

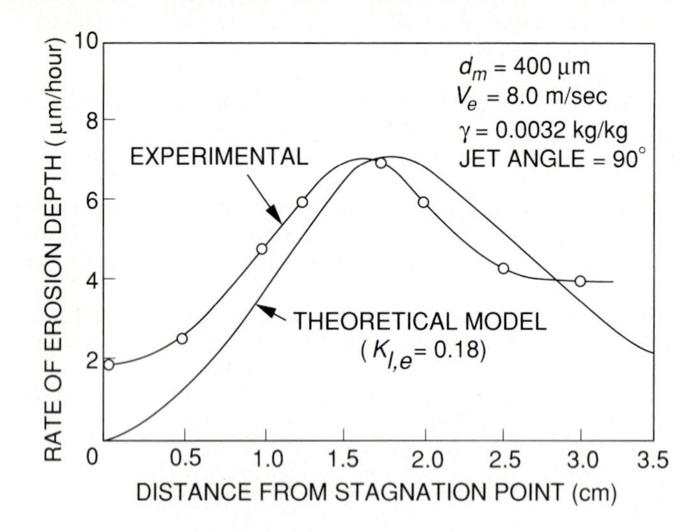

fluid mechanics of particle suspensions, i.e. small abrasives are carried in the main stream far from the stagnation point (because of the large ratio of horizontal drag force/vertical drag force); and the large abrasives have a tendency to hit the metallic plate at very close positions to the stagnation point. Thus, for the case of large abrasives, large erosion rates are obtained in a narrow band around the stagnation point; meanwhile larger spread of erosion (i.e. erosion of larger surface area of the metallic plate) is obtained for smaller abrasive sizes.

This theoretical model in common with others does not predict erosion at the stagnation point, although the experiments show noticeable erosion. This is partly due to the way we defined the mass removal in equation (6); i.e., although there is no mass removed by the "cutting process," some other effects such as low-fatigue cycle of the material and abrasive-abrasive interaction (not included in our analysis) might play an important role in erosion at the stagnation point.

Conclusion

In this study, we combined the theory of fluid mechanics of suspended particles and a simple model of weight removal of material to develop a general model of erosion of metallic materials by solid abrasives (such as sand grains).

The erosion distribution and the position of maximum erosion depth obtained in this analysis were quite consistent with the experimental measurements, except for general overestimation of the erosion rate. The introduction of an erosion coefficient (i.e. $K_{I,e}$ calculated from experiments) takes into account the effects (not considered in our analysis) of additional resistance to erosion such as surface work-hardening, surface roughness, surface geometry of cutting process, oxide layer film, secondary flow effects, and squeeze film, therefore, and knowing the properties of the liquid flow (i.e., size, density, velocity, etc.), the abrasive properties (i.e., size, density, velocity, etc.), and the properties of the metallic plate (such as hardness and density), the maximum erosion depth of the plate can be well determined for each system geometry. In this study, the treatment of erosion was carried out by mechanical considerations only. However, in some practical cases of erosion by slurries, corrosion effects might also be important.

It remains to be shown how these results can be applied to other geometries where secondary flows can be significant.

Acknowledgments

The authors wish to express their thanks to Professors Peter Huber and Forbes Dewey, Jr. for their helpful discussions and suggestions regarding this work. Technical assistance was provided by Mr. Fred "Andy" Anderson and Mr. Joseph "Tiny" Caloggero, typing by Ms. Purdy Young. Their assistance is sincerely appreciated.

Funding for this research was provided by Mobil Oil Company.

References

1. Thompson, T. L., and Aude, T. C., "Slurry Pipeline Design and Operation Pitfalls to Avoid," ASME Paper 76-PET-12, June 8, 1976.
2. Faddick, R. R., "Pipeline Wear From Abrasive Slurries," First International Conference on the Internal and External Protection of Pipes, Sept. 9–11, 1975, University of Durham.
3. Benchaita, M. T., "Erosion of Metal Pipe by Solid Particles Entrained in a Liquid," Ph.D thesis in Mechanical Engineering, Massachusetts Institute of Technology, Cambridge, Mass., Feb. 1980, pp. 87–98.
4. Mills, D., and Mason, J. S., "Learning to Live with Erosion of Bends," First International Conference on the Internal and External Protection of Pipes, Sept. 9–11, 1975, University of Durham, Paper G1.
5. Benchaita, M. T., *op. cit.,* pp. 79–81.

Writing Exercise

In order to write a report in this format, you must be fairly far along in your undergraduate studies. Choose some scientific or engineering lab research or a design project and write a progress report. Be sure to organize the section on methodology according to the nature of your research or design.

7

Innovative or Revolutionary Science and Technology

Chapters 5 and 6 presented readings and writing exercises based on traditional science and technology, the research and technological development that occupies most scientists and engineers. In traditional science and technology, knowledge develops at a steady pace, and puzzles are solved within the intellectual and theoretical framework of established disciplines.

In contrast to this steady development are the scientific and technological revolutions, which change our most basic understanding of the nature of the physical world or the best way of solving a technological problem. Galileo's revolution in astronomy, Darwin's in biology, and Einstein's in physics are examples of revolutions. Once the new theory was broadcast to the world, no one could think in the same old ways again.

Revolutions of less epic magnitude are with us today. Here are two examples of contemporary revolutions, one in archeology and anthropology, the other in physics, and astrophysics and the problem of how unpredictable behavior (chaos) can result from predictable motion.

REVOLUTIONS IN ARCHEOLOGY, PHYSICS, AND ASTROPHYSICS

William K. Stevens describes below one of the current revolutions in archeology and anthropology, which is that hierarchical societies existed much earlier than had been previously thought, and that such societies were far removed from any Edenic or pastoral model.

Prehistoric Society: A New Picture Emerges

William K. Stevens

Archeologists and anthropologists are discerning a major, previously unrecognized phase in the development of early civilization. The gradually unfolding discovery promises to revolutionize standard textbook notions about the nature of human life in the millenia just before recorded history began.

For years, the experts have believed that before 3500 or 4000 B.C., when the first cities blossomed in the Middle East, people lived pastoral lives in farming villages whose simple patterns of existence had hardly changed since people first settled down to cultivate the earth about 10,000 years ago.

"We've been thinking of these early agriculturalists as peaceful, egalitarian groups, happy in the sunshine, not terribly motivated to go off and look at other areas, really simple, self-sufficient communities," says Mary M. Voigt, an anthropologist at the University of Pennsylvania's Museum of Archeology/Anthropology.

Now, however, experts are concluding that complex, sophisticated economic and social systems developed 7,000 to 9,000 years ago—1,000 to 3,000 years before the first cities rose and humans began to write things down.

The society that developed during this period was not quite urban, the experts say. It constituted a long, stable transition—previously unknown—between earlier, simpler agricultural societies and what is called the Urban Transformation of the third and fourth millenia B.C., when urbanization, writing, and metal-making converged in one of humanity's biggest leaps forward.

The emerging new portrait of humanity in the late neolithic age is that of restless people whose trading networks spanned the known world; who had moved far beyond subsistence agriculture to become specialized producers of pottery, beads, hides, food and other goods, and who had already begun to heat and work metals.

No longer the egalitarian villagers of an earlier era, they displayed a social hierarchy, competed economically, developed the precursors of central cities with satellite villages, built walled towns, and probably made war on each other. Some may have practiced human sacrifice. Far from the bland people depicted in some anthropology texts, they painted their bodies and wore bangles.

Stevens goes on to describe the meaning of this revolution and the method by which most substantive changes of thought take place in the sciences and technology.

It has become plain to a number of archeologists and anthropologists that the model of early village life that has dominated thought on the subject for the last quarter-century is obsolete and should be discarded. This is so, they say, only partly because of new discoveries and new, science-based analyses of village artifacts. It is also an outgrowth of new ways of looking at the results of archeological excavations in the Middle East over the last three decades.

This is not a case in which a single, dramatic discovery has produced a revelatory breakthrough. Rather, as more often happens in science, it is a case in which the accumulation of exceptions to a prevailing theory seems, in time, to overwhelm and overturn the theory itself. One exception after another emerged from the Mideast digs, but for a long time they were explained away, says Dr. Voigt. "Nobody," she said, "ever really sat down and made the jump to say, 'Well, this means we have to start all over.' "

What Stevens calls "exceptions" other historians of science and technology call "anomalies." These are facts, literally irregularities, that don't fit into accepted theory. These exceptions to what is expected can sometimes be ignored; but when enough of them accumulate, and when enough professionals recognize their objective existence, a new theory often arises to accommodate them. Thomas Kuhn discusses this process in *The Structure of Scientific Revolutions* (see Chapter 7).

At a symposium called "Chaos '87: The Physics of Chaos and Systems Far From Equilibrium," held in Monterey, California, physicists and experimenters from around the world discussed the breakdown of the traditional view that the heavenly bodies operate according to a stable, clockwise, Newtonian predictability. One writer reported the discussions as follows:

Tales of Chaos: Tumbling Moons and Unstable Asteroids

James Gleick

New computer calculations, based on observations from the Voyager satellites and ground-based telescopes, show that just about every moon—except the earth's—has experienced millions of years of chaotic tumbling. . . . Unlike the familiar spin of body around an axis, . . . a tumbling moon falls end over end, twists sideways, speeds up, slows down, all the time obeying Newton's completely deterministic laws of motion, yet [defying] prediction in a way that scientists used to consider impossible. One of Saturn's moons, a 250-mile-long football-shaped body called Hyperion, appears to be tumbling now, and its unpredictability is such that even if Voyager I had measured Hyperion's position and speed to 10-digit accuracy, God's own computer would not have been able to calculate where it would be when Voyager II came by a year later.

Of this tumbling effect, which follows Newton's laws but whose plot, nevertheless, cannot be predicted, James Yorke, a mathematician at the University of Maryland, said, "Here is a departure from everything you've expected. . . . You have no special perturbation or complexity. It's just basic physics—yet it's doing something that nobody thought about." This theory applies especially to the orbits of asteroids that become "chaotic" under the gravitational influences of nearby planets. During thousands of orbits over countless years, the trajectory of the asteroid becomes increasingly off-center, causing it to tumble.

Gleick also notes the rational scepticism of scientists when new theories are first introduced and then their gradual conversion:

To many astronomers, such chaotic "bursts" of eccentricity seemed implausible. "When Wisdom [an M.I.T. professor] showed you could get this chaotic effect, it was met with considerable skepticism," Dr. Wetherill [a conference participant] said. "There was a tendency among almost everybody to say, 'Well, he just doesn't know how to integrate yet, wait till he gets a little older, he'll understand these things.' " But most are now convinced.

Gleick says that this new field, which studies how ordered systems can become chaotic (unpredictable), has great interdisciplinary application for any system displaying complex dynamical behavior. The "fantastic structures" in crystals, the form of shock waves, and the problem of tumult in fluid flows may all be affected by this

---

new approach, which adds the use of computers to the two standard methods of doing science, theory and experiment. (See also Gleick 1987.)

The energy and exultation of scientists and engineers caught up in discovery is classically exemplified by Lord Rutherford's 1911 revision of the model of the atom that J. J. Thomson had proposed in 1903. Sir Karl Popper chronicles Rutherford's response:

> Rutherford had accepted Thomson's theory according to which the positive charge must be distributed over the whole space occupied by the atom. This may be seen from his reaction to the famous experiment of Geiger and Marsden. They found that when they shot alpha particles at a very thin sheet of gold foil, a few of the alpha particles—about one in twenty thousand—were reflected by the foil, rather than merely deflected. Rutherford was incredulous. As he said later: "It was quite the most incredible event that has ever happened to me in my life. It was almost as incredible as if you fired a fifteen-inch shell at a piece of tissue paper and it came back and hit you." This remark of Rutherford's shows the utterly revolutionary character of the discovery. Rutherford realized that the experiment refuted Thomson's model of the atom, and he replaced it by his nuclear model of the atom. This was the beginning of nuclear science. Rutherford's model became widely known, even among non-physicists. (Popper 1981, 102)

Later in this chapter we'll investigate theories that attempt to account for how such scientific revolutions take place. First, though, is an introduction to a phenomenon that is almost always present in the environment of scientific and technological revolutions: the notion that how we perceive (i.e., the physical and physiological mechanisms we use to perceive) determines what we actually do perceive.

Reference List

Gleick, J. 1987. *Chaos.* New York: Viking Penguin.

Popper, K. 1981. The rationality of scientific revolutions. In *Scientific revolutions,* edited by Ian Hacking. New York: Oxford University Press.

DESCRIPTION: HOW WE PERCEIVE DETERMINES WHAT WE PERCEIVE

Animal physiologists have long been fascinated by how the mechanisms an animal uses to perceive determines what the world looks like, and so is, to that animal. For instance, bats perceive almost entirely by a sophisticated radar system: their flight is governed by complex radar that they emit, much as a jet fighter does when flying at night. The bat's world consists of sound, arranged and filtered to allow the bat to fly in total darkness.

We might think that the bat is limited by its perceptual system. No doubt it is, but its lifestyle is perfectly suited by evolution to profit from its own perceptual limitations. Humans are limited too: we can't see what is directly behind us, and our effective peripheral vision is less than 180 degrees. Many lizards can see more effectively. And, if we close one eye or lose its use through an accident, we are limited to seeing two dimensions instead of three (no depth perception).

What we know, and therefore what we think the world is like, is directly related to our powers of perception. The following selections are arranged to demonstrate how the act of augmenting our perceptual field through technology can lead to revolutions in our world view. (Similar arguments can be made in the field of the humanities, too.)

In the first selection, Harold M. Schmeck, Jr., a science writer for the *New York Times,* describes how contact X-ray microscopy enables us for the first time to see a moment in the life of a living cell.

New Tool Captures Cells Alive

Harold M. Schmeck, Jr.

Flashes of X-rays lasting only billionths of a second are giving scientists their first highly detailed looks at individual living human cells.

The images, obtained with a new technique called contact X-ray microscopy, can show internal structures as well as surface characteristics. The details are far finer in resolving power than can be obtained with the best of light microscopes. [See Figure 7-1.]

Although electron microscopes can reveal finer detail than is possible with X-ray microscopes, they cannot make images that capture an instant of life. For viewing under the electron microscope, subjects must be housed in a vacuum chamber and therefore cannot ordinarily be pictured while they are still alive. [See Figure 7-2.]

FIGURE 7-1

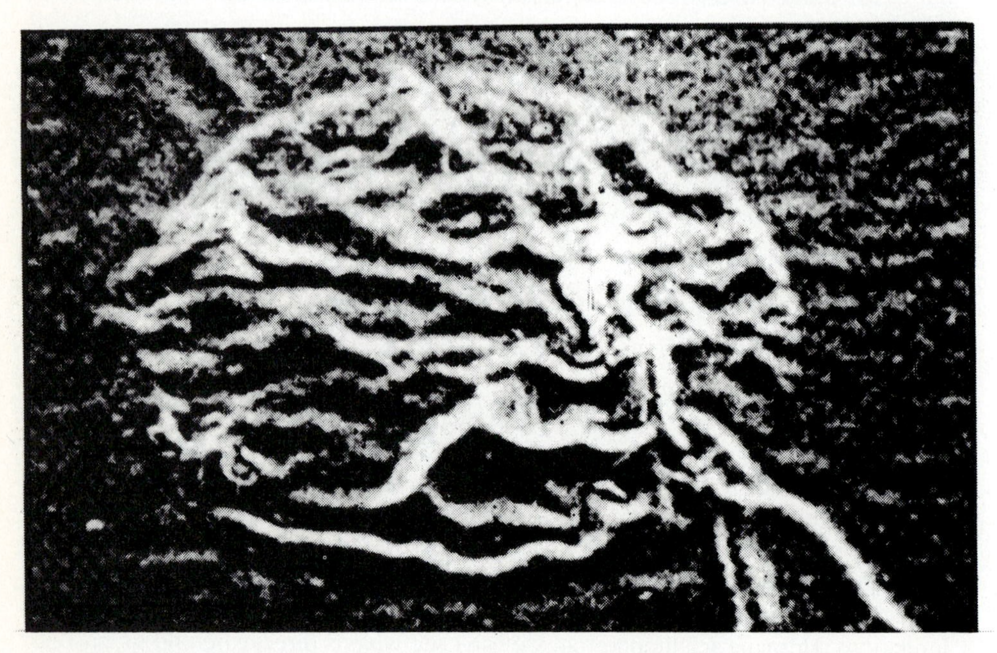

The first detailed images of a living blood platelet is obtained with X-ray microscope.

From Schmeck, H. M., Jr. 15 January 1985. New tool captures cells alive. *New York Times,* III, 1:4.

FIGURE 7-2

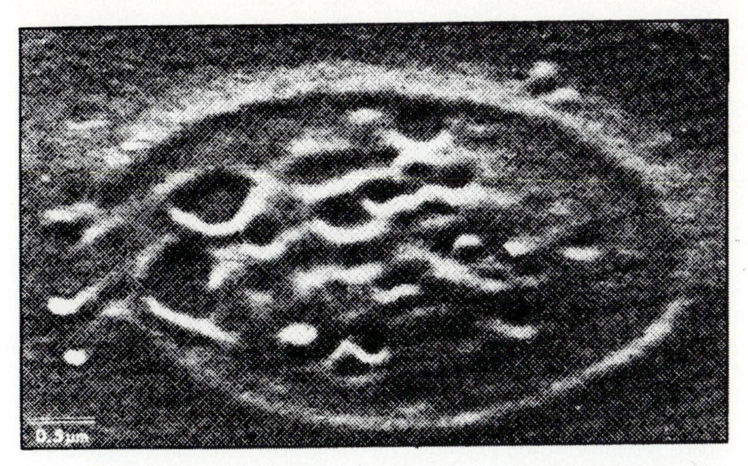

The image of a dead platelet offers far less information.

The new technique makes use of long wavelength, or "soft," X-rays that kill the specimen, but since the pulse lasts only 100 billionths of a second, the image is formed before the destructive effects occur.

"We report here what may be the first soft X-ray image of this type, that of a living human blood platelet, produced with a flash X-ray source that emits a 100-nsec pulse of soft X-rays," said the first public report of this research in the Jan. 4 issue of the journal *Science*. The abbreviation nsec means nanosecond, or billionth of a second.

Blood platelets, crucial to the healing of wounds, were the first subjects of this new kind of microscopy, but skin cells are already under study and a wide variety of other living cells and components will be examined soon.

The new microscope technology is one of many shown here in a major new exhibition on microscopy that opens today at the I.B.M. Gallery of Science and Art at Madison Avenue and 56th Street.

"We are showing microscopy from 1650 to 1990," said Dr. Cecil H. Fox, a research pathologist of the National Cancer Institute who is an adviser and major contributor to the exhibition. On display are more than 50 microscopes, including some of the earliest as well as the most modern, and many spectacular pictures that have been taken with them.

The contact X-ray microscopy technique, also known as flash X-ray microscopy, is one of the newest aspects of this ancient and important branch of art and science. The exhibit includes some of the pictures taken by this method. A film clip on view for visitors describes the technology.

While the new report on the X-ray technique describes research on platelets, the technique is considered applicable to many kinds of human cells and to the study of many important functions of life, such as cells' actions as scavengers and in secreting vital hormones and other substances.

"The images reveal details not previously seen in images of fixed or dried platelets," said the report in *Science* from a group of research workers from I.B.M.'s T.J. Watson Research Center, in Yorktown Heights, N.Y.; the National Institutes of Health, in Bethesda, Md., and New England Deaconess Hospital and Harvard Medical School in Boston.

Platelets are pancake-shaped structures, about half the size of red blood cells, that serve a vital thumb-in-the-dike function in the human circulatory system. They circulate in the blood in an inactive state until breakage or injury occurs. They become activated when they encounter a break or damage in a blood vessel. Their role in wound healing is that of clumping together quickly to stop blood flow through breaks in the injured vessels, said Dr. Mario

Baldini of Deaconess Hospital and Harvard, an expert on platelets and one of the authors of the report in Science. This stoppage comes before the actual formation of a durable blood clot. It is an early stage of wound healing.

Platelets were chosen for the new studies because they are of such great importance in understanding wound healing and probably such disease conditions as atherosclerosis.

The new method used to produce pictures of platelets in action stems partly from techniques used to produce extraordinarily small printed circuits for the computer industry, said Dr. Ralph Feder of I.B.M., first author of the report in *Science,* a pioneer in developing the contact X-ray microscopy technique.

The freshly prepared and still living platelets are exposed to the flash of X-rays from a Maxwell Laboratories Low Energy X-ray Illumination Source. A bas-relief, virtually three dimensional, image is fixed on a sheet of synthetic material sensitive to X-rays and known as an X-ray sensitive resist. This image is then observed under the electron microscope, showing details of platelet structure never obtained before. [See Figure 7-3.]

In the research, platelets have been observed under both kinds of electron microscope, transmission and scanning, and reveal somewhat different structural features under each. Some of the features shown in the pictures are still totally unknown as to function, but they are already offering new insights into the mechanisms by which platelets function.

As a key part of their function in halting bleeding, activated platelets send out projections called pseudopods and evidently tangle with each other and adhere to the blood vessel to block off blood flow. Just how the pseudopods form is still unknown, but the new X-ray microscopy shows clearly that the pseudopods are not just exterior projections, Dr. Baldini said, but have roots deep inside.

FIGURE 7-3
Photographing a Live Cell

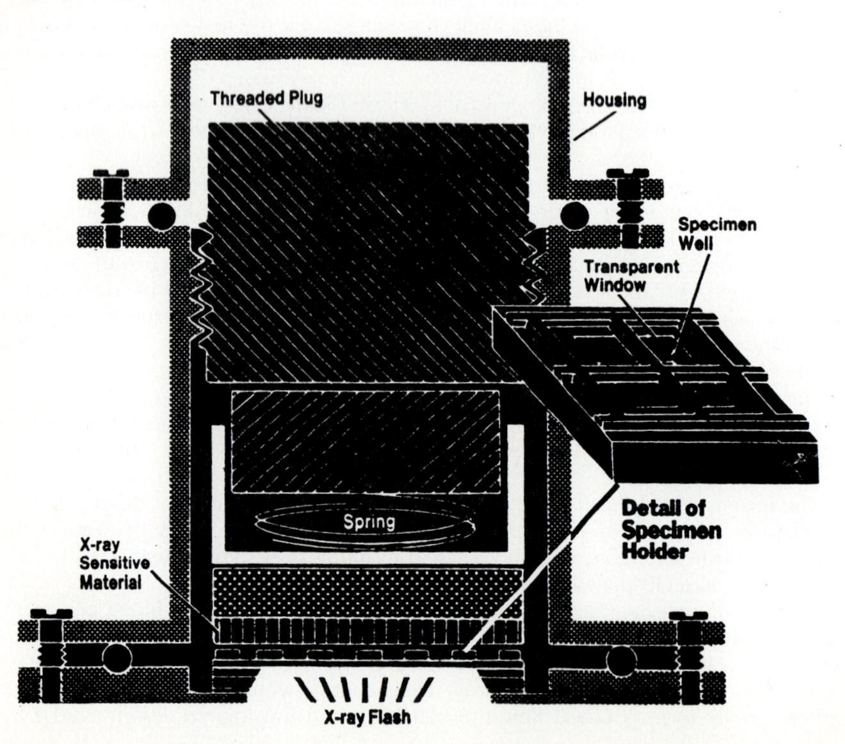

Computers and Microscopes

Other displays in the exhibition show the ability of computers paired with microscopes to improve images and to obtain information that goes far beyond the pictorial. Many years ago, Dr. Fox said, scientists developed techniques through which the dry weight of a microscope specimen could be calculated from the images that were formed under special lighting conditions. This quantitative aspect of microscopy, revealing not only the shape of a specimen, but also details of its chemistry, is an important aspect of some of the latest technologies, he said. Modern computers have made much more practical the extraction of quantitative information on weight and chemistry from suitable microscope images. . . .

Included in the historical exhibit is a replica of a small single lens device hardly larger than two joined 35 millimeter photographic slides. The microscope, of a design from the middle 1600's, was similar to those that Robert Hooke and Anton Van Leeuwenhoek used in making discoveries concerning bacteria, spermatozoa and other previously unseen things.

Combining Science and Art

Later in that century compound microscopes, using more than one lens, were produced that were themselves works of art as well as being instruments of science.

Dr. Fox said the conventional light microscope reached its highest stage of technical advancement in resolving power in the 19th century. Electron microscopes were first introduced in the 1930's. Since then a series of new technical developments has greatly expanded the kinds of information that can be obtained.

Most of the microscopes on display are from the Armed Forces Medical Museum in Washington. Many of the photographs come from the Smithsonian Institution.

While the exhibition focuses primarily on microscopy as an art and science, several high quality contemporary Zeiss light microscopes are available for use by the public.

In addition to being a display of instruments that show the progressive development of the microscope for more than three centuries, the exhibition shows hundreds of the dazzling pictures made possible by both ancient and modern microscopy, giving the visitor a real sense of the art and science through which microscopy has transformed human understanding of the world of living things.

One of the greatest revolutions in modern physics is the development of quantum mechanics, especially the indeterminacy principle first put forth by Werner Heisenberg. This principle describes one of the most profound insights about the relationship between *how* something is perceived and *what* is perceived. The next excerpt attempts to explain this very difficult principle.

Max Planck (1858–1947), German physicist, formulated the quantum theory and Planck's Constant (which describes the proportional relationship between the quantum of energy possessed by radiation and that radiation's frequency). Here he describes the indeterminacy principle.

The Philosophy of Physics

Max Planck

The quantum theory evolved originally from the radiation of light and heat; accordingly we may well begin at this point by dealing with the processes of radiation. Numerous facts allow us to regard it as proved that the energy in a beam of light of any given colour does not move in a steady continuous stream, but progresses in individual parts called photons,

---

From Planck, M. 1936. *The philosophy of physics.* Translated by W. H. Johnson, 55–60. London: George Allen & Unwin. Reprinted by permission of the publisher, Unwin Hyman Ltd.

the size of which depends exclusively upon the colour of the light; these photons fly from their source in all directions with the velocity of light and to this extent behave in accordance with Newton's emanation theory. Where the light is intense the photons follows each other so densely that they are practically equivalent to a steady continuous stream; however, as the distance from their source increases the density of the ray decreases and the photons are less close to each other, like a jet of water which grows progressively thinner until it turns into a number of individual drops of a certain magnitude. The characteristic fact is that the photons (the "drops" of energy) do not grow smaller as the energy of the ray grows less; what happens is that their magnitude remains unchanged and that they follow each other at greater intervals.

Now it is easy to see that the application of causality to these events leads us to serious difficulties. Let us take, for example, a ray of a given colour falling upon a highly-polished level sheet of glass. Part of the light will then be reflected and another part, say three times as much, will pass through the sheet. The ratio between these two parts does not depend upon the intensity of the light, or, in other words, upon the number of photons impinging on the glass. This much is shown by experience. Now if the number of impinging photons is large, e.g., a million, it is easy to state how many will be reflected and how many will penetrate: a quarter of a million will be reflected, and three-quarters of a million will penetrate. If, however, the ray of light is extremely weak, a single photon may impinge on the sheet, and then the question whether it will be reflected or will penetrate is, to say the least of it, a source of serious embarrassment. The easiest solution would be to divide it into four: but this is impossible.

But worse is to come. In the previous example we might find a way out by assuming that, while there was a temporary state of uncertainty, there might still be some hitherto unknown factor decisively influencing the photon in one sense or the other. The following case, however, seems to be entirely hopeless. It is a fact that certain colours are reflected by preference while others are allowed to penetrate by preference. When a white ray falls on the sheet the sheet looks coloured in the reflected light and also in the penetrating light. The classical wave theory of light gives an entirely satisfactory explanation of this phenomenon by saying that the light reflected at the front of the sheet interferes with that reflected at the back, so that the two reflected rays strengthen or weaken each other in accordance as the wave crest of one ray coincides with the crest or the trough of the other ray. Now the wave lengths of different colours are different, so that there are differences for the different colours, and the differences thus calculated agree exactly with actual measurements. This phenomenon too can be observed with light of the least intensity.

What happens now when a single photon impinges on the sheet? The photon must interfere with itself, since otherwise its wave length could not exert any influence. For this purpose, however, it would have to separate into parts; and this is impossible. We see thus that this view is altogether untenable.

Mechanics is in the same position as optics, as far as the quantum theory is concerned. The smallest mass points, the electrons, are in the same condition as the photons: they interfere with each other. An electron having a given velocity in this respect resembles a photon of a given power; if it impinges upon a sheet of crystal at a certain angle it is either reflected by preference or passes through by preference according to its velocity, and a complete explanation of this phenomenon in all it details is afforded by considering the wave length corresponding to its energy. The path taken by the electron when impinging upon the sheet has therefore never been calculated, and indeed it cannot be calculated.

The fundamental difficulty of determining the place of an electron moving at a certain velocity is expressed in a general manner by the uncertainty relation originally formulated by Werner Heisenberg. This relation is characteristic of quantum physics and states among other things that the measurement of an electron's velocity is inaccurate in proportion as the measurement of its position in space is accurate, and vice versa. It is not hard to discover the reason. We can determine the position of a moving electron only if we can see it and in order to see it we must illuminate it, i.e. we must allow light to fall on it. The rays falling on it impinge upon the electron and thus alter its velocity in a way which it is impossible to calculate. The more accurately we desire to determine the position of the electron, the shorter must be the light waves employed to illuminate it, the stronger will be the impact, and the greater the inaccuracy with which the velocity is determined.

This much having been discovered it is clearly impossible even in principle to transfer into the world of the senses with any desired degree of accuracy the simultaneous values of the co-ordinates and of the velocities of material points such as we find them at the core of the world image of classical physics. This impossibility makes it difficult to apply a strict causality and has led certain indeterminists to claim that the law of causality as applied to physics has been definitely refuted. On closer consideration, however, it is seen that this conclusion rests upon a confusion between the world image and the world of sense; it is at any rate premature. It is far more natural to avoid the difficulty by another method, a method which has often rendered good services in similar cases and which consists in assuming that it is meaningless, with respect to physics, to ask for the simultaneous values of the co-ordinates and of the velocities of a material point or for the path of a photon of a given colour. Evidently the law of causality cannot be blamed because it is impossible to answer a meaningless question; the blame rests with the assumptions which lead to the asking of the question, i.e. in the present case with the assumed structure of the physical world image. The classical world image has failed us and something else must be put in its place.

This has actually been done. The new world image of quantum physics is due to the desire to carry through a rigid determinism in which there is room for Planck's quantum. For this purpose the material point which had hitherto been a fundamental part of the world image had to lose this supremacy. It has been analysed into a system of material waves, and these material waves are the elements of the new world image.

Writing Exercises

The goal of these two writing assignments is for you to explore the relationship between new means of perception and the possibility of revolutionary perceptions. Since these are open-ended and exploratory writing assignments, no requirements for length or audience are specified.

1. Following the Schmeck example, write a description of an experiment in which a change in a perceptual mechanism enabled you to see something you had not seen before. The new perception you make does not, of course, have to be a real revolution; it just needs to be a new perception for you.

2. Following the example of Heisenberg's indeterminacy principle, write up an analysis of a lab you have recently completed in the physical sciences or in engineering. Base your analysis on a discussion of how the measuring or testing device you were using may have altered the material or specimen you were analyzing. Was the alteration significant? Does Heisenberg's principle have practical application other than on the subatomic scale?

EXPLANATION: THE EXTRAORDINARY BUT USUALLY UNNOTICED

As we have seen, the process of revolutionary thinking is often initiated by having our perceptions sharpened by innovative technological mechanisms. This process is also strengthened by learning to see the extraordinary where most people see only the ordinary or the unimportant. The selections in this section demonstrate this habit of mind at work.

In the following article, Walter Sullivan, a science writer for the *New York Times,* probes a question about sunshine, a phenomenon familiar to us all.

What Makes the Sun Shine? Neutrinos May Provide a Clue

Walter Sullivan

Munich—Many years ago, scientists deduced that the primary energy source of the Sun and other stars is the fusion of hydrogen nuclei. Experiments intended to verify this fundamental assumption, however, have indicated that something is seriously amiss. In theory, such a fusion reaction would bombard Earth with a stream of ghostly subatomic particles called neutrinos, yet this rich flow of neutrinos has not been observed.

A consortium of European institutions has now embarked on the most ambitious effort yet to track the elusive particles, deep in a vast cavern near Rome. If the effort is successful, scientists will be a step closer to answering one of their most puzzling questions: What makes the Sun shine?

Neutrinos are extremely difficult to detect because they pass through matter virtually unobstructed. In fact, they are thought to pass from the Sun's core to its surface, a distance of about 430,000 miles, in 2.3 seconds, while light does not diffuse to the surface for more than a million years, being constantly deflected by atoms in the dense gases in the Sun's interior. [See Figure 7-4.] After reaching the surface, however, neutrinos and light alike speed to Earth,

FIGURE 7-4
The Paths of Neutrinos and Light

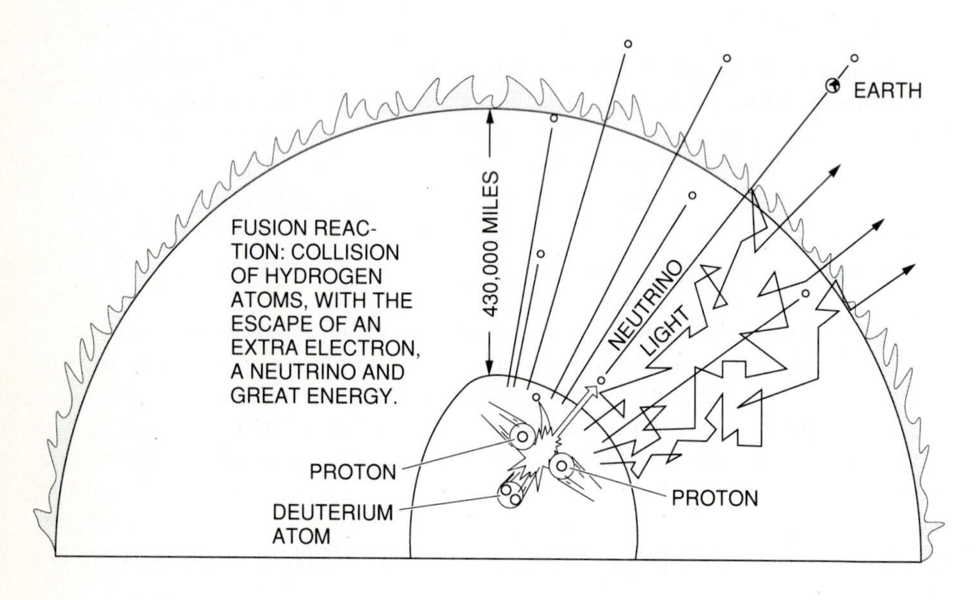

From Sullivan, W. 6 August 1985. What makes the sun shine? Neutrinos may provide a clue. *New York Times,* III, 1:1.

93 million miles away, in 8 minutes 20 seconds. Thus neutrinos provide scientists their only opportunity to "see" what is happening now in the Sun's core.

Sun May Be "Turned Off"

Even when the most elaborate current devices are used, only about five percent of the expected solar neutrinos have been found.

This has led to a variety of explanations, some of them bizarre. One suggests a temporary shutdown of the solar furnace, whose effects might not reach Earth for millions of years, causing an ice age. Another possibility is that neutrinos in flight toward Earth oscillate among three forms, only one of which, related to electrons, would be detected. In this case neutrinos would not be totally without mass, as is generally believed. Still another explanation is that our current understanding of what energizes the stars is incorrect. In the words of Dr. Till A. Kirsten of the Max Planck Institute for Nuclear Physics in Heidelberg, West Germany, "That would be rather dramatic."

30 Tons of Gallium Needed

Dr. Kirsten is spokesman for a consortium of French, West German, Italian and Israeli institutions that has been formed to find out what makes the Sun shine. Their experiment, in a manmade chamber inside the Gran Sasso tunnel east of Rome, will require 30 tons of gallium, two-thirds of the world's yearly supply.

The experiment has been blocked by the limited availability and high cost of gallium, a soft, metallic chemical element that is used almost exclusively by the electronics industry. Gallium was chosen for the tests because of its special sensitivity to the primary energy range expected for neutrinos, about 420,000 electron volts. The matter used in current tests—perchloroethylene, a cleaning fluid—is sensitive only to the most energetic neutrinos.

The chamber near Rome, 422 feet long, 60 feet wide and 60 feet high, is nearing completion, and the German participants here believe they are assured of the more than $10 million needed to purchase the gallium. The chief bottleneck for the European consortium, however, is the shortage of the element, a byproduct of aluminum production. World consumption of gallium currently matches production, and Dr. Kristen said three to four years will be needed to accumulate 30 tons once the money is available.

Two years after it goes into operation the detector should have significant results, Dr. Kristen said. This would be followed by one more year to verify the findings.

U.S.–German Venture Scrapped

The new European experiment, which arose after the collapse of negotiations for a multimillion-dollar joint venture by West Germany and the United States, is the latest attempt to detect the elusive particles.

Efforts to detect neutrinos have been carried out for many years by Dr. Raymond Davis Jr. of Brookhaven National Laboratory on Long Island using a large volume of cleaning fluid deep underground in South Dakota. He assumed that high-energy neutrinos radiated by boron 8 produced by fusion in the Sun would hit atoms of chlorine 37 in the cleaning fluid, converting it to the gas argon 37, which, because it is radioactive, can easily be measured. But to the surprise of the researchers, only about 5 percent of the expected conversions were detected.

Because the cleaning fluid tests record only the most energetic neutrinos, Dr. Davis, who is moving to Pennsylvania State University, has collaborated with Dr. Kirsten and others to develop the new gallium detection process. The gallium will be incorporated into 100 tons of liquid gallium chloride. Within this fluid, atoms of gallium will occasionally be converted by the neutrinos to germanium in the germanium chloride molecules, which then can be swept out of the tanks by helium bubbles.

Russians Plan Similar Tests

A somewhat similar Soviet test is being prepared, but physicists here doubt that it will be sensitive enough to resolve the problem.

Dr. Kirsten said he has been told that the Soviet test will use metallic gallium, of which about 20 tons is on hand. Precise detection of germanium produced in the metal will be extremely difficult, Dr. Kirsten said. For unexplained reasons, apparently related to national security, the Soviet scientists are not allowed to convert the metal to gallium chloride, although recovering the metallic form would probably be easy.

The Soviet test is to be conducted in a neutrino observatory that has been tunneled several miles into the Caucasus Mountains. The Soviet scientists reportedly also plan a test like that of Dr. Davis's, using 10 times as much cleaning fluid.

In this next selection, Charles Darwin, in one of his late writings, discusses worms, those unsung heroes of soil mechanics.

The Formation of Vegetable Mould Through the Actions of Worms with Observations on Their Habits

Charles Darwin

Worms have played a more important part in the history of the world than most persons would at first suppose. In almost all humid countries they are extraordinarily numerous, and for their size possess great muscular power. In many parts of England a weight of more than ten tons (10,516 kilogrammes) of dry earth annually passes through their bodies and is brought to the surface on each acre of land; so that the whole superficial bed of vegetable mould passes though their bodies in the course of every few years. From the collapsing of the old burrows the mould is in constant though slow movement, and the particles composing it are thus rubbed together. By these means fresh surfaces are continually exposed to the action of the carbonic acid in the soil, and of the humus-acids which appear to be still more efficient in the decomposition of rocks. The generation of the humus-acids is probably hastened during the digestion of the many half-decayed leaves which worms consume. Thus the particles of earth, forming the superficial mould, are subjected to conditions eminently favourable for their decomposition and disintegration. Moreover, the particles of the softer rocks suffer some amount of mechanical trituration in the muscular gizzards of worms, in which small stones serve as millstones.

The finely levigated castings, when brought to the surface in a moist condition, flow during rainy weather down any moderate slope; and the smaller particles are washed far down even a gently inclined surface. Castings when dry often crumble into small pellets and these are apt to roll down any sloping surface. Where the land is quite level and is covered with herbage, and where the climate is humid so that much dust cannot be blown away, it appears at first sight impossible that there should be any appreciable amount of sub-aerial denudation; but worm-castings are blown, especially whilst moist and viscid, in one uniform direction by the prevalent winds which are accompanied by rain. By these several means the superficial mould is prevented from accumulating to a great thickness; and a thick bed of mould checks in many ways the disintegration of the underlying rocks and fragments of rock.

The removal of worm-castings by the above means leads to results which are far from insignificant. It has been shown that a layer of earth, ·2 of an inch in thickness, is in many places annually brought to the surface; and if a small part of this amount flows, or rolls, or is washed, even for a short distance, down every inclined surface, or is repeatedly blown in one direction, a great effect will be produced in the course of ages. It was found by measurements

---

From Darwin, C. [1881] 1945. *The formation of vegetable mould through the actions of worms with observations on their habits,* 145–48. Reprint. London: Faber & Faber.

and calculations that on a surface with a mean inclination of 9° 26′, 2.4 cubic inches of earth which had been ejected by worms crossed, in the course of a year, a horizontal line one yard in length; so that 240 cubic inches would cross a line 100 yards in length. This latter amount in a damp state would weigh 11½ pounds. Thus a considerable weight of earth is continually moving down each side of every valley, and will in time reach its bed. Finally this earth will be transported by the streams flowing in the valleys into the ocean, the great receptacle for all matter denuded from the land. It is known from the amount of sediment annually delivered into the sea by the Mississippi, that its enormous drainage area must on an average be lowered ·00263 of an inch each year; and this would suffice in four and a half million years to lower the whole drainage area to the level of the seashore. So that, if a small fraction of the layer of fine earth, ·2 of an inch in thickness, which is annually brought to the surface by worms, is carried away, a great result cannot fail to be produced within a period which no geologist considers extremely long.

Archæologists ought to be grateful to worms, as they protect and preserve for an indefinitely long period every object, not liable to decay, which is dropped on the surface of the land, by burying it beneath their castings. Thus, also, many elegant and curious tesselated pavements and other ancient remains have been preserved; though no doubt the worms have in these cases been largely aided by earth washed and blown from the adjoining land, especially when cultivated. The old tesselated pavements have, however, often suffered by having subsided unequally from being unequally undermined by the worms. Even old massive walls may be undermined and subside; and no building is in this respect safe, unless the foundations lie 6 or 7 feet beneath the surface, at a depth at which worms cannot work. It is probable that many monoliths and some old walls have fallen down from having been undermined by worms.

Worms prepare the ground in an excellent manner for the growth of fibrous-rooted plants and for seedlings of all kinds. They periodically expose the mould to the air, and sift it so that no stones larger than the particles which they can swallow are left in it. They mingle the whole intimately together, like a gardener who prepares fine soil for his choicest plants. In this state it is well fitted to retain moisture and to absorb all soluble substances, as well as for the process of nitrification. The bones of dead animals, the harder parts of insects, the shells of land-molluscs, leaves, twigs, etc., are before long all buried beneath the accumulated castings of worms, and are thus brought in a more or less decayed state within reach of the roots of plants. Worms likewise drag an infinite number of dead leaves and other parts of plants into their burrows, partly for the sake of plugging them up and partly as food.

The leaves which are dragged into the burrows as food, after being torn into the finest shreds, partially digested, and saturated with the intestinal and urinary secretions, are commingled with much earth. This earth forms the dark coloured, rich humus which almost everywhere covers the surface of the land with a fairly well-defined layer or mantle. Hensen placed two worms in a vessel 18 inches in diameter which was filled with sand, on which fallen leaves were strewed; and these were soon dragged into their burrows to a depth of 3 inches. After about 6 weeks an almost uniform layer of sand, a centimetre (·4 inch) in thickness, was converted into humus by having passed through the alimentary canals of these two worms. It is believed by some persons that worm-burrows, which often penetrate the ground almost perpendicularly to a depth of 5 or 6 feet, materially aid in its drainage; notwithstanding that the viscid castings piled over the mouths of the burrows prevent or check the rain-water directly entering them. They allow the air to penetrate deeply into the ground. They also greatly facilitate the downward passage of roots of moderate size; and these will be nourished by the humus with which the burrows are lined. Many seeds owe their germination to having been covered by castings; and others buried to a considerable depth beneath accumulated castings lie dormant, until at some future time they are accidentally uncovered and germinate.

Worms are poorly provided with sense-organs, for they cannot be said to see, although they can just distinguish between light and darkness; they are completely deaf, and have only a feeble power of smell, the sense of touch alone is well developed. They can therefore learn but little about the outside world, and it is surprising that they should exhibit some skill in lining their burrows with their castings and with leaves, and in the case of some species in piling up their castings into tower-like constructions. But it is far more surprising that they should apparently exhibit some degrees of intelligence instead of a mere blind instinctive impulse, in their manner of plugging up the mouths of their burrows. They act in nearly the same manner as would a man, who had to close a cylindrical tube with different kinds of leaves, petioles, triangles of paper, etc., for they commonly seize such objects by their pointed ends. But with thin objects a certain number are drawn in by their broader ends. They do not act in the same unvarying manner in all cases, as do most of the lower animals; for instance, they do not drag in leaves by their foot-stalks, unless the basal part of the blade is as narrow as the apex, or narrower than it.

When we behold a wide-turf-covered expanse, we should remember that its smoothness, on which so much of its beauty depends, is mainly due to all the inequalities having been slowly levelled by worms. It is a marvellous reflection that the whole of the superficial mould over any such expanse has passed, and will again pass, every few years through the bodies of worms. The plough is one of the most ancient and most valuable of man's inventions; but long before he existed the land was in fact regularly ploughed, and still continues to be thus ploughed by earthworms. It may be doubted whether there are many other animals which have played so important a part in the history of the world, as have these lowly organized creatures. Some other animals, however, still more lowly organized, namely corals, have done far more conspicuous work in having constructed innumerable reefs and islands in the great oceans; but these are almost confined to the tropical zones.

In a vein similar to Darwin's, John Crompton describes the extraordinary engineering principles involved in the construction of a spider's web. To appreciate this description, the reader needs some knowledge of the spider's capacity for making silk. Crompton tells us that the Aranea has 6 teats, with 600 tiny taps (glands), each of which is individually controllable. The spider can express 600 streams, or can combine them in nearly limitless combinations. Aranea can produce 5 kinds of silk thread from these taps, in different sizes, textures, and strengths. Some authors have compared the strength of Aranea's silk to that of fused quartz.

The Web Weavers

John Crompton

The spokes [in the preliminary web] are marvellously equidistant considering the irregular order in which they were made; not *exactly* equidistant but very nearly so in the majority of cases. . . . The sketch has been made simpler by reducing the number of spokes. We have drawn twenty; the actual number in the web of *Aranea diadema* varies from twenty-five to thirty-five, with thirty as the average number.

. . . The centre is strengthened by a few spiral lines. . . . And here the uninitiated observer may be disappointed. Starting at A [Figure 7-5] and working even faster than before Aranea lays down a rough, wide-spaced spiral finishing at B.

Possibly the observer goes away. If he does, and decides to come back later he will find something like the state of affairs represented in [the complete Figure 7-5].

What has happened? To explain we must go back to [the initial wide-spaced spiral, A–B]. That amateurish spiral was not part of the web at all. It was scaffolding, made to be destroyed later. *We* use scaffolding when we build. So does Aranea, and for the same purpose—for a

---

From Crompton, J. [John Battersby Crompton Lamburn]. 1951. *The spider,* 22–30. Boston: Houghton Mifflin, Riverside Press. *The spider* published with six other titles by Nick Lyons Books. Reprinted by permission.

FIGURE 7-5
Construction of the Web

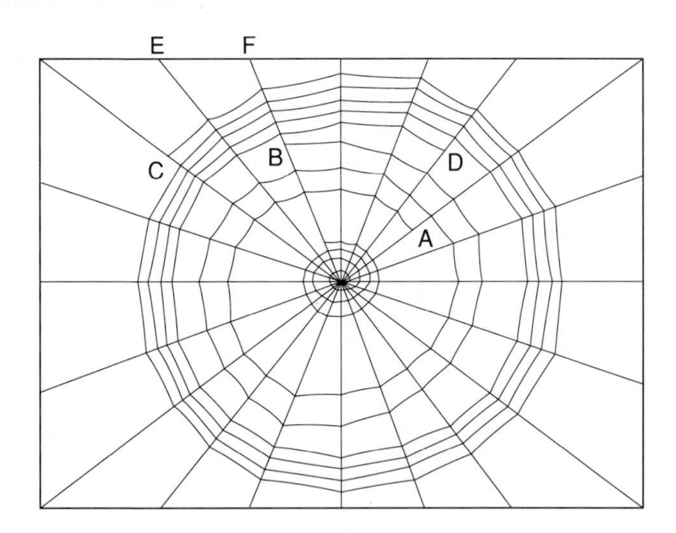

foothold; a platform from which to work, for the making of the web now becomes very tricky. A lot of labour has gone into it up to the present, but as a means of catching flies or any other game it is useless. So the spider started at C to make her real snare. From C she stretched a line to the spoke E, fastened it, then on to F, fastened it, and so on round and round in close concentric circles until she came to D, which merely happens to be the point reached when the observer came back. She will go on with those spirals until she gets near the centre, when she will stop, leaving a clear space between the last made spirals and the centre-strengthening spirals.

It will be noticed that the spiral scaffolding in the area over which she has worked has disappeared. She used it for a platform and rolled it up as she went along. By the time she comes close to the centre it will all have been rolled up and thrown away.

A change has come over Aranea in her latest task. She is moving more slowly and working more carefully. It would seem that her job has become more complicated. She pauses at every spoke and does things to the line. She has switched on to more elastic silk than she used before but it is not that that delays her, it is the fact that these lines must be made sticky. Those six hundred little pipes of hers are not *all* connected to silk factories; there is another factory inside her, a smaller concern altogether, that manufactures glue and has one of two discharge pipes of its own. So as she lays a line from one spoke to another (and in all she will have to lay down roughly 13,000 of these lines) Aranea anoints it at the same time with glue from the glue pipes.

She does more. Perhaps a line merely painted with glue would dry too quickly, or dust would render it inoperative, or perhaps glue arranged in little drops has more holding power. At any rate she arranges her glue in a number of minute droplets. Observers had known this for a long time and had studied the threads under the microscope and had found that the drops were arranged close together and all *exactly equidistant.* The spider's ability to arrange them so excited both admiration and puzzlement. Then the secret came out. If you put fluid on a violin or similar type of string and twang this string with your finger, the fluid, for some reason probably better known to you than to me, separates out into drops equally spaced. This is what Aranea does. Having laid her line between two spokes and smeared it with glue she pulls it down with a claw of one of her hind feet and lets it snap back. . . .

How did the spider learn to do this? How did she hit upon a method which would have defeated me? When I ask people they smile tolerantly and say, "Oh, instinct." If you say "instinct" you have, apparently, explained everything. The building of the spider's web, the paralysing of its prey by the hunting wasp—instinct. Instinct, and therefore nothing wonderful at all. Yet these things had to start. I know we have to use it but I hate the word instinct. We say it with such smugness. To my mind, far from explaining anything it makes the problem infinitely more complicated than it was before.

And so Aranea moves from spoke to spoke, tying, glueing, twanging, and proceeding in concentric circles towards the hub. Short of the hub, and leaving the circular space known as the "free zone" she concludes the laying of the gummed snare.

Little more remains to be done. The hub has to be attended to and a telephone installed and possibly a few more stays and supports added. At the centre . . . there is a tuft of silk, the result of all the fastenings that have been made. This is bitten away together with a further portion of the centre. The resulting hole is filled up (or not) according to the whim of the species. *Aranea diadema* fills up hers with a meshwork of irregular threads. . . .

Some, including *Meta segmentata,* the commonest of all the outdoor spiders, leave a hole in the centre; others make a platform of almost solid silk; and so on. From this and other signs the expert can tell which spider has made a web without seeing the spider at all.

What I called the "telephone," which is a good enough description for the moment, is a long line connecting the web with some retreat. It also serves as a bridge from the web to the retreat.

So there it is, the web of Aranea, a beautiful, almost invisible work of art. We admire it on dewy or frosty mornings because that is the only time we can see it but its lines then are coarsened and thickened, the delicacy has gone. It always seems a pity to me, studying one of these newly-made creations, when some clumsy daddy-long-legs or other oafish creature blunders into it and tears gaping holes. I know that the seamstress herself welcomes these vandals, but it is desecration nevertheless. It is as if an ill-bred urchin made holes in tapestry. That the urchin is going to be punished, and in the case in question punished very severely, does not lessen one's annoyance. The culprit, bound and trussed, is led off, but the hole he made remains. The tapestry is spoilt. And if you think that Aranea, is going to repair that hole you are very much mistaken. A creator, a supreme artist, does not lower herself to the extent of taking on darning. Clumsy unskilled workers like the house spiders may do it, not Aranea. She has a look at it. And then she goes away. During the day others come. There are more threshings about by foolish creatures who will not realise that they are in a snare from which escape is almost impossible. Soon there will be many holes and Aranea will take the whole thing out of its framework, and make a new one.

Sometimes she is very particular. I saw one individual discard a web and weave another after only one small hole had been made in the first. At other times she does the reverse and will suffer her web to be torn to threads without doing anything about it, but this is only when she is bloated out with food and has several roped-up prisoners awaiting slaughter in her dungeon. In the ordinary way Aranea builds a new snare once every twenty-four hours, making this snare at a given time, after dark, before dark, or before dawn, according to her species. For each species, like human builders, has, in normal circumstances, definite hours for this type of work.

The scaffolding mentioned before was necessary to give Aranea a bridge across the spokes from which to work; also to enable her to avoid treading on her own limed threads. She *is* however able to tread on the lime without getting stuck, and hoist, as it were, with her own petard; indeed she often has to tread on it, especially when dealing with ensnared insects, but she prefers not to do so unnecessarily, just as a man will often avoid mud and puddles even though wearing gumboots. The reason she is not held fast like her victims once puzzled observers. Fabre, who has solved many problems, solved this one. The secret is that the spider's feet are covered with a film of oil; they exude oil like sweat. Fabre found that the end of a glass rod, when dry, stuck to the gummed line but did not stick when lightly smeared with oil. He also found that a leg freshly taken from a garden spider did not stick unless it was left in contact with the line for a long time. A freshly taken leg, however, washed in bisulphide of carbon (which dissolves every trace of oil) stuck at once. Which seems adequately to explain things.

The glue of the spider, like our "gum," is soluble in water, and rain in any quantity washes it away. It is also acid. They say you can taste the acid if you lick the spiral part. I must admit I have licked the spiral parts and detected no taste. But it *is* acid and I sometimes wonder if it does not contain a certain amount of poison like our fly papers. The effect of this acid is well illustrated by the experience of a certain sportsman, Mr. Hart, in America. Mr. Hart was out shooting and his path led him through some woods. Here he encountered numerous spider webs of the sort that tore my helmet off in Africa. They are not pleasant things to force one's way through, so Mr. Hart put his shot-gun in front of him to break and ward off the webs. Soon the barrels were festooned with trailing threads, and four hours later he found that these strands—especially where they had lain thickest—had destroyed the "blueing" of the barrels.

I have protested against overuse of the word "instinct." The making of the web, however, *is* now (apart possibly from the selection of a site) entirely instinctive. Remote and clever ancestors found out how to do it, not at once but by a series of improvements, and possibly taught the art to their children just as the carnivorae teach their children to hunt. That spiders to-day, after their children have achieved a certain age, only regard them as something to eat has no bearing on the question. Because an animal behaves in a certain way now it does not follow that it behaved in the same way a million years ago. Indeed the reverse is the case, for nothing stands still. Even now the mother spider absolutely dotes on her *very* young. That doting fondness may have been carried to a much later stage in ancestors, so that parent spiders then *may* have taught their young to weave. They may have been more social too. At least families may have lived together. Then competition came in. Spiders began to cover the earth; food became short. The mother found that her own young were taking food *she* might have had and which she needed. She chased them off. Later on in the scale of time she found them good for food. And so on until to-day each spider lives alone and the young take good care to leave their mother before the present short period of love has ended. The more primitive types of present-day spiders are however very similar in appearance to their remote ancestors of the carboniferous period, so that there are no scientific grounds for supposing that the habits of spiders generally have greatly changed.

We know that the ability to make a web is inborn in the spider, chiefly because the babies make webs also—and adorable little webs they are; perfect miniatures no longer than a postage stamp and correct in almost every detail, though not possessing all the features of the web of the adult. Watching the making of them under a lens is more fascinating than watching the work of their mothers; the field is so small that it is easier to keep it under observation.

The young of most species display asexual traits; girls and boys have similar voices, girls will climb trees and boys may play with dolls. So it is not surprising to learn that immature male spiders make little webs almost as good, at first, as those of their sisters, but when they become adult they cast away, together with their last skin, what they doubtless consider childish things. They are adult only after the final moult and then all ability to weave leaves them. They become purely rovers, lascivious roués, following and accosting females. We will waste however no sympathy on the female on this account for she is more than able to look after herself.

Closer to home, David Bodanis enlightens us about margarine, in a fascinating book about the unexpected in our daily life, *The Secret House.*

Morning

David Bodanis

Over the toast comes the margarine, nice gloppy melting chunks, spread fearlessly, applied thick, dripping in and demanding more. It used to be butter on the toast, but since all the warnings about cholesterol and heart attacks butter has been banned from this sensible

---

From Bodanis, D. 1986. *The secret house: 24 hours in the strange and unexpected world in which we spend our nights and days,* 35–37. New York: Simon & Schuster.

household. Only this lighter, thinner, fresh and non-cardiac wholly vegetarian margarine is being used. Or rather, the users think it's lighter, thinner, fresher, etc. A look at the way it's manufactured would suggest otherwise.

Margarine is made from fat. It was first invented in response to an award offered by the French Emperor, Napoleon III, after the urban rebellions of 1848 to find a cheap source of fat for the working classes who could not afford butter. Today there's soya fat in margarine, also the fat you get from squished herrings, and about 20 per cent of the total is beef fat or even nice old-fashioned lard—pig's fat. All these fats are mixed together and melted, and if you think molten pig's fat smells bad, you should wait till you've had the misfortune to walk through a factory where it's being stirred in with boiling herring and other fats. The whole mess is so repulsive, so clearly distasteful and unmarketable (it comes out colored grey on top of it all) that before anything else is done it has to be funneled into even larger deodorizing vats to try to get rid of the stink.

What comes out of the deodorizer, while at least it can be approached without gagging, is still not quite the tempting substance commercial margarine is supposed to be. It's grey, it's sticky, and it's far, far too chunky. The fats that were temporarily boiled apart in the deodorizer can't restrain themselves for long, and have clotted back into large, unattractive lumps. Those lumps have to go.

The grey gloop is poured into another vat, metal shavings are clunked in before it, then the vat is screwed shut and high-pressure hydrogen gas is sprayed in. The fats are boiled and compressed in there, they react with the nickel and hydrogen, and when finally the ordeal is over and the top taken off, the lumps are gone, squeezed clear out in all the ruckus.

There's more. Beef dripping, lard and herring fat don't cost very much, but if at this stage they could be diluted with something even cheaper, even more easy to get in quantity than pig's fat, then the cost of producing the margarine would drop even lower. This extra substance is waiting in another vat in the factory, right next to the one where the de-lumping took place. It's milk—of a sort.

By government regulations there are two main grades of milk in most countries: Grade A, which is fresh, checked and suitable for drinking; and a lesser Grade B, which ordinary consumers don't usually see, but which being slightly older, or having a bit more bacteria in it than is best, is used for condensed milk, commercial cakes, and baby milk mixtures. The milk that's waiting to be mixed in with the fat in the margarine factory is the second kind, or even one grade below that. It's not fresh; in fact it's going sour. Even though it's already been pasteurized once, the factory engineers have to give it another pasteurizing heat treatment to get some of the worse stuff out of it. After that it's strained, filtered and poured in with the waiting fat.

This presents a problem. Oil and water don't mix (think of French salad dressing), and the lard and fish fat coming in through one spout are highly oily, while the sour milk pouring in from the other spout is that 88 per cent water. To ensure a match, other substances have to be sprayed into the vat where they're heading. Soap-like emulsifiers are squeezed in to foam around each drop of sour milk water and so keep it from joining up with the other milky water there in the vat. Then lots of starch is poured in to make the combined mix even more gooey, and to see that things don't slip back.

It seems it would take a genius to make a palatable product out of this soapy and starch-full mixture of grey sour milk and animal fat. Luckily, unsung geniuses aplenty labor in these margarine factories. First some color is added, something to cover over the vile grey. Normal yellow dyes wouldn't work, because they grey is so deep that it would keep on poking through. Extra-strong dyes based on sulfur-refined coal tars are used instead.

Then a stifling strong flavorer is mixed in to make it taste like something other than the miscellaneous lard, other fats and old milk it is. Then vitamins—because all this processing has made it nutritiously almost valueless. The result of all these labours is then compressed, cooled, scraped, cut into long blocks, cut into smaller blocks, and finally dropped into plastic tubs.

There's something else. Back at the beginning a little sunflower oil is sometimes mixed in. That's not because it's especially needed, and usually not because there's enough to make any difference, but rather so the designers can have something suggestive of sun-kissed meadows and open spaces to print on the cover. Even the French Academy of Sciences that sponsored the original stuff had their doubts. The award for the first margarine was granted in 1869, but eleven years later the Academy decreed that it couldn't be used in government cafeterias: it was, they said, too revolting on the palate.

With that toastly topping the breakfast preparations are over. Now it's time to consume, to chew, chomp, swallow and chew again. Only it must be done quickly, feverishly, for the clock in the kitchen is nearing eight and there is still much to do. The filling and warming substances enveloping the insides are one thing: now it's time to return to the bedroom, shake off the dressing gown, and do something about enveloping the outsides.

The final selection for this section comes from one of the most innovative books in twentieth-century biology: *On Growth and Form,* by D'Arcy Wentworth Thompson. Sir Peter Medawar, the eminent twentieth-century biologist, has called Thompson's work "beyond comparison the finest work of literature in all the annals of science that have been recorded in the English tongue" (in Gleick 1987, 200).

Thompson's great insight was to apply mathematics to biological forms, thus elucidating their structure. As you'll see in the next section, Arthur Koestler would call this "bisociation," in which one discipline is used to shed new light on another.

Thompson's two-volume work covers natural forms from the cells of a beehive to the mechanical structure of animal skeletal frameworks (where he applies principles from structural engineering). Though often overlooked by contemporary authors, Thompson's work has provided the foundation for the new field of biomechanics.

In the following selection, Thompson describes his extraordinary finding that the spiral of the *Nautilus* and of other similar molluscan shells increases in regular logarithmic increments.

The Equiangular Spiral of the Nautilus Shell

D'Arcy Wentworth Thompson

So far as we have now gone, we have studied the elementary properties of the equiangular spiral, including its fundamental property of *continued similarity;* and we have accordingly learned that the shell or the horn tends *necessarily* to assume the form of this mathematical figure, because in these structures growth proceeds by successive increments which are always similar in form, similarly situated, and of constant relative magnitude one to another. Our chief objects in enquiring further into the mathematical properties of the equiangular spiral will be: (1) to find means of confirming and verifying the fact that the shell (or other organic curve) is actually an equiangular spiral; (2) to learn how, by the properties of the curve, we may further extend our knowledge or simplify our descriptions of the shell; and (3) to understand the factors by which the characteristic form of any particular equiangular spiral is determined, and so to comprehend the nature of the specific or generic differences between one spiral shell and another.

Of the elementary properties of the equiangular spiral the following are those which we may most easily investigate in the concrete case of the molluscan shell: (1) that the polar radii whose vectorial angles are in arithmetical progression are themselves in geometrical progression; hence (2) that the vectorial angles are proportional to the *logarithms* of the

---

From Thompson, Sir. D'A. W. [1917] 1963. *On growth and form.* 2d ed., 769–74. Reprint. London: Cambridge University Press. Published by the Syndics of the Cambridge University Press. Reprinted by permission of the publisher.

FIGURE 7-6

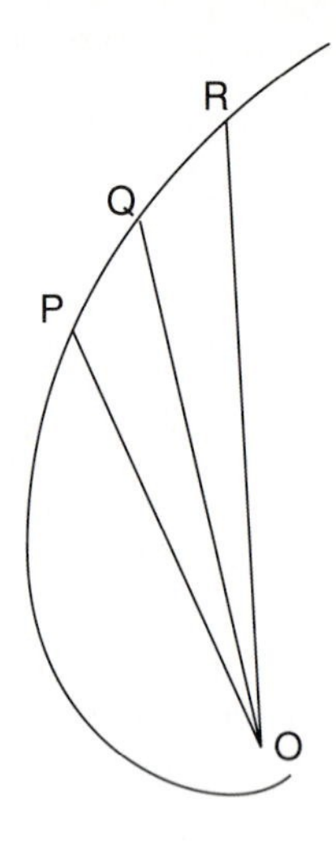

corresponding radii; and (3) that the tangent at any point of an equiangular spiral makes a constant angle (called the *angle of the spiral*) with the polar radius vector.

The first of these propositions may be written in a simpler form, as follows: radii which form equal angles about the pole of the equiangular spiral are themselves continued proportionals. That is to say, in [Figure 7-6], when the angle *ROQ* is equal to the angle *QOP,* then *OP:OQ::OQ:OR.*

A particular case of this proposition is when the equal angles are each angles of 360°: that is to say when in each case the radius vector makes a complete revolution, and when, therefore, *P, Q* and *R* all lie upon the same radius.

It was by observing with the help of very careful measurement this continued proportionality, that Moseley was enabled to verify his first assumpüon, based on the general appearance of the shell, that the shell of *Nautilus* was actually an equiangular spiral, and this demonstration he was soon afterwards in a position to generalise by extending it to all spiral Ammonitoid and Gastropod mollusca. For, taking a median transverse section of a *Nautilus pompilius,* and carefully measuring the successive breadths of the whorls (from the dark line which marks what was originally the outer surface, before it was covered up by fresh deposits on the part of the growing and advancing shell), Moseley found that "the distance of any two of its whorls measured upon a radius vector is one-third that of the two next whorls measured upon the same radius vector. Thus in [Figure 7-7], *ab* is one-third of *bc, de* of *ef, gh* of *hi,* and *kl* of *lm.* The curve is therefore an equiangular spiral."

The numerical ratio in the case of the *Nautilus* happens to be one of unusual simplicity. Let us take, with Moseley, a somewhat more complicated example.

From the apex of a large *Turritella (Turbo) duplicata* [Figure 7-8] a line was drawn across its whorls, and their widths were measured upon it in succession, beginning with the last but one. The measurements were, as before, made with a fine pair of compasses and a diagonal scale. The sight was assisted by a magnifying glass. In a parallel column to the following admeasurements are the terms of a geometric progression, whose fist term is the width of the widest whorl measured, and whose common ratio is 1·1804.

The close coincidence between the observed and the calculated figures is very remarkable, and is amply sufficient to justify the conclusion that we are here dealing with a true logarithmic spiral.

FIGURE 7-7

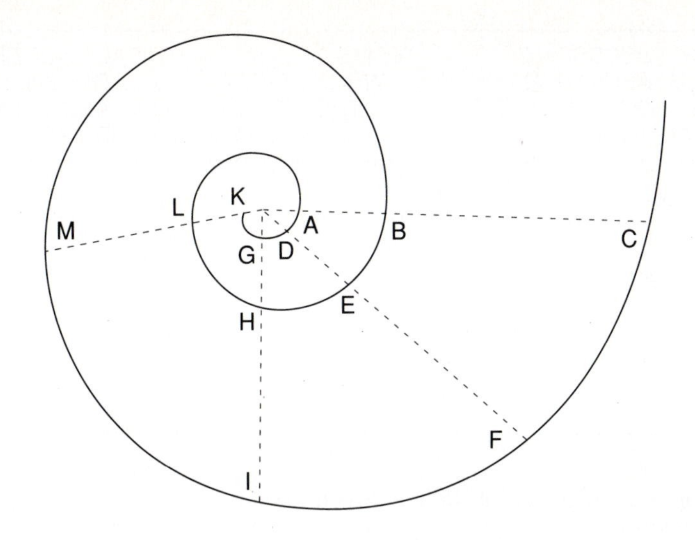

FIGURE 7-8
Turritella duplicata

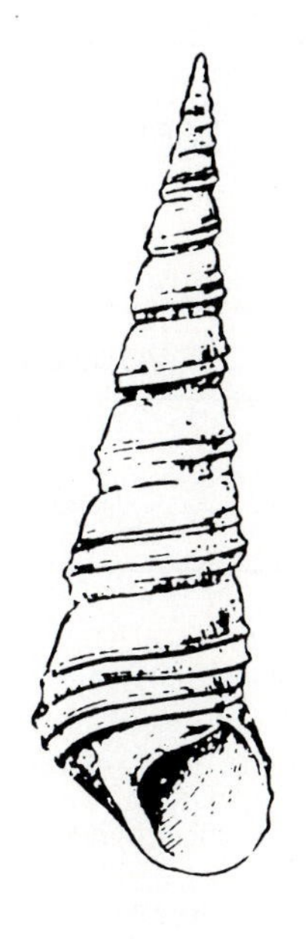

Turritella Duplicata

WIDTHS OF SUCCESSIVE WHORLS, MEASURED IN INCHES AND PARTS OF AN INCH	TERMS OF A GEOMETRICAL PROGRESSION, WHOSE FIRST TERM IS THE WIDTH OF THE WIDEST WHORL, AND WHOSE COMMON RATIO IS 1·1804
1·31	1·310
1·12	1·110
0·94	0·940
0·80	0·797
0·67	0·675
0·57	0·572
0·48	0·484
0·41	0·410

Nevertheless, in order to verify his conclusion still further, and to get partially rid of the inaccuracies due to successive small measurements, Moseley proceeded to investigate the same shell, measuring not single whorls but groups of whorls taken several at a time: making use of the following property of a geometrical progression, that "if μ represent the ratio of the sum of every even number *(m)* of its terms to the sum of half that number of terms, then the common ratio *(r)* of the series is represented by the formula

$$r=(\mu-1)^{\frac{2}{m}}."$$

Accordingly, Moseley made the following measurements, beginning from the second and third whorls respectively:

WIDTH OF		
SIX WHORLS	THREE WHORLS	RATIO μ
5·37	2·03	2·645
4·55	1·72	2·645
FOUR WHORLS	TWO WHORLS	
4·15	1·74	2·385
3·52	1·47	2·394

"By the ratios of the two first admeasurements, the formula gives

$$r=(1{\cdot}645)^{\frac{1}{3}}=1{\cdot}1804.$$

By the mean of the ratios deduced from the second two admeasurements, it gives

$$r=(1{\cdot}389)^{\frac{1}{2}}=1{\cdot}1806.$$

"It is scarcely possible to imagine a more accurate verification than is deduced from these larger admeasurements, and we may with safety annex to the species *Turbo duplicatus* the characteristic number 1·18."

By similar and equally concordant observations, Moseley found for *Turbo phasianus* the characteristic ratio, 1·75; and for *Buccinum subulatum* that of 1·13.

From the measurements of *Turritella duplicata* (on p. 772), it is perhaps worth while to illustrate the logarithmic statement of the same thing: that is to say, the elementary fact, or corollary, that if the successive radii be in geometric progression, their logarithms will differ from one another by a constant amount.

Turritella Duplicata

WIDTHS OF SUCCESSIVE WHORLS	LOGARITHMS OF DO.		DIFFERENCES OF LOGARITHMS	RATIOS OF SUCCESSIVE WIDTHS
131	2·11727		—	—
112	2·04922		0·06805	1·170
94	1·97313		0·07609	1·191
80	1·90309		0·07004	1·175
67	1·82607		0·07702	1·194
57	1·75587		0·07020	1·175
48	1·68124		0·07463	1·188
41	1·61278		0·06846	1·171
		Mean	0·07207	1·1806

And 0·07207 is the logarithm of 1·1805.

Lastly, we may if we please, in this simple case, reduce the whole matter to arithmetic, and, dividing the width of each whorl by that of the next, see that these quotients are nearly identical, and that their mean value, or common ratio, is precisely that which we have already found.

We may shew, in the same simple fashion, by measurements of *Terebra,* . . . how the relative widths of successive whorls fall into a geometric progression, the criterion of a logarithmic spiral.

Writing Exercise

Following these fascinating examples, describe a common object or mechanism that has extraordinary properties. Or, following Thompson, use a mathematical or geometric overlay to elucidate an unexpected property of a physical object (living or inert).

Since this is an exploratory assignment, no length or audience is specified.

Reference List

Gleick, J. 1987. *Chaos.* New York: Viking Penguin.

REVOLUTIONARY CONCEPTS: THREE THEORIES OF SCIENTIFIC DISCOVERY AND TECHNOLOGICAL INNOVATION

Inspired by the profound revolutions in science and technology since the mid–nineteenth century, philosophers and theoreticians of the history of science and technology have tried to explain how such revolutions take place: how they begin, develop, and mature. There are three main schools of thought on this matter, all identified by their originator.

1. Arthur Koestler: sudden illumination or insight.
2. Thomas Kuhn: paradigm shift.
3. Herbert Simon: problem solving.

Arthur Koestler's magisterial work, *The Act of Creation,* analyzes the creative act in all fields of human endeavor. Koestler thinks that creativity—whether in science, technology, visual arts, poetry, or whatever field—conforms to a basic pattern: creative people first immerse themselves in a problem or confusion and then, mysteriously, solve the problem by relating two previously unrelated methods of thinking or feeling, or by overlaying the methodology of one field with that of another. He calls this "bisociation," which is closely associated with metaphoric thinking. Here is a simple example of this process.

Let's say that you are doing field research for a course in the behavioral psychology of preschoolers. Your professor sends you to a local day-care center, with the assignment of observing and trying to make sense of (trying to define and explain) how children use toys. You soon observe a repetitive series of actions. One or two children will tend to oversee, distribute, and keep track of the toys, often keeping their preferred ones, if they can, for themselves. When a new preschooler shows up, these leaders will hoard the toys, keeping them from the new arrival. Only gradually will the new child be allowed to have toys, and he or she will usually first be given the least valued ones. Eventually, as the child becomes part of the group, he or she will get more and better toys. Soon this child will become part of the group. This process may be ameliorated by the school's adult overseer, who will insist from time to time that the children share, though most of the children object.

At the same time, you are having two other experiences. You have a puppy at home and are training it as a house dog. To make it comfortable, you have given it teething toys and pillows. While these were, at first, all over the house, your puppy has decided to carry them all to his new "spot," the landing of your stairs to the second floor. Your puppy now naps in this place and keeps his toys there. If his toys are moved around the house, he moves many of them back.

You are also reading a book by Robert Ardrey, called *The Territorial Imperative,* for a sociology class. Its main thesis concerns the seemingly in-born urge that humans and many other animals have to mark off their territories and accumulate their possessions within it.

These two experiences, you suddenly see, can be used to explain the behavior of your preschoolers—behavior that the school's overseer has called antisocial. You begin to realize that the child hoarding his toys is following the territorial imperative, as a means to keeping valued objects for himself and his group. Instead of being antisocial, the child is actually preserving the quality of life for his own group.

In making this connection, you have used the definition of an abstract concept (Ardrey's idea of the territorial imperative) to explain a heretofore unexplained set of events (the child hoarding his toys), or to make clear the principle behind a set of events that is taken for granted. You have done so by relating a familiar but unexplained action (hoarding) to an unfamiliar or new set of information (the hypothesis of the territorial imperative). And you have extended the power of your explanation by seeing that your puppy is following the same biologically encoded urge.

You have increased the order of your world view by assimilating one more fact. You can now use your theory to explain, or find an ordering pattern in, other behavior, such as the way countries make and protect borders and the way bureaucrats make and protect their pet projects.

Using theories from one field to illuminate another is the part of the discovery process of Arthur Koestler's bisociation theory. Koestler explains, in his *The act of creation,* that in the "basic, bisociative pattern of the creative, synthesis" is characterized by "the sudden interlocking of two previously unrelated skills, or matrices of thought" (1969, 121). He calls such synthesis a "Eureka act—the sudden shaking together of two previously unconnected matrices. . . . If all goes well that single, explosive contact will lead to a lasting fusion of the two matrices—a new synthesis will emerge, a further advance in the mental evolution will have been achieved" (212).

A sense of mystery surrounds the creative act, and a sense of wonder is the fruit of such intuition. Koestler contrasts the popular idea of scientists with what he believes to be the truth.

Thinking Aside

Arthur Koestler

In the popular imagination these men of science appear as sober ice-cold logicians, electronic brains mounted on dry sticks. But if one were shown an anthology of typical extracts from their letters and autobiographies with no names mentioned, and then asked to guess their profession, the likeliest answer would be: a bunch of poets or musicians of a rather romantically naïve kind. The themes that reverberate through their intimate writings are: the belittling of logic and deductive reasoning (except for verification after the act); horror of the one-track mind; distrust of too much consistency ("One should carry one's theories lightly," wrote Titchener); scepticism regarding all-too-conscious thinking ("It seems to me that what you call full consciousness is a limit case which can never be fully accomplished. This seems to me connected with the fact called the narrowness of consciousness *Enge des Bewusstseins,"*—Einstein). This sceptical reserve is compensated by trust in intuition and in unconscious guidance by quasi-religious or by aesthetic sensibilities. "I cannot believe that God plays dice with the world," Einstein repeated on several occasions, rejecting the tendency in modern physics to replace causality by statistical probabilities. "There is a scientific taste just as there is a literary or artistic one," wrote Renan. Hadamard emphasized that the mathematician is in most cases unable to foresee whether a tentative line of attack will be successful; but he has a "sense of beauty that can inform us, and I cannot see anything else allowing us

---

Excerpts from Koestler, A. 1969. *The act of creation.* New York: Dell. Reprinted by permission of Sterling Lord Literistic, Inc.

to foresee. This is undoubtedly the way the Greek geometers thought when they investigated the ellipse, because there is no other conceivable way." Poincaré was equally specific: "It may be surprising to see emotional sensibility invoked *a propos* of mathematical demonstrations which, it would seem, can interest only the intellect. This would be to forget the feeling of mathematical beauty, of the harmony of numbers and forms, of geometric elegance. This is a true aesthetic feeling that all real mathematicians know. The useful combinations [of ideas] are precisely the most beautiful, I mean those best able to charm this special sensibility." Max Planck, the father of quantum theory, wrote in his autobiography that the pioneer scientist must have "a vivid intuitive imagination for new ideas not generated by deduction, but by *artistically* creative imagination." The quotations could be continued indefinitely, yet I cannot recall any explicit statement to the contrary by any eminent mathematician or physicist.

Here, then, is the apparent paradox. A branch of knowledge which operates predominantly with abstract symbols, whose entire rationale and credo are objectivity, verifiability, logicality, turns out to be dependent on mental processes which are subjective, irrational, and verifiable only after the event. (146–47)

This vision of the creative scientist and engineer leads Koestler to a theory of the evolution of creative ideas in science and engineering.

The Evolution of Ideas

Arthur Koestler

There is a theory, put forward by George Sarton, and held to be self-evident by many scientists, which says, broadly speaking, that the history of science is the only history which displays a cumulative progress of knowledge; that, accordingly, the progress of science is the only yardstick by which we can measure the progress of mankind; and moreover, that the word "progress" itself has no clearly defined meaning in any field of activity—except the field of science.

This is the kind of pronouncement where it is advisable to hold one's breath and count to ten before expressing indignant protest or smug agreement, according to one's allegiance to eggheads or engineers. Personally I believe that there is a grain of truth in Sarton's proposition—but no more than that.

Separations and Reintegrations

There are certain analogies between the characteristic stages in the history of an individual discovery, and the historical development of a branch of science as a whole. Thus a 'blocked matrix' in the individual mind reflects some kind of impasse into which a science has manœuvred itself. The "period of incubation," with its frustrations, tensions, random tries, and false inspirations, corresponds to the critical periods of "fertile anarchy" which recur, from time to time, in the history of every science. These crises have, as we saw, a destructive and a constructive aspect. In the case of the individual scientist, they involve a temporary retreat to some more primitive form of ideation—innocence regained through the sacrifice of hard-won intellectual positions and established beliefs; in the case of a branch of science taken as a whole, the crisis manifests itself in a relaxation of the rigid rules of the game, a thawing of the collective matrix, the breakdown of mental habits and absolute frontiers—a process of *reculer pour mieux sauter* on an historic scale. The Eureka act proper, the moment of truth experienced by the creative individual, is paralleled on the collective plane by the emergence, out of the scattered fragments, of a new synthesis, brought about by a quick succession of individual discoveries—where, characteristically, the same discovery is often made by several individuals at the same time. . . .

The last stage—verification, elaboration, consolidation—is by far the least spectacular, the most exacting, and occupies the longest periods of time both in the life of the individual and in the historical evolution of science. Copernicus picked up the ancient Pythagorean teaching of the sun as the centre of all planetary motions when he was a student in Renaissance Italy (where the idea was much discussed at the time), and spent the rest of his life elaborating it into a system. Darwin hit on the idea of evolution by natural selection at the age of twenty-nine; the remaining forty-four years of his life were devoted to its corroboration and exposition. Pasteur's life reads like a story divided into several chapters. Each chapter represents a period which he devoted to one field of research; at the beginning of each period stands the publication of a short preliminary note which contained the basic discovery in a nutshell; then followed ten or fifteen years of elaboration, consolidation, clarification.

The collective advances of science as a whole, and of each of its specialized branches, show the same alternation between relatively brief eruptions which lead to the conquest of new frontiers, and long periods of consolidation. In the case of the individual, this protracted chore has its natural limits at three score years and ten, or thereabouts; but on the historical stage, the assimilation, consolidation, interpretation, and elaboration of a once revolutionary discovery may go on for generations, and even centuries. The new territory opened up by the impetuous advance of a few geniuses, acting as a spearhead, is subsequently occupied by the solid phalanxes of mediocrity; and soon the revolution turns into a new orthodoxy, with its unavoidable symptoms of one-sidedness, over-specialization, loss of contact with other provinces of knowledge, and ultimately, estrangement from reality. We see this happening—unavoidably, it seems—at various times in the history of various sciences. The emergent orthodoxy hardens into a "closed system" of thought, unwilling or unable to assimilate new empirical data or to adjust itself to significant changes in other fields of knowledge; sooner or later the matrix is blocked, a new crisis arises, leading to a new synthesis, and the cycle starts again.

This does not mean, of course, that science does not advance; only that it advances in a jerky, unpredictable, "unscientific" way. Although "in the year 1500 Europe knew less than Archimedes who died in the year 212 B.C.," it would nevertheless be foolish to deny that today we know considerably more than Archimedes. And I mean by that not only the fantastic and threatening achievements of applied science which have transformed this planet to a point where it is becoming increasingly uninhabitable; but that we also know more than Archimedes in other more worthwhile ways, by having gained deeper insights into the structure of the universe, from the spiral nebulae to the acid molecules which govern heredity.

But these insights were not gained by the steady advance of science along a straight line. Mental evolution is a continuation of biological evolution, and in various respects resembles its crooked ways. "Evolution is known to be a wasteful, fumbling process characterized by sudden mutations of unknown cause, by the slow grinding of selection, and by the dead-ends of over-specialization and loss of adaptability. "Progress" can by definition never go wrong; evolution constantly does; and so does the evolution of ideas, including those of "exact science." New ideas are thrown up spontaneously like mutations; the vast majority of them are useless, the equivalent of biological freaks without survival-value. There is a constant struggle for survival between competing theories in every branch of the history of thought. When we call ideas "fertile" or "sterile," we are unconsciously guided by biological analogy. . . .

"Moreover, there occur in biological evolution periods of crisis and transition when there is a rapid, almost explosive, branching out in all directions, often resulting in a radical change in the dominant trend of development. After these stages of "adaptative radiations," when the species is plastic and malleable, there usually follow periods of stabilization and specialization along the new lines—which again often lead into dead ends of rigid over-specialization."

But there the analogy ends. The branching of the evolutionist's tree of life is a one-way process; giraffes and whales do not bisociate to give rise to a new synthesis. The evolution of ideas, on the other hand, is a tale of ever-repeated differentiation, specialization and reintegrations on a higher level; a progression from primordial unity through variety to more complex patterns of unity-in-variety. (224–26)

Koestler's three classic illustrations of this process are of the printing press (Gutenberg), the concept of gravity (Kepler and Newton), and natural selection (Darwin). Here is his excerpt about the printing press, an engineering marvel that changed the course of Western history.

The Printing Press

Arthur Koestler

At the dawn of the fifteenth century printing was no longer a novelty in Europe. Printing from wooden blocks on vellum, silk, and cloth apparently started in the twelfth century, and printing on paper was widely practised in the second half of the fourteenth. The blocks were engraved in relief with pictures or text or both, then thoroughly wetted with a brown distemper-like substance; a sheet of damp paper was laid on the block and the back of the paper was rubbed with a so-called *frotton*—a dabber or burnisher—until an impression of the carved relief was transferred to it. Each sheet could be printed on only one side by this method, but the blank backs of the sheets could be pasted together and then gathered into quires and bound in the same manner as manuscript-books. These "block books" or *xylographs* circulated already in considerable numbers during Gutenberg's youth.

He was born in 1398 at Mainz and was really called Gensfleisch, meaning gooseflesh, but preferred to adopt the name of his mother's birthplace. The story of his life is obscure, highlighted by a succession of lawsuits against money-lenders and other printers; his claim to priority is the subject of a century-old controversy. But there exists a series of letters to a correspondent, Frère Cordelier, which has an authentic ring and gives a graphic description of the manner in which Gutenberg arrived at his invention. Whether others, such as Costa of Haarlem, made the same invention at the same time or before Gutenberg is, from our point of view, irrelevant.

Oddly enough, the starting point of Gutenberg's invention was not the block-books—he does not seem to have been acquainted with them—but playing-cards. In his first letter to Cordelier he wrote:

> *For a month my head has been working; a Minerva, fully armed, must issue from my brain. . . . You have seen, as I have, playing-cards and pictures of saints. . . . These cards and pictures are engraved on small pieces of wood, and below the pictures there are words and entire lines also engraved. . . . A thick ink is applied to the engraving; and upon this a leaf of paper, slightly damp, is placed; then this wood, this ink, this paper is rubbed and rubbed until the back of the paper is polished. This paper is then taken off and you see on it the picture just as if the design had been traced upon it, and the words as if they had been written; the ink applied to the engraving has become attached to the paper, attracted by its softness and by its moisture. . . .*
>
> *Well, what has been done for a few words, for a few lines, I must succeed in doing for large pages of writing, for large leaves covered entirely on both sides, for whole books, for the first of all books, the Bible. . . .*
>
> *How? It is useless to think of engraving on pieces of wood the whole thirteen hundred pages. . . .*
>
> *What am I to do? I do not know: but I know what I want to do: I wish to manifold the Bible, I wish to have the copies ready for the pilgrimage to Aix la Chapelle.*

Here, then, we have matrix or skill No. I: the printing from woodblocks by means of rubbing.

In the letters which follow we see him desperately searching for a simpler method to replace the laborious carving of letters in wood:

> *Every coin begins with a punch. The punch is a little rod of steel, one end of which is engraved with the shape of one letter, several letters, all the signs which are seen in relief on a coin. The punch is moistened and driven into a piece of steel, which becomes the "hollow" or "stamp." It is into these coin-stamps, moistened in their turn, that are placed the little discs of gold, to be converted into coins, by a powerful blow.*

This is the first intimation of the method of type-casting. It leads Gutenberg, by way of analogy, to the *seal:* "When you apply to the vellum or paper the seal of your community, everything has been said, everything is done, everything is there. Do you not see that you can repeat as many times as necessary the seal covered with signs and characters?"

Yet all this is insufficient. He may cast letters in the form of coins, or seals, instead of engraving the wood, yet they will never make a clear print by the clumsy rubbing method; so long as his search remains confined to this one and only traditional method of making an "imprint," the problem remains blocked. To solve it, an entirely different kind of skill must be brought in. He tries this and that; he thinks of everything under the sun: it is the period of incubation. When the favourable opportunity at last offers itself he is ready for it:

> *I took part in the wine harvest. I watched the wine flowing, and going back from the effect to the cause, I studied the power of this press which nothing can resist. . . .*

At this moment it occurs to him that the same, steady pressure might be applied by a seal or coin—preferably of lead, which is easy to cast—on paper, and that owing to the pressure, the lead would leave a trace on the paper—Eureka!

> *. . . A simple substitution which is a ray of light. . . . To work then! God has revealed to me the secret that I demanded of Him. . . . I have had a large quantity of lead brought to my house and that is the pen with which I shall write.*

"The ray of light" was the bisociation of wine-press and seal—which, added together, become the letter-press. The wine-press has been lifted out of its context, the mushy pulp, the flowing red liquid, the jolly revelry—as Sultan's branch was wrenched out of the context of the tree—and connected with the stamping of vellum with a seal. From now onward these separate skills, which previously had been as different as the butcher's, the baker's, and the candlestick-maker's, will appear integrated in a single, complex matrix:

> *One must strike, cast, make a form like the seal of your community; a mould such as that used for casting your pewter cups; letters in relief like those on your coins, and the punch for producing them like your foot when it multiplies its print. There is the Bible! (121–24)*

Koestler's notion that all scientific and technological evolution goes through the three stages of discovery, consolidation, and rigidity prepares us for the second theory of scientific discovery.

Thomas S. Kuhn, arguably this century's most famous science historian, concentrates more than Koestler on the intellectual rigors of scientific discovery. For Kuhn—writing in *The Structure of Scientific Revolutions*—scientists, engineers, and the world in which they work all share a common paradigm, a set of "universally recognized scientific achievements that for a time provide model problems and solutions to a community of practitioners" (1970, viii). Such a common paradigm often gains the status of a world view, as, for instance, the world view of atomic physics holds sway for us.

Under such paradigms, scientists and engineers do their work. Kuhn calls such work "normal science, the activity in which most scientists inevitably spend almost all their time. [It] is predicated on the assumption that the scientific community knows what the world is like" (5). As a conservative activity, normal science and research is "a strenuous and devoted attempt to force nature into the conceptual boxes supplied by professional education" (5). As a tradition-bound activity, normal science "often suppresses fundamental novelties because they are necessarily subversive of its basic commitments" (5).

Kuhn believes that revolutions begin when such novelties, called anomalies, accumulate experimentally and when they cannot, "despite repeated effort, be aligned with professional expectation" (6). When anomalies reach what might be thought of as a critical mass, scientists and engineers are forced to abandon traditional theories of normal science and to create a new paradigm (a hypothesis on the way to becoming a theory) that will accommodate the new data. Such a paradigm shift constitutes a scientific revolution. Kuhn's example of the discovery of oxygen shows how a paradigm shift comes about.

Anomaly and the Emergence of Scientific Discoveries

Thomas Kuhn

Normal science, the puzzle-solving activity we have just examined, is a highly cumulative enterprise, eminently successful in its aim, the steady extension of the scope and precision of scientific knowledge. In all these respects it fits with great precision the most usual image of scientific work. Yet one standard product of the scientific enterprise is missing. Normal science does not aim at novelties of fact or theory and, when successful, finds none. New and unsuspected phenomena are, however, repeatedly uncovered by scientific research, and radical new theories have again and again been invented by scientists. History even suggests that the scientific enterprise has developed a uniquely powerful technique for producing surprises of this sort. If this characteristic of science is to be reconciled with what has already been said, then research under a paradigm must be a particularly effective way of inducing paradigm change. That is what fundamental novelties of fact and theory do. Produced inadvertently by a game played under one set of rules, their assimilation requires the elaboration of another set. After they have become parts of science, the enterprise, at least of those specialists in whose particular field the novelties lie, is never quite the same again.

We must now ask how changes of this sort can come about, considering first discoveries, or novelties of fact, and then inventions, or novelties of theory. That distinction between discovery and invention or between fact and theory will, however, immediately prove to be exceedingly artificial. Its artificiality is an important clue to several of this essay's main theses. Examining selected discoveries in the rest of this section, we shall quickly find that they are not isolated events but extended episodes with a regularly recurrent structure. Discovery commences with the awareness of anomaly, i.e., with the recognition that nature has somehow violated the paradigm-induced expectations that govern normal science. It then continues with a more or less extended exploration of the area of anomaly. And it closes only when the paradigm theory has been adjusted so that the anomalous has become the expected. Assimilating a new sort of fact demands a more than additive adjustment of theory, and until that adjustment is completed—until the scientist has learned to see nature in a different way—the new fact is not quite a scientific fact at all.

To see how closely factual and theoretical novelty are intertwined in scientific discovery examine a particularly famous example, the discovery of oxygen. At least three different men have a legitimate claim to it, and several other chemists must, in the early 1770's, have had enriched air in a laboratory vessel without knowing it. The progress of normal science, in this case of pneumatic chemistry, prepared the way to a breakthrough quite thoroughly. The earliest of the claimants to prepare a relatively pure sample of the gas was the Swedish apothecary, C. W. Scheele. We may, however, ignore his work since it was not published until oxygen's discovery had repeatedly been announced elsewhere and thus had no effect upon

---

From Kuhn, T. S. 1970. *The structure of scientific revolutions.* 2d ed, 52–56. Chicago: University of Chicago Press.

the historical pattern that most concerns us here. The second in time to establish a claim was the British scientist and divine, Joseph Priestley, who collected the gas released by heated red oxide of mercury as one item in a prolonged normal investigation of the "airs" evolved by a large number of solid substances. In 1774 he identified the gas thus produced as nitrous oxide and in 1775, led by further tests, as common air with less than its usual quantity of phlogiston. The third claimant, Lavoisier, started the work that led him to oxygen after Priestley's experiments of 1774 and possibly as the result of a hint from Priestley. Early in 1775 Lavoisier reported that the gas obtained by heating the red oxide of mercury was "air itself entire without alteration [except that] . . . it comes out more pure, more respirable." By 1777, probably with the assistance of a second hint from Priestley, Lavoisier had concluded that the gas was a distinct species, one of the two main constituents of the atmosphere, a conclusion that Priestley was never able to accept.

This pattern of discovery raises a question that can be asked about every novel phenomenon that has ever entered the consciousness of scientists. Was it Priestley or Lavoisier, if either, who first discovered oxygen? In any case, when was oxygen discovered? In that form the question could be asked even if only one claimant had existed. As a ruling about priority and date, an answer does not at all concern us. Nevertheless, an attempt to produce one will illuminate the nature of discovery, because there is no answer of the kind that is sought. Discovery is not the sort of process about which the question is appropriately asked. The fact that it is asked—the priority for oxygen has repeatedly been contested since the 1780's—is a symptom of something askew in the image of science that gives discovery so fundamental a role. Look once more at our example. Priestley's claim to the discovery of oxygen is based upon his priority in isolating a gas that was later recognized as a distinct species. But Priestley's sample was not pure, and, if holding impure oxygen in one's hands is to discover it, that had been done by everyone who ever bottled atmospheric air. Besides, if Priestley was the discoverer, when was the discovery made? In 1774 he thought he had obtained nitrous oxide, a species he already knew; in 1775 he saw the gas as dephlogisticated air, which is still not oxygen or even, for phlogistic chemists, a quite unexpected sort of gas. Lavoisier's claim may be stronger, but it presents the same problems. If we refuse the palm to Priestley, we cannot award it to Lavoisier for the work of 1775 which led him to identify the gas as the "air itself entire." Presumably we wait for the work of 1776 and 1777 which led Lavoisier to see not merely the gas but what the gas was. Yet even this award could be questioned, for in 1777 and to the end of his life Lavoisier insisted that oxygen was an atomic "principle of acidity" and that oxygen gas was formed only when that "principle" united with caloric, the matter of heat. Shall we therefore say that oxygen had not yet been discovered in 1777? Some may be tempted to do so. But the principle of acidity was not banished from chemistry until after 1810, and caloric lingered until the 1860's. Oxygen had become a standard chemical substance before either of those dates.

Clearly we need a new vocabulary and concepts for analyzing events like the discovery of oxygen. Though undoubtedly correct, the sentence, "Oxygen was discovered," misleads by suggesting that discovering something is a single simple act assimilable to our usual (and also questionable) concept of seeing. That is why we so readily assume that discovering, like seeing or touching, should be unequivocally attributable to an individual and to a moment in time. But the latter attribution is always impossible, and the former often is as well. Ignoring Scheele, we can safely say that oxygen had not been discovered before 1774, and we would probably also say that it had been discovered by 1777 or shortly thereafter. But within those limits or others like them, any attempt to date the discovery must inevitably be arbitrary because discovering a new sort of phenomenon is necessarily a complex event, one which involves recognizing both *that* something is and *what* it is. Note, for example, that if oxygen were dephlogisticated air for us, we should insist without hesitation that Priestley had discovered it, though we would still not know quite when. But if both observation and conceptualization, fact and assimilation to theory, are inseparably linked in discovery, then discovery is a process and must take time. Only when all the relevant conceptual categories are prepared in advance, in which case the phenomenon would not be of a new sort, can discovering *that* and discovering *what* occur effortlessly, together, and in an instant.

Grant now that discovery involves an extended, though not necessarily long, process of conceptual assimilation. Can we also say that it involves a change in paradigm? To that question, no general answer can yet be given, but in this case at least, the answer must be yes. What Lavoisier announced in his papers from 1777 on was not so much the discovery of oxygen as the oxygen theory of combustion. That theory was the keystone for a reformulation of chemistry so vast that it is usually called the chemical revolution. Indeed, if the discovery of oxygen had not been an intimate part of the emergence of a new paradigm for chemistry, the question of priority from which we began would never have seemed so important. In this case as in others, the value placed upon a new phenomenon and thus upon its discoverer varies with our estimate of the extent to which the phenomenon violated paradigm-induced anticipations. Notice, however, since it will be important later, that the discovery of oxygen was not by itself the cause of the change in chemical theory. Long before he played any part in the discovery of the new gas, Lavoisier was convinced both that something was wrong with the phlogiston theory and that burning bodies absorbed some part of the atmosphere. That much he had recorded in a sealed note deposited with the Secretary of the French Academy in 1772. What the work on oxygen did was to give much additional form and structure to Lavoisier's earlier sense that something was amiss. It told him a thing he was already prepared to discover—the nature of the substance that combustion removes from the atmosphere. That advance awareness of difficulties must be a significant part of what enabled Lavoisier to see in experiments like Priestley's a gas that Priestley had been unable to see there himself. Conversely, the fact that a major paradigm revision was needed to see what Lavoisier saw must be the principal reason why Priestley was, to the end of his long life, unable to see it.

Ever since the initial publication of Kuhn's work in 1962, philosophers of the history of science and technology have been debating to what extent Kuhn's theories hold true for all such revolutions. Although their final verdict is still out, Kuhn's theories have launched a whole new branch of academic study into the origins and development of scientific discovery and technical innovation. In the latter case, for instance, much evidence has been found that high-level tinkering—a kind of extended, intensive play—is responsible for many of the best technological innovations. Such seems to be the case, for instance, in the work of Thomas Edison.

The major new theory that challenges those of Koestler and Kuhn comes from the area of artificial intelligence. Herbert A. Simon and his colleagues have written computer programs designed to generate nontrivial scientific discoveries. To replace Koestler's mysterious category of insight and Kuhn's problematic paradigm, these programs, as detailed in *Scientific Discovery* (1987), assume that "scientific discovery is problem solving writ large" (6).

The search for a problem solution is not carried out by random trial and error, but is selective. It is guided in the direction of a *goal situation* (or symbolic expressions describing a goal) by rules of thumb, called *heuristics.* Heuristics make use of information extracted from the problem definitions and the states already explored in the problem space to identify promising paths for search. (8)

In order to create such programs, Simon and his associates have analyzed classic examples in the history of scientific discovery and have concluded that a problem–solution format can be used to understand and describe scientific discovery.

We believe that science is also poetry, and—perhaps even more heretical—that discovery has its reasons, as poetry does. However romantic and heroic we find the moment of discovery to be, we cannot believe either that the events leading up to that moment are entirely random and chaotic or that they require genius that can be understood only by congenial minds. We believe that finding order in the world must itself be a process impregnated with purpose and reason. We believe that the process of discovery can be described and modeled, and that there are better and worse routes to discovery—more and less efficient paths. (3–4)

In the following excerpt, Simon and his colleagues exemplify their method by describing Max Planck's discovery of the law of blackbody radiation.

Scientific Discovery

Pat Langley, Herbert A. Simon, Gary L. Bradshaw, and Jan M. Zytkow

What is a Normative Theory of Discovery?

As we have already seen, a "normative theory of discovery" cannot mean a set of rules for deriving theories conclusively from observations. Instead, a normative theory of discovery would be a set of criteria for judging the efficacy and the efficiency of processes used to discover scientific theories. Presumably the criteria can be derived from the goals of the scientific activity. That is to say, a normative theory rests on contingent propositions such as "If process X is to be efficacious for attaining goal Y, then it should have properties A, B, and C." (See Simon 1973a.) Given such norms, we would be justified in saying that a person who adhered to them would be a better scientist ("better" meaning "more likely to make discoveries").

The idea that there can be a normative theory of discovery is no more surprising than the idea that there can be a normative theory of surgery. Some surgeons do better work than others, presumably because they have better heuristics and techniques (some in the form of conscious principles and problem-solving methods, some in the form of abilities to recognize the critical features of situations, some in the form of tools and instruments, and some in the form of practiced motor skills). The combination of all these heuristics and techniques makes skill in surgery more than a matter of chance, but not a matter of certainty in any particular operation. Patients who might have been saved do die, even in the hands of the most skillful surgeons.

Some scientists, too, do better work than others and make more important discoveries. It seems reasonable to presume that the superior scientists have more effective methodological principles and problem-solving methods, are better able to recognize critical features in data and theoretical formulations, have better laboratory and computing instruments, and are more skillful in the laboratory than their less successful colleagues. Of course, even the finest scientists often fail to solve problems to which they have addressed themselves; Einstein's unsuccessful search for a unified field theory comes immediately to mind, as does the failure of Lorentz and Poincaré to find a wholly satisfactory theory of special relativity.

Skill in surgery may be specialized to particular classes of operations. The superiority of a brain surgeon may rest on heuristics specialized to the anatomy of the human head, and may provide him with no special ability with organ transplants. Similarly, the heuristics possessed by a particular scientist may provide him with a comparative advantage only in some relatively limited domain of science. It is even possible that there is no "scientific method" at all, but only special methods for doing research involving gene transplants, or experiments on nuclear magnetic resonance, or theoretical work in particle physics. The processes of experimentation and theory construction may be quite different in physics, geology, economics, and psychology.

The work reported in this book derives from the hypothesis that, if there is no single "scientific method," there are at least a number of scientific methods that are broadly

From Langley, P., H. A. Simon, G. L. Bradshaw, and J. M. Zytkow. 1987. *Scientific discovery: Computational explorations of the creative processes,* 45–54. Cambridge: The MIT Press.

applicable over many domains of science. This is not to deny that there are also special methods, some of them applicable to only very narrow realm, nor is it to claim that general methods are more important for carrying out scientific work successfully than are special ones. Our only claims are that relatively general methods do play (and have historically played) a significant role in scientific discovery, that such methods are numerous, that they can be identified, and that their effects can be studied.

Cognitive-science research on human problem solving has revealed that humans, in solving all kinds of problems, use both special and general methods. Methods adapted to a particular domain, when available, may be far more powerful than general methods that make no use of the knowledge and structure of the domain. However, general methods, which will usually be weak methods, may be the only ones at hand on the frontiers of knowledge, where few relevant special techniques are yet available. Thus, our task here is to elucidate the role of weak methods in scientific discovery, where they can be expected to play a major role in creative scientific work.

The standard against which weak methods must be judged is the limiting case of random search. Scientific discoveries seldom, if ever, emerge from random, trial-and-error search; the spaces to be searched are far too large for that. Rationality for a scientist consists in using the best means he has available—the best heuristics—for narrowing the search down to manageable proportions (sometimes at the cost of ruling out good solution candidates). If the scientist's tools are weak (perhaps because of the novelty of the problem), a great deal of residual search may still be required; but we must regard such a search process as rational if it employs all the heuristics that are known to be applicable to the domain. This is the concept of rationality that is relevant to the creative process and to problem solving in general, and it is with this kind of rationality that a normative theory of creativity and scientific discovery is concerned.

An Example: Blackbody Radiation

Our way of conceptualizing the discovery process can perhaps best be explained with the help of concrete examples. The two examples we shall employ lie beyond anything we have simulated with our computer programs. We chose them because they represent two of the most remarkable events in the history of physics, so that the encomium of "creativity" cannot be denied to them. If discoveries of the magnitude considered here can be explained in terms of understandable cognitive processes, even though they have not yet been simulated, they should give us some basis for approaching the modeling through simulation of other scientific discoveries of the first magnitude.

The first example is Max Planck's discovery of the law of blackbody radiation, a discovery that opened the way to the development of the quantum theory and provided the initial formulation of that theory. The example is really a whole set of examples, for a complex set of events was involved. At least one of these events can be explained with very good resolution in terms of steps each taking only a few hours; others require much more time.

In the case of blackbody radiation, the problem was to characterize the equilibrium distribution, by wavelength, of radiating energy inside a perfectly reflecting cavity. The goal was to find a function expressing the relative intensity of radiation at each wavelength as a function of that wavelength and the temperature of the cavity. The problem emerged in the middle of the nineteenth century and received considerable attention from Gustav Kirchhoff and other leading physicists from that time until the first decades of the twentieth century. Steady advances in experimental instruments over the last quarter of the nineteenth century permitted increasingly accurate measurements of the intensity function over broader and broader ranges of temperatures and wavelengths. Thus, theorists concerned with the problem were confronted with increasingly stringent tests of their hypotheses, which culminated in 1899–1900 in the availability of data spanning a wide range of conditions.

Planck, a theoretical physicist, devoted a major part of his attention and activity to the blackbody problem over the years 1895–1900. Another German theoretical physicist, Wilhelm Wien, also played an important role in the events we shall describe. Of course, we cannot tell the whole story here; it has been the subject of several substantial books as well as a number of briefer accounts. The facts on which we shall draw are well established in this literature; see Klein 1962 and Kuhn 1978.

By the early 1890s, a good deal of theoretical work had been done on the blackbody problem, bringing to bear upon it the theories of electromagnetism, thermodynamics, and mechanics of that era. However, the theory had not developed to the point of yielding deductive derivations of a complete formula for blackbody radiation. Meanwhile, several empirical formulas ("curve-fitting exercises") had been proposed to account of the available empirical data. In particular, Wien had constructed (in 1896) an incomplete and not entirely satisfactory explanation of a formula first proposed on wholly empirical grounds by Friedrich Paschen. Wien's formula, as it was generally called, provided an excellent fit to the data then available. It was of the form

$$I(\nu)=\frac{A}{\exp(k\nu/T)}, \tag{2.1}$$

where $I(v)$ is the intensity of radiation of frequency v, T is the temperature of the cavity, and A and k are constants. Discrepancies between the formula and data for low temperatures and high frequencies (short wavelengths) were seldom more than a few percent. Data were unavailable for high temperatures and low frequencies. Wien's formula gained wide acceptance as a correct description of the phenomena and as the basis for a physical explanation of them.

In 1899, Planck succeeded, or thought he had succeeded, in deriving Wien's law deductively from the principles of classical physics. Planck believed that his demonstration showed that the law was unique—that no alternative law was compatible with classical principles. By "deriving the law from the principles of classical physics" we mean that Planck formulated a physical model to describe the black body and its cavity, and described the behavior of the model in classical terms.

Here we see already, in coarse grain, some of the kinds of developmental steps with which we shall be concerned throughout this book. In the period up to 1899 there were theoretical steps (by Kirchhoff, Wien, Planck, and others) connecting blackbody radiation to established physical theory; there were experimental steps, improving and extending the measurements; and there were curve-fitting steps, seeking to provide parsimonious descriptions of the data. As each step of whichever kind was taken, it provided a new set of initial conditions for the next act of theorizing, experimenting, or curve fitting. If there was an autonomous first mover in this system, it was the experimentation, which largely responded to new advances in instrumentation and which was not directly affected by theoretical progress. Nevertheless, the growing richness and interest of the theory undoubtedly contributed major motivation to the vigor of the experimental activities and the desire to extend the range of the empirical data. The theoretical work provided a rationale for the experimentation without much influencing its specific course.

Neither Wien's law nor Planck's derivation of it, completed in 1899 and published early in 1900, survived very long the empirical tests provided by new experiments with higher temperatures and lower frequencies. By the summer of 1900, in the new range of observation, discrepancies had been discovered between data and formula that could not easily be explained away as errors of measurement. On the afternoon of Sunday, October 7, Heinrich Rubens (one of the experimentalists in Berlin working on blackbody radiation) and his wife called on the Plancks. Rubens described to Planck new findings that clearly violated Wien's law. In fact, he told Planck, these findings demonstrated that, for small values of v/T, I was linear in the reciprocal, T/v; that is, $I(v)=AT/kv$. Before he went to bed that night, Planck had conjectured a new formula to replace Wien's law, and had mailed a postcard to Rubens describing it. The new formula was closely similar to Wien's. Planck had simply replaced $\exp(kv/T)$ in the former by $\exp(kv/T)-1$, to obtain

$$I(\nu)=\frac{A}{\exp(k\nu/T)-1}. \tag{2.2}$$

This is the law that has been known ever since as Planck's quantum-theoretical law of blackbody radiation. Planck announced it on October 19, explaining that he had arrived at it by constructing "completely arbitrary expressions for the entropy which although they are more complicated than Wien's expression [i.e., the expression consistent with Wien's formula] still seem to satisfy just as completely all requirements of the thermodynamic and electromagnetic theory" and pointing out, with some note of apology, that his earlier demonstration of the "necessity" of Wien's law had obviously been erroneous. He also identified the probable locus of the error in his reasoning: the expression for entropy, which he had postulated with very little physical motivation.

We will return to the crucial day of October 7, 1900, but before doing so we will sketch out the rest of the story. Planck was now faced with the task of providing a physical explanation for the new law. He was able to construct and announce one by December. The explanation rested on the nonclassical quantization of energy; however, as Klein (1962), Kuhn (1978), and others have shown, the derivation was sufficiently complex and difficult that neither Planck nor other physicists were aware of its deadly consequences for classical physics until about 1905 or 1906, when these were pointed out by Albert Einstein and Paul Ehrenfest.

Notice that in this last phase, the final months of 1900, Planck was working backward from a known result, a formula induced directly from the data, and was aiming at providing a physical "story" that would explain why the formula was just so and not otherwise. This was exactly what he had done in 1899 for Wien's law. By the verdict of history, the 1899 story was invalid, the 1900 story valid (or essentially so; we do not tell it today in exactly the same way). The history of events would seem to imply that, whatever the facts (or whatever they were thought to be at a given moment of time), a physicist of sufficient expertise and ingenuity (but perhaps only one at Planck's level) might find a rationalization that would fit these facts to theory. We are reminded again of Poincaré's dictum that "hypotheses are what we lack the least." There is no reason to believe, and no evidence, that the reasoning processes Planck applied to his task of 1899 were any different from those he applied in 1900. The goal was different, however—a different formula to which he had to build a bridge from classical theory.

A crucial step in Planck's 1900 derivation employed probabilities. The particular probability model he selected (involving indistinguishable electrons) had no special physical motivation except that it led to the right, and already known, formula. In fact, had Planck used the standard (and more defensible) Boltzmann statistics, he would have been led to the "classical" Rayleigh-Jeans law of blackbody radiation, with $I(v)$ linear in T/v, which, like Wien's law, holds only asymptotically for one range of v/T. (The correct law, given by Planck's formula, is in fact an interpolation between Wien's formula, which fits large values of v/T, and that of Rayleigh and Jeans, which fits small values.)

Planck's theoretical derivation, therefore, was in considerable measure driven by the data (or, more exact, by a formula obtained inductively from the data). The solution to the problem that Planck posed for himself had to satisfy constraints imposed by the data as well as constraints imposed by classical theory. The latter were sufficiently ill defined that it was possible to violate them inadvertently and thereby find a solution under the driving force of the given empirical formula.

We go back now to October 7, 1900. What happened during the hours in which Planck was seeking an alternative, consistent with the data, to Wien's formula? There are two possible answers to this question, but they both lead to the same conclusion: The problem was solved by finding a formula that interpolated smoothly between the two known limiting functions, Wien's formula for high values of the v/T ratio and the linear (in T/v) formula communicated to Planck by Rubens for low values of v/T. In his October 19 paper, Planck describes one path toward this interpolation without asserting quite explicitly that it was the actual path of his discovery. There was also a second path, and it seems even simpler and more obvious.

First, consider the path mentioned by Planck (see Kangro 1972). From the radiation formula (either Wien's or the linear one), and with the help of the then-accepted physical theory of the phenomena, it is easy to calculate the second derivative, d^2S/dU^2, where S is the entropy of the system and U is its energy. S and U are connected with T by the relation $dS/dU = 1/T$. Planck was already familiar with this derivation in connection with Wien's law, and the same derivation can be carried through directly for the linear formula. We obtain

$$\frac{d^2S}{dU^2} = \frac{A}{U} \tag{2.3}$$

for Wien's law and

$$\frac{d^2S}{dU^2} = \frac{A}{U^2} \tag{2.4}$$

for the linear intensity law.

Now, Planck goes on to say in his October 19 paper (Kangro 1970, p. 36):

I finally started to construct completely arbitrary expressions for the entropy which although they are more complicated than Wien's expression still seem to satisfy just as completely all requirements of the thermodynamic and electromagnetic theory.

I was especially attracted by one of the expressions thus constructed which is nearly as simple as Wien's expression and which would deserve to be investigated since Wien's expression is not sufficient to cover all observations. We get this expression by putting

$$d^2S/dU^2 = \alpha/[U(\beta + U)].$$

It is by far the simplest of all expressions which lead to S *as a logarithmic function of* U. . . .

If we accept this account as an actual description of the process that Planck used to solve his problem, then we see that he searched among an undefined class of functions until he found a "simple" one that had the desired limiting properties. We do not have any very good measure of how hard or how easy a search this is, except that Planck completed it in a few hours.

There is an alternative path to the goal that makes use of the intensity laws without calculating the derivatives of the entropies. It consists in looking directly for a "simple" interpolating function between Wien's law, for high v/T, and the law that is linear in T/v, for low v/T. Consider Wien's law in the form $I = A/e^x$, where x is essentially v/T. To find the behavior of the function in the neighborhood of $x = 0$ (low v/T), we expand e^x into a Taylor series, obtaining the familiar expression

$$e^x = 1 + x + x^2/2! + \ldots, \tag{2.5}$$

whence, subtracting unity from both sides and neglecting higher-order terms, we obtain

$$e^x - 1 = x. \tag{2.6}$$

Hence, if we replace e^x in Wien's law by $(e^x - 1)$, we obtain a function that, by equation 2.6, is approximately linear in x for small x, but which goes asymptotically to e^x as x (that is, v/T) grows large. Thus, we have obtained a new function, $I = A/[\exp(v/T) - 1]$, which interpolates between Wien's law for large v/T and the linear law for small v/T and which is precisely Planck's law of radiation as stated in equation 2.2.

In the spirit of casual empiricism, we sought to assess just how difficult this derivation is, and whether it is likely to occur to a skilled applied mathematician within a moderate span of time. We gave the interpolation problem, without explaining its physical origins, to eight colleagues who are professional physical scientists and applied mathematicians. Five gave the same answer that Planck arrived at, each in under two minutes. Three of the five used exactly the process described above (expansion of e^x into series); the other two used different, but equally simple, procedures. None of them recognized the problem as related to blackbody radiation or thought of Planck's law until they had derived it.

Given this evidence, we must conclude that, for someone possessing the standard heuristics of an applied mathematician (all of which were known in 1900), even in the absence of the physical cues that were available to Planck, no extensive search is required to solve the problem. Planck's success in solving it within a few hours should occasion no surprise. We do not know, of course, whether Planck took the route just described, or the one he himself described, or some other route. We know only that no unusual processes, but only standard processes used by and generally available to theoreticians in the physical sciences, need be postulated in order to explain the actual discovery.

It would be hard to argue that either Planck's behavior or the behavior of our colleagues when they were presented with Planck's problem was a matter of chance or of pure "inspiration." Rather, they quite patently represent applications of these entirely sensible heuristic principles:

- Given two distinct functions that fit the data excellently in two different regions, look for an interpolating function that approximates the given functions in the limit in the appropriate regions.
- To examine the limiting behavior of a function, express it as a Taylor series around the limit point.

These procedures will not always work—they are weak methods, but scientists who apply them will more often be successful in finding laws that fit data well than scientists who do not. Why, then, didn't others discover Planck's law in 1900 or earlier? Planck was one of a small number of theoreticians interested in this problem, and one of an even smaller number who were conversant with the most recent data. Moreover, the refutation by these data of Planck's previous "explanation" of Wien's law gave him strong motivation to put the problem at the top of his agenda. The "accident" happened to a thoroughly prepared and motivated mind.

Planck's construction in December 1900 of a physical explanation or rationalization of the new formula also is attributable to something other than inspiration or intuition. His previous derivation of Wien's law told him at just what point his theory needed correction, and his new knowledge of the "correct" law provided a criterion that any new explanation must satisfy: It must lead to the new law. Planck applied the heuristic of working backward from a known result, generating subgoals as he went, and examined alternative probability models until he found one (a very unorthodox one, although he was not aware of that fact at the time) that yielded the desired result. Knowing what answer he had to reach greatly limited the size of his search space and informed him immediately when he had succeeded.

This example should make clear the sense in which we think normative statements can be made about the discovery process. Such statements will take the form of descriptions of good heuristics, or of evaluative statements about the relative merits of different heuristics or other methods. The evaluative statements can be generated, in turn, either by examining historical evidence of discovery or failure of discovery or by constructing computer programs that incorporate the heuristics and then testing the efficacy of the computer programs as machines for making discoveries.

The next selections describe scientific revolutions. As you read them, try to decide whether the theories of Koestler, Kuhn, or Simon best fit the facts.

No modern scientific revolution has affected the public consciousness more than the theory that man descended from a lower form of life. Charles Darwin (1809–1882) described this theory in two works: *On the Origin of Species by Means of Natural Selection* (1859) and *The Descent of Man* (originally published in 1871). The following selection comes from the latter work.

On the Affinities and Genealogy of Man

Charles Darwin

Even if it be granted that the difference between man and his nearest allies is as great in corporeal structure as some naturalists maintain, and although we must grant that the differ-

---

From Darwin, C. 1874. *The descent of man, and selection in relation to sex.* 2d ed., revised. New York: A. L. Burt.

ence between them is immense in mental power, yet the facts given in the earlier chapters appear to declare, in the plainest manner, that man is descended from some lower form, notwithstanding that connecting-links have not hitherto been discovered.

Man is liable to numerous, slight, and diversified variations, which are induced by the same general causes, are governed and transmitted in accordance with the same general laws, as in the lower animals. Man has multiplied so rapidly, that he has necessarily been exposed to struggle for existence, and consequently to natural selection. He has given rise to many races, some of which differ so much from each other, that they have often been ranked by naturalists as distinct species. His body is constructed on the same homological plan as that of other mammals. He passes through the same phases of embryological development. He retains many rudimentary and useless structures, which no doubt were once serviceable. Characters occasionally make their re-appearance in him, which we have reason to believe were possessed by his early progenitors. If the origin of man had been wholly different from that of all other animals, these various appearances would be mere empty deceptions; but such an admission is incredible. These appearances, on the other hand, are intelligible, at least to a large extent, if man is the co-descendant with other mammals of some unknown and lower form.

Some naturalists, from being deeply impressed with the mental and spiritual powers of man, have divided the whole organic world into three kingdoms, the Human, the Animal, and the Vegetable, thus giving to man a separate kingdom. Spiritual powers cannot be compared or classed by the naturalist: but he may endeavour to shew, as I have done, that the mental faculties of man and the lower animals do not differ in kind, although immensely in degree. A difference in degree, however great, does not justify us in placing man in a distinct kingdom, as will perhaps be best illustrated by comparing the mental powers of two insects, namely, a coccus or scale-insect and an ant, which undoubtedly belong to the same class. The difference is here greater than, though of a somewhat different kind from, that between man and the highest mammal. The female coccus, whilst young, attaches itself by its proboscis to a plant; sucks the sap, but never moves again; is fertilised and lays eggs; and this is its whole history. On the other hand, to describe the habits and mental powers of worker-ants, would require, as Pierre Huber has shewn, a large volume; I may, however, briefly specify a few points. Ants certainly communicate information to each other, and several unite for the same work, or for games of play. They recognise their fellow-ants after months of absence, and feel sympathy for each other. They build great edifices, keep them clean, close the doors in the evening, and post sentries. They make roads as well as tunnels under rivers, and temporary bridges over them, by clinging together. They collect food for the community, and when an object, too large for entrance, is brought to the nest, they enlarge the door, and afterwards build it up again. They store up seeds, of which they prevent the germination, and which, if damp, are brought up to the surface to dry. They keep aphides and other insects as milch-cows. They go out to battle in regular bands, and freely sacrifice their lives for the common weal. They emigrate according to a preconcerted plan. They capture slaves. They move the eggs of their aphides, as well as their own eggs and cocoons, into warm parts of the nest, in order that they may be quickly hatched; and endless similar facts could be given. On the whole, the difference in mental power between an ant and a coccus is immense; yet no one has ever dreamed of placing these insects in distinct classes, much less in distinct kingdoms. No doubt the difference is bridged over by other insects; and this is not the case with man and the higher apes. But we have every reason to believe that the breaks in the series are simply the results of many forms having become extinct.

Professor Owen, relying chiefly on the structure of the brain, has divided the mammalian series into four sub-classes. One of these he devotes to man; in another he places both the Marsupials and the Monotremata; so that he makes man as distinct from all other mammals as are these two latter groups conjoined. This view has not been accepted, as far as I am aware, by any naturalist capable of forming an independent judgment and therefore need not here be further considered.

We can understand why a classification founded on any single character or organ—even an organ so wonderfully complex and important as the brain—or on the high development of the mental faculties, is almost sure to prove unsatisfactory. This principle has indeed been tried with hymenopterous insects; but when thus classed by their habits or instincts, the arrangement proved thoroughly artificial. Classifications may, of course, be based on any character whatever, as on size, colour, or the element inhabited; but naturalists have long felt a profound conviction that there is a natural system. This system, it is now generally admitted, must be, as far as possible, genealogical in arrangement,—that is the co-descendants of the same form must be kept together in one group, apart from the co-descendants of any other form; but if the parent-forms are related, so will be their descendants, and the two groups together will form a larger group. The amount of difference between the several groups—that is the amount of modification which each has undergone—is expressed by such terms as genera, families, orders, and classes. As we have no record of the lines of descent, the pedigree can be discovered only by observing the degrees of resemblance between the beings which are to be classed. For this object numerous points of resemblance are of much more importance than the amount of similarity or dissimilarity in a few points. If two languages were found to resemble each other in a multitude of words and points of construction, they would be universally recognised as having sprung from a common source, notwithstanding that they differed greatly in some few words or points of construction. But with organic beings the points of resemblance must not consist of adaptations to similar habits of life: two animals may, for instance, have had their whole frames modified for living in the water, and yet they will not be brought any nearer to each other in the natural system. Hence we can see how it is that resemblances in several unimportant structures, in useless and rudimentary organs, or not now functionally active, or in an embryological condition, are by far the most serviceable for classification; for they can hardly be due to adaptations within a late period; and thus they reveal the old lines of descent or of true affinity.

We can further see why a great amount of modification in some one character ought not to lead us to separate widely any two organisms. A part which already differs much from the same part in other allied forms has already, according to the theory of evolution, varied much; consequently it would (as long as the organism remained exposed to the same exciting conditions) be liable to further variations of the same kind; and these, if beneficial, would be preserved, and thus be continually augmented. In many cases the continued development of a part, for instance, of the beak of a bird, or of the teeth of a mammal, would not aid the species in gaining its food, or for any other object; but with man we can see no definite limit to the continued development of the brain and mental faculties, as far as advantage is concerned. Therefore in determining the position of man in the natural or genealogical system, the extreme development of his brain ought not to outweigh a multitude of resemblances in other less important or quite unimportant points.

The greater number of naturalists who have taken into consideration the whole structure of man, including his mental faculties, have followed Blumenbach and Cuvier, and have placed man in a separate Order, under the title of the Bimana, and therefore on an equality with the orders of the Quadrumana, Carnivora, &c. Recently many of our best naturalists have recurred to the view first propounded by Linnæus, so remarkable for his sagacity, and have placed man in the same Order with the Quadrumana, under the title of the Primates. The justice of this conclusion will be admitted: for in the first place, we must bear in mind the comparative insignificance for classification of the great development of the brain in man, and that the strongly-marked differences between the skulls of man and the Quadrumana (lately insisted upon by Bischoff, Aeby, and others) apparently follow from their differently developed brains. In the second place, we must remember that nearly all the other and more important differences between man and the quadrumana are manifestly adaptive in their nature, and relate chiefly to the erect position of man; such as the structure of his hand, foot, and pelvis, the curvature of his spine, and the position of his head. The family of Seals offers a good illustration of the small importance of adaptive characters for classification. These animals differ from all other Carnivora in the form of their bodies and in the structure of their limbs, far more than does man from the higher apes; yet in most systems, from that of Cuvier to the most recent one by Mr. Flower, seals are ranked as a mere family in the Order of the Carnivora. If man had not been his own classifier, he would never have thought of founding a separate order for his own reception.

It would be beyond my limits, and quite beyond my knowledge, even to name the innumerable points of structure in which man agrees with the other Primates. Our great anatomist and philosopher, Prof. Huxley, has fully discussed this subject, and concludes that man in all parts of his organization differs less from the higher apes, then these do from the lower members of the same group. Consequently there "is no justification for placing man in a distinct order."

Lower Stages in the Genealogy of Man

We have thus far endeavoured rudely to trace the genealogy of the Vertebrata by the aid of their mutual affinities. We will now look to man as he exists; and we shall, I think, be able partially to restore the structure of our early progenitors, during successive periods, but not in due order of time. This can be effected by means of the rudiments which man still retains, by the characters which occasionally make their appearance in him through reversion, and by the aid of the principles of morphology and embryology. The various facts, to which I shall here allude, have been given in the previous chapters.

The early progenitors of man must have been once covered with hair, both sexes having beards; their ears were probably pointed, and capable of movement; and their bodies were provided with a tail, having the proper muscles. Their limbs and bodies were also acted on by many muscles which now only occasionally reappear, but are normally present in the Quadrumana. At this or some earlier period, the great artery and nerve of the humerus ran through a supra-condyloid foramen. The intestine gave forth a much larger diverticulum or cæcum than that now existing. The foot was then prehensile, judging from the condition of the great toe in the fœtus; and our progenitors, no doubt, were arboreal in their habits, and frequented some warm, forest clad land. The males had great canine teeth, which served them as formidable weapons. At a much earlier period the uterus was double; the excreta were voided through a cloaca; and the eye was protected by a third eyelid or nictitating membrane. At a still earlier period the progenitors of man must have been aquatic in their habits; for morphology plainly tells us that our lungs consist of a modified swim-bladder, which once served as a float. The clefts on the neck in the embryo of man show where the branchiæ once existed. In the lunar or weekly recurrent periods of some of our functions we apparently still retain traces of our primordial birthplace, a shore washed by the tides. At about this same early period the true kidneys were replaced by the corpora wolffiana. The heart existed as a simple pulsating vessel; and the chorda dorsalis took the place of a vertebral column. These early ancestors of man, thus seen in the dim recesses of time, must have been as simply, or even still more simply organised than the lancelet or amphioxus.

There is one other point deserving a fuller notice. It has long been known that in the vertebrate kingdom one sex bears rudiments of various accessory parts, appertaining to the reproductive system, which properly belongs to the opposite sex; and it has now been ascertained that at a very early embryonic period both sexes possess true male and female glands. Hence some remote progenitor of the whole vertebrate kingdom appears to have been hermaphrodite or androgynous. But here we encounter a singular difficulty. In the mammalian class the males possess rudiments of a uterus with the adjacent passage, in their vesiculæ prostaticæ; they bear also rudiments of mammæ, and some male Marsupials have traces of a marsupial sack. Other analogous facts could be added. Are we, then, to suppose that some extremely ancient mammal continued androgynous, after it had acquired the chief distinctions of its class, and therefore after it had diverged from the lower classes of the vertebrate kingdom? This seems very improbable, for we have to look to fishes, the lowest of all the classes, to find any still existent androgynous forms. That various accessory parts, proper to each sex, are found in a rudimentary condition in the opposite sex, may be explained by such organs having been gradually acquired by the one sex, and then transmitted in a more or less imperfect state to the other. When we treat of sexual selection, we shall meet with innumerable instances of this form of transmission,—as in the case of the spurs, plumes, and brilliant colours, acquired for battle or ornament by male birds, and inherited by the females in an imperfect or rudimentary condition.

The possession by male mammals of functionally imperfect mammary organs is, in some respects, especially curious. The Monotremata have the proper milk-secreting glands with orifices, but no nipples; and as these animals stand at the very base of the mammalian series, it is probable that the progenitors of the class also had milk-secreting glands, but no nipples. This conclusion is supported by what is known of their manner of development; for Professor Turner informs me, on the authority of Kölliker and Langer, that in the embryo the mammary glands can be distinctly traced before the nipples are in the least visible; and the development of successive parts in the individual generally represents and accords with the development of successive beings in the same line of descent. The Marsupials differ from the Monotremata by possessing nipples; so that probably these organs were first acquired by the Marsupials, after they had diverged from, and risen above, the Monotremata, and were then transmitted to the placental mammals. No one will suppose that the Marsupials still remained androgynous, after they had approximately acquired their present structure. How then are we to account for male mammals possessing mammæ? It is possible that they were first developed in the females and then transferred to the males, but from what follows this is hardly probable.

It may be suggested, as another view, that long after the progenitors of the whole mammalian class had ceased to be androgynous, both sexes yielded milk, and thus nourished their young; and in the case of the Marsupials, that both sexes carried their young in marsupial sacks. This will not appear altogether improbable, if we reflect that the males of existing syngnathous fishes receive the eggs of the females in their abdominal pouches, hatch them, and afterwards, as some believe, nourish the young;—that certain other male fishes hatch the eggs within their mouths or branchial cavities;—that certain male toads take the chaplets of eggs from the females, and wind them round their own thighs, keeping them there until the tadpoles are born;—that certain male birds undertake the whole duty of incubation, and that male pigeons, as well as the females, feed their nestlings with a secretion from their crops. But the above suggestion first occurred to me from mammary glands of male mammals being so much more perfectly developed than the rudiments of the other accessory reproductive parts, which are found in the one sex though proper to the other. The mammary glands and nipples, as they exist in male mammals, can indeed hardly be called rudimentary; they are merely not fully developed, and not functionally active. They are sympathetically affected under the influence of certain diseases, like the same organs in the female. They often secrete a few drops of milk at birth and at puberty: this latter fact occurred in the curious case before referred to, where a young man possessed two pairs of mammæ. In man and some other male mammals these organs have been known occasionally to become so well developed during maturity as to yield a fair supply of milk. Now if we suppose that during a former prolonged period male mammals aided the females in nursing their offspring, and that afterwards from some cause (as from the production of a smaller number of young) the males ceased to give this aid, disuse of the organs during maturity would lead to their becoming inactive; and from two well-known principles of inheritance, this state of inactivity would probably be transmitted to the males at the corresponding age of maturity. But at an earlier age these organs would be left unaffected, so that they would be almost equally well developed in the young of both sexes.

Conclusion

Von Baer has defined advancement or progress in the organic scale better than any one else, as resting on the amount of differentiation and specialisation of the several parts of a being,—when arrived at maturity, as I should be inclined to add. Now as organisms have become slowly adapted to diversified lines of life by means of natural selection, their parts will have become more and more differentiated and specialised for various functions from the advantage gained by the division of physiological labour. The same part appears often to have been modified first for one purpose, and then long afterwards for some other and quite distinct purpose; and thus all the parts are rendered more and more complex. But each organism still retains the general type of structure of the progenitor from which it was aboriginally derived. In accordance with this view it seems, if we turn to geological evidence, that organisation on the whole has advanced throughout the world by slow and interrupted steps. In the great kingdom of the Vertebrata it has culminated in man. It must not, however, be supposed that groups of organic beings are always supplanted, and disappear as soon as they have given birth to other and more perfect groups. The latter, though victorious over their predecessors, may not have become better adapted for all places in the economy of nature. Some old forms appear to have survived from inhabiting protected sites, where they have not been exposed to very severe competition; and these often aid us in constructing our genealogies, by giving us a fair idea of former and lost populations. But we must not fall into the error of looking at the existing members of any lowly-organised group as perfect representatives of their ancient predecessors.

The most ancient progenitors in the kingdom of the Vertebrata, at which we are able to obtain an obscure glance, apparently consisted of a group of marine animals, resembling the larvæ of existing Ascidians. These animals probably gave rise to a group of fishes, as lowly organised as the lancelet; and from these the Ganoids, and other fishes like the Lepidosiren, must have been developed. From such fish a very small advance would carry us on to the Amphibians. We have seen that birds and reptiles were once intimately connected together; and the Monotremata now connect mammals with reptiles in a slight degree. But no one can at present say by what line of descent the three higher and related classes, namely, mammals, birds, and reptiles, were derived from the two lower vertebrate classes, namely, amphibians and fishes. In the class of mammals the steps are not difficult to conceive which led from the ancient Monotremata to the ancient Marsupials; and from these to the early progenitors of the placental mammals. We may thus ascend to the Lemuridæ; and the interval is not very wide from these to the Simiadæ. The Simiadæ then branched off into two great stems, the New World and Old World monkeys; and from the latter, at a remote period, Man, the wonder and glory of the Universe, proceeded.

Thus we have given to man a pedigree of prodigious length, but not, it may be said, of noble quality. The world, it has often been remarked, appears as if it had long been preparing for the advent of man: and this, in one sense is strictly true, for he owes his birth to a long line of progenitors. If any single link in this chain had never existed, man would not have been exactly what he now is. Unless we wilfully close our eyes, we may, with our present knowledge, approximately recognise our parentage; nor need we feel ashamed of it. The most humble organism is something much higher than the inorganic dust under our feet; and no one with an unbiased mind can study any living creature, however humble, without being struck with enthusiasm at its marvellous structure and properties.

Marja Sklodowska was born in Warsaw, Poland, in 1867. The University of Warsaw refused her admission in 1891 because she was a woman, so she entered the Sorbonne, earning there her Master's and Doctor's degrees in physics. She was awarded the Nobel Prize in physics (1903) with her husband, Pierre Curie, and in chemistry (1911). Marie Curie discovered the radioactive properties of radium, thorium, and polonium; she isolated radium from pitchblend (uranium ore) and determined its atomic weight. After Pierre died in a street accident, Marie Curie succeeded to his professorship in physics at the Sorbonne, becoming the first woman faculty member in the 650 years of the university's existence. Marie Curie served with heroic distinction in the First World War, establishing X-ray units and training X-ray technicians at the front. Tragically, she died of complications from her exposure to radiation. The following excerpt comes from her doctoral thesis.

Radioactive Substances

Marie Curie

Communication of Radio-Activity to Substances Initially Inactive.

During the course of our researches on radio-active substances M. Curie and I have observed that every substance which remains for some time in the vicinity of a radium salt becomes itself radio-active. In our first publication on this subject, we confined ourselves to proving that the radio-activity thus acquired by substances initially inactive is not due to the transference of radio-active particles to the surface of these substances. This is proved beyond dispute by all the experiments which will be here described, and by the laws according to which the radio-activity excited in naturally inactive bodies disappears when the latter are removed from the influence of radium.

We have given the name of *induced radio-activity* to the new phenomenon thus discovered.

In the same publication, we indicated the essential characteristics of induced radio-activity. We excited screens of different substances by placing them in the neighbourhood of solid radium salts, and we investigated the radio-activity of these screens by the electrical method. We observed the following facts:—

1. The activity of a screen exposed to the action of radium increases with the time of exposure, approaching to a definite limit according to an asymptotic law.
2. The activity of a screen which has been excited by the action of radium, and which is afterwards withdrawn from its action, disappears in a few days. This induced activity approaches zero as a function of the time, following an asymptotic law.
3. Other things being equal, the radio-activity induced by the same radium product upon different screens is independent of the nature of the screen. Glass, paper, metals, all acquire the same degree of activity.
4. The radio-activity induced in one screen by differing radium products has a limiting value which rises with the increased activity of the product.

Shortly afterwards, Mr. Rutherford published a research, which showed that compounds of thorium are capable of producing the phenomenon of induced radio-activity. Mr. Rutherford discovered for this phenomenon the same laws as those just enunciated, besides this additional important fact, that bodies charged with negative electricity become more active than others. Mr. Rutherford also observed that air passed over thorium oxide preserves a notable conductivity for about ten minutes. Air in this condition communicates induced radio-activity to inactive substances, especially to those negatively charged. Mr. Rutherford explains his experiences by the supposition that compounds of thorium, particularly the oxide, give rise to a *radio-active emanation* capable of being carried by air currents and charged with positive electricity. This emanation would be the origin of induced radio-activity. M. Dorn has repeated, with salts of barium containing radium, the experiments of Mr. Rutherford with thorium oxide.

M. Debierne has shown that actinium causes, to a marked degree, induced activity of bodies placed in its vicinity. As in the case of thorium, there is a considerable carriage of activity by air currents.

Induced radio-activity has various aspects, and irregular results are obtained when the activity of a substance in the neighbourhood of radium is excited in free air. MM. Currie and Debierne have observed, however, that the phenomenon is quite regular when operating in a closed vessel; they therefore investigated induced activity in a closed space. . . .

Part played by Gases in the Phenomena of Induced Radio-activity. The gases present in an enclosure containing a solid salt or a solution of a salt of radium are radio-active. This

---

From Curie, M. [1903] 1983. Radioactive substances, 69–72, 92–94. Doctoral thesis, reprint. New York: Philosophical Library. Reprinted by permission of the Philosophical Library.

radio-activity persists when the gas is drawn off with a tube and collected in a test-tube. The sides of the test-tube become themselves radio-active, and the glass of the test-tube is luminous in the dark. The activity and luminosity of the test-tube finally completely disappear, but very gradually, and a month afterwards radio-activity may still be detected.

Since the beginning of our researches, M. Curie and I have, by heating pitchblende, extracted a strongly radio-active gas, but, as in the preceding experiment, the activity of this gas finally completely vanished.

We could discern no new ray in the spectrum of this gas; this was therefore not a case of a new radio-active gas, and we understood later that it was the phenomenon of induced radio-activity.

Thus, for thorium, radium, and actinium induced radio-activity is progressively propagated through the gases from the radiating body to the walls of the enclosure containing it, and the exciting principle is carried away with the gas itself, when the latter is extracted from the enclosure.

When the radio-activity of radium compounds is measured by the electrical method by means of the apparatus, . . . the air between the plates is itself radio-active; however, on passing a current of air between the plates, there is no observable lowering of the intensity of the current, which proves that the radio-activity distributed in the space between the plates is of little account in comparison with that of the radium itself in the solid state.

It is quite otherwise with thorium. The irregularities which I observed in determining the radio-activity of the thorium compounds arose from the fact that at this point I was working with a condenser open to the air; the least air current caused a considerable change in the intensity of the current, because the radio-activity dispersed in the space in the vicinity of the thorium is considerable as compared with the radio-activity of the substance.

This effect is still more marked in the case of actinium. A very active compound of actinium appears much less active when a current of air is passed over the substance.

The radio-active energy is therefore contained in the gas in a special form. Mr. Rutherford suggests that radio-active bodies generate an *emanation* or gaseous material which carries the radio-activity. In the opinion of M. Curie and myself, the generation of a gas by radium is a supposition which is not so far justified. We consider the emanation as radio-active energy stored up in the gas in a form hitherto unknown. . . .

Nature and Cause of the Phenomena of Radio-Activity.

From the beginning of research upon the radio-active bodies, and when the properties of these bodies were yet hardly known, the spontaneity of their radiation presented itself as a problem having the greatest interest for physicists. Today we have advanced considerably in the understanding of radio-active bodies, and are able to isolate one of very great power, viz., radium. With the object of making use of the remarkable properties of radium, a profound investigation of the rays emitted by radio-active bodies is indispensable; the various groups of rays under investigation present points of similarity with the groups of rays existing in Crookes tubes: cathode rays, Röngten rays, canal rays. The same groups of rays are found in the secondary radiation produced by Röntgen rays, and in the radiation of bodies which have acquired radio-activity by induction.

But if the nature of the radiation is actually better known, the cause of this spontaneous radiation remains a mystery, and the phenomena always presents itself to us as a profound and wonderful enigma.

The spontaneously radio-active bodies, and in the first place radium, are sources of energy. The evolution of energy, to which they give rise, is manifested by Becquerel radiation, by chemical and luminous effects, and by the continuous generation of heat.

The question often arises as to whether energy is created within the radio-active bodies themselves, or whether it is borrowed by them from external sources. No one of the numerous hypotheses arising from these two points of view has yet received experimental confirmation.

The radio-active energy may be assumed to have been initially accumulated and then gradually dissipated, as happens in the case of long continued phosphorescence. We imagine the evolution of radio-active energy to correspond to a transformation of the nature of the atom of the active body; the fact of the continuous generation of heat by radium speaks in favour of this hypothesis. The transformation may be assumed to be accompanied by a loss of weight and by an emission of material particles constituting the radiation. The source of energy may yet be sought in the energy of gravitation. Finally, we may imagine that space is constantly traversed by radiations yet unknown, which are arrested in their course by radio-active bodies and transformed into radio-active energy.

Many reasons are adduced for and against these different views, and most often attempts at experimental verifications of the conclusions drawn from these hypotheses have given negative results. The radio-active energy of uranium and radium apparently neither becomes exhausted nor varies appreciably with lapse of time. Demarcay examined spectroscopically a specimen of pure radium chloride after a five months' interval, and observed no change in the spectrum. The principal barium line, which was visible in the spectrum indicating the presence of a trace of barium, had not increased in intensity during the interval, showing therefore that there was no transformation of radium into barium to an appreciable extent.

The variations of weight announced by M. Heydweiller in radium compounds cannot yet be looked upon as established facts.

Elster and Geitel found that the radio-activity of uranium is not affected at the bottom of a mine-shaft 850 m. deep; a layer of earth of this thickness would therefore not affect the hypothetical primary radiation which would be excited by the radio-activity of uranium.

We have determined the radio-activity of uranium at midday and at midnight, thinking that if the hypothetical primary radiation had its origin in the sun it would be partly absorbed in traversing the earth. The experiment showed no difference in the two determinations.

Conclusions

I will define, in conclusion, the part I have personally taken in the researches upon radio-active bodies.

I have investigated the radio-activity of uranium compounds. I have examined other bodies for the existence of radio-activity, and found the property to be possessed by thorium compounds. I have made clear the atomic character of the radio-activity of the compounds of uranium and thorium.

I have conducted a research upon radio-active substances other than uranium and thorium. To this end I investigated a large number of substances by an accurate electrometric method, and I discovered that certain minerals posses activity which is not to be accounted for by their content of uranium and thorium.

From this I concluded that these minerals must contain a radio-active body different from uranium and thorium, and more strongly radio-active than the latter metals.

In conjunction with M. Curie, and subsequently with MM. Curie and Bémont, I was able to extract from pitch-blende two strongly radio-active bodies—polonium and radium.

I have been continuously engaged upon the chemical examination and preparation of these substances. I effected the fractionations necessary to the concentration of radium, and I succeeded in isolating pure radium chloride. Concurrently with this work, I made several atomic weight determinations with a very small quantity of material, and was finally able to determine the atomic weight of radium with a very fair degree of accuracy. The work has proved *that radium is a new chemical element.* Thus the new method of investigating new chemical elements, established by M. Curie and myself, based upon radio-activity, is fully justified.

I have investigated the law of absorption of polonium rays, and of the absorbable rays of radium, and have demonstrated that this law of absorption is peculiar and different from the known laws of other radiations.

I have investigated the variation of activity of radium salts, the effect of solution and of heating, and the renewal of activity with time, after solution or after heating.

In conjunction with M. Curie, I have examined different effects produced by the new radio-active substances (electric, photographic, fluorescent, luminous colourations, &c.).

In conjunction with M. Curie, I have established the fact that radium gives rise to rays charged with negative electricity.

Our researches upon the new radio-active bodies have given rise to a scientific movement, and have been the starting-point of numerous researches in connection with new radio-active substances, and with the investigation of the radiation of the known radio-active bodies.

Writing Exercises

1. Read further in the works of Koestler, Kuhn, and Simon. Write an analysis of a scientific discovery or technological innovation in your field, and apply to it one of these three theories.

2. Using the theory of Koestler, Kuhn, or Simon, write a solution to a hypothetical problem in your field of study.

Reference List

Langley, P., H. A. Simon, G. L. Bradshaw, and J. M. Zytkow. 1987. *Scientific discovery: Computational explorations of the creative processes.* Cambridge: MIT Press.

8

Exploration in Science and Technology

One of our most extraordinary features as homo sapiens ("human that knows") is our capacity to seek knowledge. We are curious, and we explore our environment to understand and control it for our own purposes.

As future scientists and engineers, you should be familiar with the actual process of great innovations; these usually involve a deep knowledge of science, technology, and engineering, as well as creative hunches, many trials, and the freedom and time and tenacity to work through to a solution.

One of the most amazing and intriguing technological innovations since the Civil War is the electric light bulb. The following selection gives a fascinating narrative of how this profound technological innovation actually came about.

Thomas Alva Edison (1847–1931) was America's greatest inventor. His 1,093 patents overshadow even the great Ben Franklin. By the time he began work on the electric light bulb in 1878, he had already captured the world's attention and some of its wealth by improving the stock ticker (used at stock exchanges) and the machines involved in telegraphy. After he filed a patent in 1878 for an incandescent lamp with a platinum filament, utility companies and bankers (among them, J. P. Morgan) backed Edison financially so he could expand his laboratory at Menlo Park, New Jersey, which employed more than 90 workers at peak times. For the next four years he devoted most of his time to perfecting the light bulb.

Two of the major problems Edison and his assistants faced were the material to be used for the filament of the bulb and the insulation material for such a filament. Edison and his colleagues tested at least 1,600 different materials for the filament (including platinum, nickel, and carbon), eventually finding that burned cardboard filaments were better than those made from metals, but that Japanese bamboo was the best of all: this finding came after testing 6,000 vegetable fibers!

The following excerpt narrates part of this complex story; it offers a case study that exemplifies the trial-and-error nature of most discovery processes.

The Triumph of Carbon

Robert Friedel and Paul Israel

On October 19, following several days of difficulties with the vacuum pump, Upton wrote in his laboratory notes that he had returned to carbon: "A stick of carbon brought up in a vacuum to 40 candles (say). Mr. B. trying to make a spiral of carbon. Grease in the pump. Gauge broken. Trying to make carbon spirals." A couple of days later, on the 21st, Upton added to these remarks, "Stick of carbon about .020 and 1/2 inch long gave cold 4 ohms, incandescent 2.3 ohms, very good light. Pt. wires melted." This same day, "Mr. B"—Batchelor—resumed the record of his own carbon-lamp experiments, proceeding toward the actual breakthrough by his customary thorough and careful steps. He first described a series of attempts at making spirals from the tar and lampblack putty. The putty itself was of no use in a light, since it contained a number of compounds that either evaporated or melted at high temperatures. The putty spirals had therefore to be carbonized—baked in an airless container until volatile compounds were driven off and only a skeleton of pure carbon remained. The carbonizing stage, however, was always the stumbling block in the forming of usable tar-lampblack spirals, and Batchelor proceeded to investigate the nature of the problem. When he tried to carbonize pieces of the putty in a glass tube, he was able to observe the hot putty giving off a "yellow oily liquid" which he guessed was "benzole" (i.e., benzine) or a similar substance. This, he surmised, was a cause of the difficulty in successfully carbonizing spirals.

Making spirals was still a key goal to Batchelor. Indeed, his notes at this point dwelt on the problem to such a degree that they hardly betrayed how close he was to a major breakthrough:

> *Electric Light Carbon Wire—*
> *A spiral wound round a paper cone no matter how thin always breaks, because it contracts so much. If the heating is done slowly this is modified but with the present proportion of tar and lampblack it will always break. . . . One of the great difficulties is to keep the spiral in position whilst you carbonize it. This might be remedied to a great extent by using a hollow sleeve winding the spiral inside with something to hold the ends whilst they are being fastened to the leading wires.*

October 21, as far as the laboratory record reveals, came to an end without the dramatic success that subsequent accounts of the electric light's invention attributed to it.

October 22, however, saw something distinctly different. Batchelor's notes of that morning do not suggest great drama, but the activity at Menlo Park took a definite turn, as all the laboratory accounts make clear. In his notebook Batchelor wrote:

> *Carbon Spirals—9 A.M.*
> *We made some very interesting experiments on straight carbons made from cotton thread so.*
>
> *We took a piece of 6 cord thread No. 24ˢ which is about 13 thousandths [of an inch] in thickness and after fastening to Pt wires we carbonized it in a closed chamber. We put in a bulb and in vacuo it gave a light equal to about ½ candle 18 cells carbon [battery], it had resistance of 113 ohms at starting & afterward went up to 140—probably due to vibration.*

The note does not say how long this feeble light lasted, but on the same day Upton described similar trials with slightly different details:

From Friedel, R., and P. Israel. 1986. *Edison's electric light: Biography of an invention.* New Brunswick, N.J.: Rutgers University Press. (The first paragraph is adapted from an article by Marshall Fox in the New York *Herald,* December 21, 1879.)

> *Carbon spirals and threads. Trying to make a lamp of a carbonized thread. 100 ohms can be made from an inch of .010 inch thread. A thread with 45 ohms resistance when cold was brought up in a high vacuum to 4 candles for two or three hours and then the resistance seemed to concentrate in one spot. Resistance cold 800 ohms.*

These observations carry no hint of triumph, no inventor's "eureka!" to set them off, but unquestionably signify the crucial transition of the electric-light search.

Batchelor continued his experiments that day in his usual methodical fashion, constructing a series of lamps with various types of carbon filaments:

> *1-of vulcanized fibre*
> *2-Thread rubbed with tarred lampblack*
> *3-Soft paper*
> *4-Fish line*
> *5-Fine thread plaited together 6 strands*
> *6-Soft paper saturated with tar*
> *7-Tar & Lampblack with half its bulk of finely divided lime work[ed] down to .020—straight one ½ inch*
> *8-200^{is} 6 cord 8 strand*
> *9-20^{s} coats 6 cord—no coating of any kind*
> *10-Cardboard*
> *11-Cotton soaked in tar (boiling) & put on*

Batchelor's notes reveal that he was seeking not only the best material but also the best configuration for a carbon filament; the need was still strongly felt to put relatively long lengths of filament into a small space. The results of his tests confirmed the significance of the behavior of the carbonized thread. Number 2 (thread rubbed with lampblack and tar) "gave an elegant light equal to 22 candles" when powered by an 18-cell battery. Number 3 "came up to 1½ gas jets," but went out due to a short circuit in the lead-in wires. It was, in fact, number 9—simple uncoated cotton thread—that provided the real triumph. In the middle of the night—1:30 by Batchelor's account—the bulb with the simple carbonized thread was put on eighteen cells and kept on. How much light it ultimately yielded is not stated; Batchelor says only that he brought it up initially to "½ candle." This, however, would have been too faint to have excited interest, so it is likely that the full eighteen cells made the lamp much brighter. After number 9 had burned for 13½ hours, more cells were added at 3:00 the following afternoon. It continued to burn for an hour longer, yielding a light equivalent to 3 gas jets (at least 30 candles), when the glass bulb overheated and cracked. None of Batchelor's other experiments had results equal to those with this simple carbonized thread.

The Menlo Park response to the success of the cotton-thread lamp, however, was not to treat it as a finished invention, but rather as the beginning of a new experimental path. Edison, of course, appreciated success as quickly as anyone, but the problems posed by the incredibly fragile carbonized thread filament were large indeed, and the stubbornly held belief in the need for a spiral or coiled filament provided little encouragement for the continued use of short, brittle lengths of thread like October 22's number 9. While Upton embarked on measurements to determine the resistance and power requirements of the carbonized-thread lamps, Batchelor spent the last week of October carbonizing a long list of other materials, including celluloid, wood shavings from boxwood, spruce, hickory, baywood, cedar, rosewood, and maple, punk, cork, flax, coconut hair and shell, and a variety of papers. In addition to carbonizing experiments, Batchelor continued to test the series of various lamps he had started on October 22, a series that eventually reached number 260. His tests were not completed until more than two months later. During these two months, Upton and Batchelor, with help from Kruesi, Jehl, and others, devoted an enormous amount of effort to perfecting the carbon lamp and devising means for producing and using it. The transition of the Menlo Park operation from research in quest of a feasible lamp to development of a practical and marketable product was remarkably rapid and smooth, reflecting the basic Edison attitude that inventions were worth something only when they were usable and saleable. This complete changeover, more than anything else, signified the realization of Edison and his men that they finally had an important invention on their hands. Nonetheless, they also saw that there was still some distance to go between what they had accomplished and the system they envisioned as a successful rival to gaslight. November was a time of alternating optimism and frustration, succinctly expressed in Francis Upton's regular letters home. On November 2, he wrote:

> *The electric light is coming up. We have had a fine burner made of a piece of carbonized thread which gave a light of two or three gas jets. Mr. Edison now proposes to give an exhibition of some lamps in actual operation. There is some talk if he can show a number of lamps of organizing a large company with three or five millions capital to push the matter through. I have been offered $1,000 for five shares of my stock. . . . Edison says the stock is worth a thousand dollars a share or more, yet he is always sanguine and his valuations are on his hopes more than his realities.*

The next week, however, Upton reported, "The Electric Light seems to be a continued trouble for as yet we cannot make what we want and see the untold millions roll upon Menlo Park that my hopes want to see." Finally, a week later, on November 16, Upton's letter home reflected complete confidence in their achievement:

> *Just at the present I am very much elated at the prospects of the Electric Light. During the past week Mr. Edison has succeeded in obtaining the first lamp that answers the purpose we have wished. It is cheap—much more so that we even hoped to have."*

He then went on to describe the cardboard filament lamp in all its elegant simplicity.

Another indicator of the attitude at Menlo Park toward the carbon lamp breakthrough was the patent application for an "electric lamp" filed November 4, 1879 (and eventually granted as U.S. Patent 223,898 on January 27, 1880). Since this was to be the key patent in the Edison system, it merits a careful look. In the beginning, Edison spells out clearly the distinction between his carbon lamp and all others:

> *The object of this invention is to produce electric lamps giving light by incandescence, which lamps shall have high resistance, so as to allow of the practical subdivision of electric light. The invention consists in a light-giving body of carbon wire or sheets coiled or arranged in such a manner as to offer great resistance to the passage of the electric current, and at the same time present but a slight surface from which radiation can take place. The invention further consists in placing such burner of great resistance in a nearly-perfect vacuum, to prevent oxidation and injury to the conductor by the atmosphere.*

After the fashion of such things, the patent then details the making of the carbon filament (this is where the term is introduced) from a variety of materials, conceding the extreme fragility of the finished product. The final form of the filament is invariably specified as a spiral or coil, and the patent drawing depicts such a form. It is highly unlikely, however, that a successful lamp was ever made at the time with a carbon spiral since the carbonized materials were simply too brittle to allow such shapes. The patent also specifies joining the carbon filament to platinum lead-in wires using a lampblack-and-tar putty, rather than clamps. This, too, was apparently wishful thinking, for all lamps made in this period, and for months afterward, required tiny platinum clamps to secure the filament [Figure 8-1]. One other feature for which Edison claimed patent protection, and which was to be surprisingly significant in ensuing years, was the use of a sealed enclosure entirely of glass, dispensing with troublesome metal-glass connections except for the tiny platinum lead-in wires. The basic incandescent lamp patent was only one of literally hundreds that Edison received for his work in electric light and power, but it was unquestionably the most significant, economically and conceptually.

FIGURE 8-1
Carbon Filament Lamp, November 1879

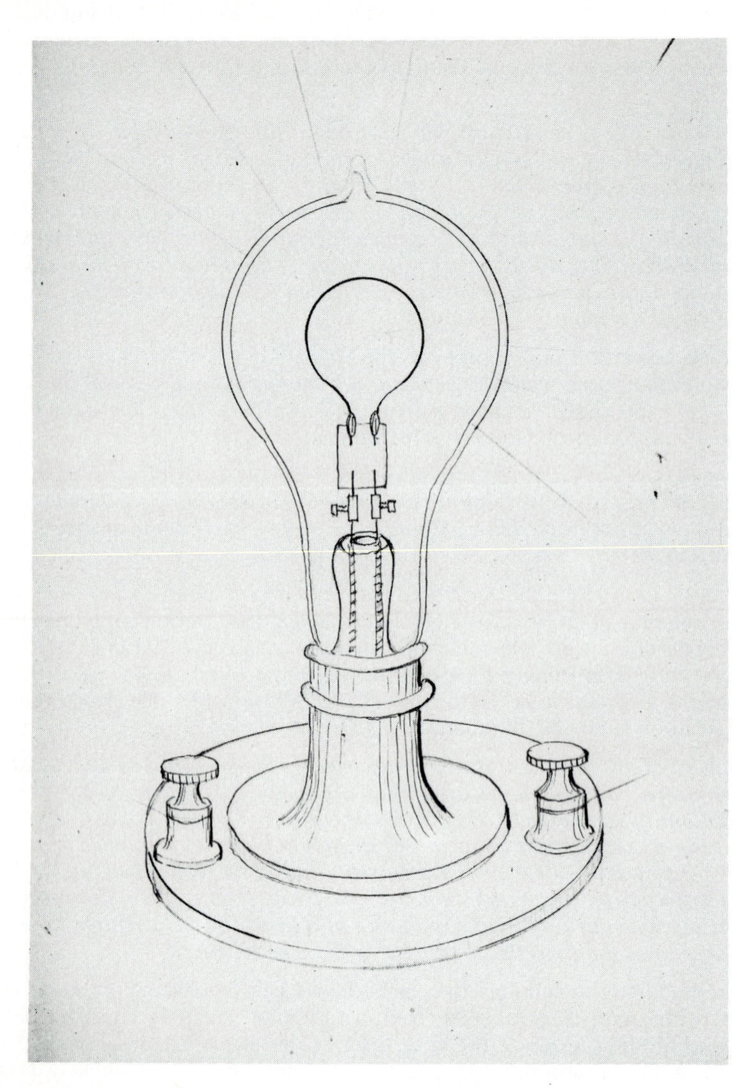

Samuel D. Mott, Edison's patent draftsman, drew this lamp showing the method then used to clamp the horseshoe-shaped filament to the lead-in wires.

The patent application submitted November 4 did not so much describe what had actually been made at Menlo Park as what Edison and his colleagues knew should be made. The process of learning what *could* be made occupied most of the remainder of 1879, as Edison put the Menlo Park team to work not only to make practical lamps but also to construct enough of a complete lighting system to convince a by now sceptical public and an ever more wary financial community. The attitude that guided the Menlo Park workers is well reflected in another of Upton's letters home:

The Electric Light is slowly advancing from the last big step. We now know we have something and that is what we [did] not know until last week. We can compete with gas in a great many ways now though not as completely as we wish, yet there seems to be nothing to prevent our getting a perfect burner that shall do as well as gas. Time and cost will prove what we have to be good or bad. (100–7)

Following this success, the search for the ideal carbon form continued.

Despite the continuation of older projects and the persistent intrusion of new ones, the laboratory in 1880 revolved around the electric light and its support system. By late spring the most important problems had defined themselves: (1) how to make the light bulb sturdier and more reliable, (2) how to make bulb production cheaper and suitable for factory operations, (3) how to determine the quantitative requirements of anticipated system installations, and (4) how to make the generator more reliable and more efficient, especially with large loads. Efforts to perfect the bulb continued to focus on finding an improved filament material. Mott's journal relates the progress of the Menlo Park experimenters as they attempted approach after approach in search for the ideal carbon form:

(April 8) Those at work on the lamp and on carbons (Batchelor, Force, Mr. Edison, Flammer and some of the men in Shop) greatly interested in the efforts to devise suitable means and devices for reducing woods to sufficient small dimensions for carbonizing, and have been trying several different devices, Ott being of the opinion that a very fine keen saw will leave the wood smoother and in better shape for the carbons.

(April 29) Wood loop cut from the thin worked holly milled by Force and cut after manner and in same former used for cardboard, carbonized by Van Cleve, were measured and put in lamps ready for pump, resistence 125 and 194 ohms.

(May 14) Carbonization. Several moulds of Bast fibers were carefully prepared and formed around wood for carbonization, but the wood proved very detrimental, every one having been broken in the moulds during the process. Van Cleve is preparing some more for trial.

(May 20) Carbonization. Van Cleve carbonized three moulds of bent wooden loops by securing the strips in slotted nickel plates; he got them out very nicely and in good shape. Bast fiber. Four of the Bast fiber lamps were measured and tested with current of 103 volts they gave from 30 to 32 candles and about six per horse power. They were connected to main wires in Laboratory and during the first hour three of them broke in the clamps and glass but the fiber in each instance remained in globe unbroken. Showing the fiber to make strong carbon but difficult to form good contact with.

These are only samplings of the day-by-day entries by Mott on the often frustrating hunt for a more durable and more reliable filament material.

Besides a continuous record of the search for a better filament, Mott's journal also provides a much clearer picture of the different activities pursued as part of the quest. For example, as each general kind of material was considered and tested for suitability as a filament source, special tools had to be devised to properly shape it. For weeks, Mott wrote of work done to devise a wood-milling machine for preparing wooden loops. On May 21 he noted the result:

Wood Milling. Dean is jubilant over his success today in working the cam milling machine with complete success and getting out about 100 loops of box & holly in excellent shape and in several cases sawing them so perfectly that the whole five loops were left joined at the thick ends, although the machine have been worked for some time with indifferent success. Today is the first Dean has felt satisfied with its working.

Other work related to the filament search involved developing better molds for the carbonization process, methods for extracting gum and resin from wood prior to carbonization, and devising other shaping tools. Mott confirmed that the number of different materials tested was not very large. By late spring, tests were limited to a few kinds of woods and bast fibers. While the experiments on woods continued through the spring, the usefulness of naturally fibrous material became more evident as time went on, and bast and other fibrous substances were the subject of the bulk of the lamp experiments by summer. One of the primary advantages of fibers was their superior stability after carbonization. According to Mott (June 12): "Mr. Edison observed that the Bast fibers shrink in carbonizing about 17 per cent, against 33% shrinkage in paper, woods, &c" It was natural, therefore, that further experiments should focus on other fibrous materials.

Bast fiber received most of the laboratory's attention during June, but persistent problems were encountered in the connections between carbonized fibers and clamps. On June 25 Mott reported that some loops were cut out of osier willow and palm leaf samples but turned out to be very fragile after carbonizing. A few days later some palm leaf loops were carbonized successfully, but the first lamps made from "palmetto" were not very promising. More were tried over the next week without markedly greater success. Finally, upon returning from a short Fourth of July vacation, Mott wrote that "A collection of Bamboo Reed and choice Bast have been obtained and some loops cut out but none yet put in the lamps to test (July 8)." On July 10, he reported the first tests of bamboo, but nothing spectacular in the results: "Bamboo cut from top or outside edge of fan. Resistance cold 188 ohms at 16 C. 114 ohms and gave 8.6 per horsepower." Performance of the bamboo in a lamp was somewhat like that of the bast fiber: (July 12) "A bamboo lamp tested Saturday (July 10) was put at 44 candles this morning, got very blue at clamp and lasted 1 hour 15 minutes." Bamboo was to be the climax of the search for a sturdier, more reliable filament material for commercial use, but its debut in the Menlo Park lab, while promising, was no more exciting than that of dozens of materials before it.

A few more weeks of experiments were required to determine that bamboo was indeed the material sought after. It is instructive to follow Mott's recounting of how this conclusion was reached. Soon after the initial tests, Batchelor, who was in charge of the carbonization experiments, figured out the shrinkage of the bamboo after carbonization (20 percent) and gave instructions for making a former for bamboo loops. By this time the laboratory had devised a means for speeding up the crucial life tests for new filament materials. Instead of being run at standard operating voltage to yield around 16 candlepower (the desired light output for commercial use), lamps were put at higher voltages and tested at outputs of 40 candles or more. This procedure reduced testing time from hours or days to minutes. Mott describes one early bamboo test: (July 17) "Bamboo. Carbon of Bamboo with slight notch cut in one side set burning at about forty candles; in about five minutes the clamp on one side melted down forming a globule on the end of the wire and destroying the carbon. Lamp Number 1277 Book No. 57 Page 159." Further tests indicated the real superiority of bamboo: (July 21) "Average test on Lamps: From the lamps so far tested at 44 candles, the average life was taken and found to be, for Bast 6 Minutes, Calcutta Bamboo 17 Minutes and paper about three minutes. The Bamboo carbons were in many cases imperfect which has probably reduced the average for them, but which will be undoubtedly raised when proper precaution and care is used in selecting only perfectly cut and prepared carbons." At about this same time efforts were made to acquire a better quality of bamboo, with further gratifying results: (July 19) "Bamboo Pure. Some Pure Bamboo (genuine) was obtained and given to Bradley from which to cut some loops. The genuine is very fine grained and works nicer than the other." This "genuine" bamboo was later referred to as "Japanese Bamboo" and did indeed prove to be superior to the "Calcutta Bamboo" they had previously used. By August it was clear that they had what they wanted, Mott reporting: (August 2) "6$^{in.}$ Bamboo. Lamp burned 3 hrs. 24 mins. at 71 candles and gave nearly 7 per h.p.—the best lamp ever yet made here from vegetable carbon." From this day on, Mott made no more references to experiments on filament materials; "carbons," "fibers," and "loops" always referred to bamboo, and the tools devised for shaping them were made with that material in mind. (154–57)

Mass producing the loops presented still more problems.

. . . Soon after the adoption of bamboo as the basic filament material, it was observed that many bamboo loops emerged badly bent from the carbonization process. At first, Batchelor found a clever way of straightening them after insertion into the bulbs; Mott described it this way:

> *(Aug. 4) Straightening carbons. Some of the longer carbons have at time bent over to one side vary much after they had been placed in the lamps and heated. To straighten them the lamp is placed between the poles of an electric magnet and the current turned on on the lamp. One pole attracts while the other repels the charged carbon and lamp is placed between the poles in such relation that the polar action is utilized to straighten the carbon through the glass globe. I find that this "little dodge" has been worked for some time but today is the first I have caught them at it.*

This expedient was, however, not suitable for the factory. Batchelor continued attacking the problem, and Mott's description of how the solution finally emerged gives another revealing glimpse of the Menlo Park style:

> *(Aug. 12) Carbonization. Mr. Batchelor's experiments with different heats did not solve the problem or reveal the cause of the loops bending over. A thorough discussion to night suggested the theory that it might be due to the fact that the way the present loops are cut and carbonized the pith and outside of the Bamboo always comes on the side of the loop, and to test the theory whether the pith side, being more loose and porous than the outside did not shrink more from the effects of the heat, than the more firm and compact outside. It was determined to change a carbon forming mould so that the widened ends might stand at right angles with the face and thus bring the pith either on inside or outside of the loop. The mould was then arranged and between twelve o'clock and morning Mr. Batchelor got out three or four, had them put in lamps exhausted and heated very high and to the gratification of all the loops remained erect and they justly feel that the problem is solved and that carbonization is now worked to a fine art.*

Batchelor's own laboratory notes give another perspective on how this problem was attacked:

> *The cause of the bending over of the loop after it is heated in vacuo I thought was due to insufficient heating in carbonization but after a series of experiments to determine that point we came to the conclusion that whether heated slightly or to a high temperature some of each bent whilst others kept straight.*
>
> *We then remembered that some bamboo fibres which were 4" long and of which we made a great number almost all kept straight. We also remembered that almost all of these were put in the clamps edgeways instead of flatways. This led us to see that the way Bradley cut them from the cane, and the bending them flatways afterwards, would have the* "pith side" *on one face and the "hard-shell" side on the other face[;] unequal shrinkage of course must occur on two such faces and cause the bending.*
>
> *We now made a mould for carbonizing that would hold the fibre edgeways so. . . . From this mould we tried some on the pumps and they not only were perfectly flat themselves but did not change their upright position with the most intense heat we could get on them.*

The problem of the bent bamboo loops was just one of many small (but often crucial) details that had to be worked out as lamp manufacture was systematized. Mott's reference to the "thorough discussion" that followed Batchelor's initial difficulties provides a good instance of the cooperative team effort that was as characteristic of the laboratory as the independent and friendly competition illustrated by the work of Jehl, Boehm, and others on vacuum pump design. It is well to note that Edison's participation is not mentioned by either Mott or

Batchelor (indeed, Mott records that Edison was absent all day), for, in dealing with such details, the laboratory could function very well without Edison, particularly when the details were being handled by Batchelor, Upton, or another member of the team with substantial responsibilities. Batchelor's notes offer a revealing picture of the kind of reasoning applied in laboratory discussions—an empirical and yet clearly deductive approach that is the mainstay of any establishment devoted to solving technical problems. The "perspiration" that Edison said was ninety-nine percent of inventive genius was effective because it was guided by alert and perceptive thinking. (164–67)

Writing Exercise

Using exploration methods similar to those of Edison, think of a mechanical device that the world is ready for but doesn't yet have. Describe the device and its usefulness. Then use your description to write a promotional brochure or the kind of fold-out, multipage advertisement that might appear in a glossy magazine.

9

Reflecting About Science and Technology

When their daily work is done, many scientists and engineers often turn their pens to reflections about science and technology. For instance, what are the social, political, and personal impacts of technological innovations and scientific theories? How can scientific concepts be applied to the seemingly inchoate nature of human life? How can scientific and technological achievements be evaluated?

Why must technology be kept close to the human scale? For instance, of what significance is it that humans are building computers that are "smarter" than they are—smarter in the sense that these computers can solve problems that humans either cannot solve or ones would require enormous amounts of time to solve?

Aside from communicating scientific and technical information to others, how is writing—the act of writing—important in itself to scientists and engineers?

The changes in the undergraduate engineering curriculum at the Massachusetts Institute of Technology underscore this kind of reflective thinking, in which science and technology are placed within the broader perspective of the humanities, fine arts, and social sciences. As reported in the *New York Times* (1 June 1987), undergraduates at M.I.T. are now required to "pursue more systematic study of the arts, humanities, and social sciences." Paul E. Gray, president of M.I.T., explained the reasons for the revision: "A professional engineer can no longer be narrowly focused on technical interests. He lives and operates in a social system, and he needs to understand cultural and human values. Humanities courses cannot be viewed merely as frosting on the cake."

The following selections reflect the urge to build bridges of understanding between humanistic thinking and scientific and technological thinking. They take us into the meditative world of scientists and engineers, revealing their human side and the human side of science and technology as well.

SPECULATING ON THE IMPACT OF TECHNOLOGY

One irony of technological development is that it makes us more powerful in one sense and more weak in another. The automobile enables us to travel long distances relatively quickly and comfortably. But as it makes us used to and therefore dependent on its luxury, it enslaves us because most of us cannot fix it when it breaks down. Thus, we come to believe that we need something whose engineering structure we do not understand and whose long-term consequences we tend to ignore. Was it really so bad just walking or riding in a horse-drawn cart? Were humans really less happy without cars?

The following selections from contemporary writers pursue the extensions of these kinds of questions. Witold Rybczynski questions the value of advanced technology and says that the field is now polarized between two extremes: the boosters of technology and the obstructionists. William H. Calvin, a neurobiologist, echoes Rybczynski's concern about the ecological impacts of technology as part of his narrative describing a rafting trip down the Colorado River and through the Grand Canyon.

The Prosthetic God

Witold Rybczynski

> *Man has, as it were, become a kind of prosthetic god. When he puts on all his auxiliary organs he is truly magnificent; but those organs have not grown on to him and they still give him much trouble at times.*
>
> —Sigmund Freud, *Civilization and Its Discontents*

Consider man, the prosthetic god. Not being able to run very fast or for very long, he has grafted on to himself additional feet, until he can travel farther and faster than any other animal, and not only on land but also on and under water and in the air. He can reinforce his eyes with glasses, telescopes, and microscopes. Thanks to orbiting satellites, he can, without displacing himself, count wildebeest in the African veldt, or missile silos outside Novosibirsk. Lacking the dolphin's ability to communicate great distances, he amplifies his voice with the aid of radio waves. In addition to his genetic code, which he shares with all other living things, he has acquired a perpetually growing communal memory in the shape of the written word, the photograph, and the recording. Everything we know now, we know forever.

These auxiliary parts not only extend man's natural abilities but change them beyond recognition. The book extends not only our memory, but also, says Argentine author Jorge Luis Borges, our imagination. The brain is supercharged by the computer; perhaps, one day, the two may become indistinguishable. Thanks also to the computer, all the books of the world can be assembled into one great Alexandrian library—no bigger than a small bedroom—through which a person may stroll, so to speak, finding in five minutes the one book he previously could not have found in a lifetime's searching. What is more, these extensions are detachable. The prosthetic eyes and ears can be sent far into the solar system to look at the other side of the moon, or on a deep interior voyage to see something even more marvelous, the human embryo.

Technology, as Freud drolly observed, is really a set of artificial organs, extensions of our natural ones. He understood, as many still do not, that the relationship between ourselves and our tools is often blurred, and frequently intimate.

Tools are not unique to the human; a number of animals and birds use tools both in hunting and in nest building. Whether man thinks because he uses tools, or vice versa, the fact remains that from the beginning of recorded history tools and technology have been part and parcel of man's essence. The prehistoric men and women who killed with a club, instead of with bare hands, or who, lacking protective pelts, covered themselves with animal skins,

---

From Rybczynski, W. 1983. The prosthetic god. In *Taming the tiger: The struggle to control technology,* 3–11. New York: Viking Press.

may have been imitating the sea otter, which cracks clams on a stone it picks up from the seabed, or the squirrel, which burrows under a pile of leaves to hibernate. It is even possible that the first projectile may not have been thrown by a man but by an ape. But it is the evolution and invention of improved tools that distinguish men from animals. It is human logic that leads from the projectile to the slingshot, from the Roman catapult to the ancient ballista, and, with the invention of gunpowder, to the cannon, which is, after all, merely a device to throw a stone a still greater distance.

But using technology to overcome man's biological limitations has not been altogether a painless experience. The artificial limbs, to continue Freud's metaphor, itch. Sometimes the scar will not heal; inflammation and infection set in. Occasionally the body rejects the implanted organ, and, if it is not quickly removed, serious damage can result. The patient barely recovers before another *assemblage* is attempted. The surgeons—the scientists and engineers—frequently neglect to tell the patient what the new organ is for, which is sometimes unpleasant and sometimes terrifying. Unmentioned also, until too late, are the accompanying side effects of this hazardous prosthesis. And always that itch.

Little wonder that the patient sometimes rebels. Do the surgeons really know what they are doing? he asks. Will the new appurtenance by compatible with the rest of the anatomy? Is it really needed, or will it do more harm than good? It is easy to overextend a metaphor, but it is a fact that growing numbers of people are beginning to question the wisdom of continued technological progress. Perhaps it would be better to remain where we are, or even backtrack a little. Much of this growing distrust of technology is based on the belief that technological development is ungovernable and ultimately outside of human control.

If the invention of fire represents one of the first instances of man's control over the forces of nature, as Friedrich Engels claimed, then the first man-caused forest fire must surely be one of the earliest examples of technology out of control. One can imagine that even then, in the murky shadows of prehistory, our ancestors must have had a lively discussion about the propriety of the continued use of this new discovery, so wonderful yet so awesome. The optimists obviously won out, as optimists usually do.

That technology has almost always had unintended side effects has often been cited by critics as a symptom of man's inability to control his inventions. Yet paradoxically, it is precisely the unpredictability of technology that has proved to be its most endearing characteristic. For instance, when John Gorrie, a Florida doctor, patented an ice-making machine in 1851, he intended it as a device to cool the rooms of his yellow fever patients. Like many doctors at the time, Gorrie mistakenly believed that yellow fever was due to miasma, or "bad air." His ice-making machine thus had little effect on curing this disease, but it was a very important contribution to the development of quite another technology—refrigeration.

The popular story of the absent-minded scientist who confusedly mixes two chemicals to produce an important discovery is not uncommon. The unpredictability of technological invention is illustrated by the case of Henri Moissan, a French chemist who was trying to manufacture artificial diamonds. In the process, he discovered, quite accidentally, the compound calcium carbide. Calcium carbide did not make diamonds, but when combined with water it did produce a burnable gas—acetylene—which became the most common form of domestic lighting until the invention of the electric light bulb. The accidents continued. A German chemist, Fritz Klatte, took up Moissan's invention and tried to develop an aircraft dope using acetylene. In 1912, one of the many unsuccessful combinations he tried turned out to be the world's first plastic—vinyl chloride.

In 1886, an Atlanta pharmacist concocted a compound which he claimed would whiten teeth, cleanse the mouth, harden and beautify the gums and relieve mental and physical exhaustion. John Styth Pemberton was the unrenowned inventor of Globe of Flower Cough Syrup, Triplex Liver Pills, and Indian Queen Hair Dye, and his latest patent medicine was equally unsuccessful. He sold it to another druggist, who marketed the odd mixture of coca leaves and cola nuts as a soft drink, under the same name that Pemberton had given it—Coca-Cola. Thus are the foundations for multinational corporations laid.

Of course the *negative* unforeseen effects of technological innovation can be frightening. One of the most notorious contemporary examples concerns the Aswan High Dam in Egypt, the enormous Soviet-aided irrigation and hydroelectric project that was to be the linchpin of Egyptian rural development. Instead, it is turning out to have the opposite effect. The dam, which is located in the extreme south of the country, prevents nutrient-rich silt from the Upper Nile from reaching the lower-river areas in the north. As a result, the seasonal cycle of fertilization, which has been the mainstay of Egyptian agriculture for centuries, has been interrupted. Although flooding still occurs, the level of nutrients in the otherwise sandy soil has dropped drastically, and the reduction in food production, coupled with Egypt's rising population, has created severe food shortages. Another ecological megadisaster may occur in Brazil, where, some scientists believe, the large-scale lumber industry could denude the jungle to the point that the reduction in vegetation, and hence in oxygen production, will affect the biosphere on a global scale.

These two examples give the impression that it is the scale of technology that is at fault; but even "small" technological innovations can have a large environmental impact. The innocuous flush toilet is a case in point. The first American urban sewerage system was built in Brooklyn in 1857, and represented at the time a hygienic and effective solution to the unsanitary cesspools and drainage ditches that had come to characterize the urban environment in the nineteenth century. Today the city of New York dumps five million tons of sewage sludge about twelve miles offshore in the Atlantic Ocean. The liquid residue that is left after so-called sewage treatment contains vast quantities of bacteria and concentrations of various heavy metals, such as zinc, chromium, and lead. In the last five years, New Jersey and Long Island beaches have been periodically shut down as the "foreign substance" (actually very American) washes ashore. The effect on marine life will take longer to be felt but could be more serious than it is on human life.

It sometimes seems that every step forward in technological development is (potentially, at least) accompanied by one step backward in terms of environmental or personal safety. The aerosol, for instance, turns out to emit fluorocarbons whose increased presence in the atmosphere depletes the ozone layer and may have severe long-term effects on the quantity of ultraviolet rays that reach the surface of the earth. Similarly, asbestos, a life-saving, fireproof material, is also a vicious carcinogen. DDT, which was developed to control malarial mosquitoes, is turning up in the food chain in dangerous concentrations, and as a result it has been banned in many countries.

It is technological "errors" such as these that give rise to the nagging feeling that the proud locomotive of progress is really an ominous juggernaut, and that our imminent demise, like that of the followers of the god Jagannath, who threw themselves beneath the wheels of his great ceremonial chariot, will be equally poignant for being self-inflicted.

Our fear of the machine is heightened by the feeling that the twentieth century is the first to experience the unintended side effects of technology. It may be a consolation to learn that humanity has experienced the impact of technological change no less mordantly in earlier times. The historian Lynn White, Jr., has documented in *Medieval Religion and Technology* how the adoption of the fireplace and chimney disturbed everyday life. Until the eleventh century, heating had been provided by open fireplaces. There were no mantels or even chimneys, and smoke was simply allowed to filter out through the thatch or shingles of the roof. Thus the entire household—lord, freeman, and servant—lived, ate, and slept (in curtained compartments) in the great hall, gathering for warmth around the central, open, cooking fire. With the development of the fireplace and chimney, it became possible to heat individual rooms on different floors throughout the house. As a result, the forced egalitarianism of the early Middle Ages began to give way to a more individual way of living. White quotes an eleventh-century observer, who wrote, "Now every rich man eats by himself in a private parlor to be rid of poor men, or in a chamber with a chimney, and leaves the great hall." Thus the chimney contributed in no small part to the upstairs-downstairs stratification that would last into the twentieth century.

White also describes the origin of another invention of the Middle Ages: the national flag. He traces it to the pennons that knights affixed to their lances. These heraldic colors, which in the case of the king became the national colors, had a utilitarian purpose quite apart from their symbolic function. They were intended to prevent the lance from transfixing a foe during a mounted charge, for if the lance became embedded too deeply in the body of one's unfortunate opponent, there was a danger that one might not be able to pull it out and would be left weaponless on the field of battle. The cloth pennon also had the advantage that, unlike a solid crosspiece, it could not get caught in the enemy's armor but could always be ripped free. These "grimly functional rags" were first used by the mounted soldiers of Central Asia, who simply tied a piece of cloth or horsehair to the ends of their spears. The barbarian invaders worried about piercing their opponents because they used a device unknown to the Europeans—the stirrup. The stirrup, which was quickly adopted by the Franks, transmitted the entire weight of both horse and rider to the lance point, giving the term "shock combat" its true meaning.

I should hasten to point out that White is no technological determinist. His point is that technology creates the opportunity for social change, not that it is always invented in response to specific problems. Indeed, sometimes innovation seems haphazard or simply accidental. The technology of artillery, for instance, was introduced into Europe in the thirteenth century. The first cannon were made out of bronze, and the expansion of their manufacture relied on a highly developed bronze-casting industry. This industry happened to be present in Europe at the time, although it had been perfected in producing a very dissimilar article—bronze church bells.

While technological developments have often pursued their own, sometimes random, logic, they have always been accompanied by overt attempts at exercising human control. In 1139, Pope Innocent II banned the crossbow as too cruel a weapon (though its use against the Moors—infidels—was permitted). The ban was not successful, in part, one guesses, because its aim was as much political as humanitarian, since it attempted to redress the balance between the mounted aristocrat and the commoner crossbowman. In another vein, Edward III of England, in order to promote the use of the Welsh longbow (the most devastating weapon of its day), prohibited on pain of death the practice of all sports *except* archery.

There are many examples of technological control in the workplace. In 1559, Queen Elizabeth I of England refused to grant a royal patent for a knitting machine on the grounds that it would deprive too many of her subjects of employment. Charles II, acting in the same spirit, gave a charter to the hosiers' guild which would protect them against mechanization until the nineteenth century.

Not all technological control was exercised by benign monarchs. Frequently workers who have felt themselves threatened by technical innovation have agitated for these innovations to be curtailed. Often this agitation has taken the form of violence directed against the guilty machine. In the nineteenth century, Belgian weavers took to "accidentally" dropping their heavy wooden clogs (or *sabots*) into the delicate mechanism of the loom, giving rise to the expression "sabotage." Probably the most well-known example of machine-wrecking in the workplace occurred during the Luddite rebellion in England. . . . Although Luddism was by no means unique, it has become a generic description for all violent reactions to mechanization.

It may seem surprising that the issue of technology control should lead to violence. On the one hand, the choice of a technology is the choice of a particular set of values, and thus the debate about nuclear energy, or small-scale industry, or chemical agriculture, is really a debate about human values masquerading as a discussion of technical options. On the other hand, as history shows us, technology frequently contains the seeds of disruptive and threatening change. Did the medieval homeowner realize what he was getting into when he started building fireplaces? Not that we have more foresight today, but we are probably more skeptical about the side effects of technological innovation.

What distinguishes the current debate about controlling technology is that the two sides are so far apart. It seems to be less a question of what to do, than of whether to do anything at all. Instead of a discussion of how to control technology there is polarization. On the one hand are the "Boosters," for whom all technological change represents progress, and who see any attempt to control the technical future as a loss of nerve. On the other hand are the "Obstructionists," for whom *any* technical innovation is a threat, and for whom controlling technology means gradually reducing it, or at least doing so whenever possible. While the

Boosters reflect a nineteenth-century optimism about the machine, the Obstructionists represent a more recent, and more novel view.

Dams, Irrigation, and the Salinity Factor

William Calvin

> *Most men, it seems to me, do not care for Nature and would sell their share in all her beauty. . . . It is for the reason that some do not care . . . that we need to continue to protect all from the vandalism of a few.*
>
> —Henry David Thoreau, *The Journals*

> *We fear the cold and the things we do not understand. But most of all we fear the doings of the heedless ones among ourselves.*
>
> —An Inuit Shaman, *quoted by an early Arctic explorer*

The scars on both sides of the river are man-made, a jarring note in this wilderness of running water. They are test holes drilled back into the limestone, extending several city blocks deep inside the Muav and the Redwall cliffs.

Surveyors' splashy marks dot the cliffs like graffiti. A tall, unnaturally steep pile of rubble can be seen below each hole, deposited so recently that the weather and river have not had time to spread out the pile. An abandoned barge lies half-sunk near the base of one pile. There used to be scaffolding and a tramline down the cliff to these sites, since removed by the Park Service.

This is not an old mine; the limestone here is rather too common for that. These test holes were the first stage in the construction of a large dam. Right here. The dam-builders planned to back up the Colorado 54 miles, back past Lee's Ferry to the base of the Glen Canyon Dam, burying all the places we've seen so far under a huge lake. That's why all the new government buildings back at Lee's Ferry are incongruously located high above river level—they thought the low-lying sites near the boat launch would soon be flooded. Thus the car campground is perched in a windswept location, and has had to be provided with sheet metal windbreaks to supplement the wilderness experience. Not only would Lee's Ferry have been flooded but also the nautiloids, Redwall Cavern, South Canyon, the Anasazi ruins and petroglyphs, Silver Grotto, North Canyon, and all the rest. As if burying Glen Canyon weren't enough.

The Glen Canyon Dam, just upstream from Lee's Ferry, actually did back up the Colorado to drown a 186-mile region every bit as spectacular as this section of Marble Canyon. There was no organized opposition to the flooding of Glen Canyon, perhaps because few realized how beautiful the area was. Lake Powell, a drag strip for speedboats, now covers those canyons and gradually fills them with red mud.

Marble Canyon Dam was defeated in the sixties by the energetic efforts of Martin Litton, owner of one of the river companies, and the Sierra Club, then led by David Brower. Plus thousands of other people. The government dam-builders tried to promote the advantages of another new lake, saying that "people will like to sightsee by speedboat." I can still remember the Sierra Club's full-page newspaper ads headlined: "SHOULD WE FLOOD THE SISTINE CHAPEL SO TOURISTS CAN GET NEARER THE CEILING?" While the government cancelled the Sierra Club's tax-deductible status over that fight, the conservationists finally won the battle.

From Calvin, W. 1986. Dams, irrigation, and the salinity factor. In *The river that flows uphill: A journey from the big bang to the big brain,* 71–76. San Francisco: Sierra Club Books. Reprinted with permission of Macmillan Publishing Company.

The Arizona and Utah politicians are endlessly fond of building dams with the federal taxpayer's money. The official Arizona state road maps still show a dam site inside Grand Canyon National Park at Mile 238 on the Colorado, variously named Bridge Canyon Dam or Hualapai Dam, showing that the fight to prevent the Grand Canyon itself from being flooded isn't over yet. Bridge Canyon Dam was officially cancelled by the U.S. government only in 1984. The "flooding easement" on the park and Indian lands was finally allowed to expire—but not without the dam-builders making a final effort to promote the flooding of the lower part of the Grand Canyon, scarring up the Canyon with access roads, and stringing it with high-voltage powerlines. But the Arizona politicians are still spending money on studies to promote it.

Unfortunately, the plans and the test-hole data for such projects get filed away, perhaps to be used in the future. And the local politicians are always trying to get Congress to give them money to build yet another dam. It's a tradition in this part of the world, one which has been repeatedly successful, this waiting game of exploiters. All it takes is a recession, and long-term preservation goals can be overridden by a short-term drive for profits and jobs. Those who would preserve the wilderness have to fight repeated battles with each new generation of politicians to keep the exploiters out. To lose once is unthinkable.

Even national park status seems no obstacle to the dam-builders and their powerful supporters. The badly overcrowded Yosemite National Park in California's Sierra Nevada Mountains used to include a second similarly spectacular valley just to the north, named Hetch Hetchy. Since 1920 it has been drowned, a Yosemite we'll never see. That generation probably rationalized flooding Hetch Hetchy by saying that one valley remained and that was enough. Today, the map handed out to the Yosemite visitors ought to remind people about why the valley is now so badly overcrowded, but it doesn't. You'd think that one such mistake would have been a sufficient lesson.

A giant cookie-cutter has been at work on the landscape. That's what Marsha reported seeing on her flight out to Phoenix. Great circular spots of green amidst the brown.

"Those are irrigated fields," Ben explained. "From giant lawnsprinklers that march over the fields, swiveling around a central pumphouse."

"But those cookies were giants!" exclaimed Marsha. "Someone on the plane told me that they could be seen from the moon more easily than the Great Wall of China!"

"The lawn-sprinklers are giants too. They put out a stream like a fire hose. It'd knock you flat on your back if you wandered into it," explained Ben. "Those sprinklers are several stories high."

"Is there a white ring of salt around the cookies yet?" asked one of the boatmen whose name I haven't learned yet. "There soon will be, you know."

The salt of the earth can be the downfall of civilizations. The boatmen are ardent ecologists. They speak of the shortsightedness of our irrigation practices as well as the lost beauty of the landscape. Just fly over the world's largest irrigation system in the Indus Valley of Pakistan, they say, and you'll think there is snow down there. It is salt, encrusted atop the soil—ruined soil in which crops will no longer grow. Short-term gain yielded long-term disaster. And one can see that same white salt buildup along stream beds in the United States, from Colorado to California.

The reason for this problem is familiar to anyone who knows a little about drilling wells for water. There is a water table, even beneath a desert, below which water sinks no further, thanks to a resistant rock layer such as shale. Above the porous layer that contains the water supply is likely to be another layer of rather salty rock, perhaps a sandstone or siltstone that was compacted beneath an ancient ocean.

As water drains through soil it picks up a bit of salt on the way. Where rain usually falls, the salt gets washed out over the millennia, carried underground along the water table. But if an unnatural amount of water comes down through the soil in an area that doesn't normally get much rain—as in the irrigated Southwest—there is lots of salt left to pick up. The well-water in the area of Phoenix, Arizona, is already starting to go salty from all the irrigation. By the time the Colorado River water reaches the Mexican border, it has been through the soil repeatedly, and has become so salty that the United States was obligated to build a desalination plant for what's left of the Colorado River (the Mexicans aren't left much) in order to live up to a treaty with Mexico which guaranteed the quality of the water delivered to that nation.

Such technological "quick-fixes" draw our attention away from the real danger: that poorly managed irrigation will raise salt to the soil's surface. After all, irrigation converts local land with less than 20 centimeters (8 inches) of rainfall a year into wellwatered farmland. And all that rock and soil between the water table and the surface becomes wet but then, when the irrigation stops and the surface starts drying, water is wicked back upward, carrying salt with it. The salt emerges slowly but irresistibly, not unlike the way manganese is drawn out of the limestone around here to create desert varnish. Eventually the topsoil becomes salty and infertile. A white "irrigation varnish," just like the white ring that builds up around the tops of the pots used for houseplants, becomes the harbinger of stunted growth and crop failure.

Or the salt washes out downhill somewhere and contaminates the streams there. If a valley that lacks natural drainage—such as California's great central agricultural valley, the San Joaquin—is irrigated with even slightly salty water, the salt continues to build up over the years as the water evaporates. The water that does run off is so salty that evaporation ponds containing it may be a health hazard because of heavy metals leached out of the earth: in the San Joaquin, waterfowl signaled the problem with a high rate of birth defects and fetal mortality.

The politicians and farmers down here in Arizona and Utah know most of these things, but short-term money talks louder. They assume that by the time the long-term arrives, science and technology will have discovered a way to bail them out, again with the taxpayer's money.

History, too, speaks of the dangers of soil salinity, as revealed by irrigation canals over 6,000 years old that have been found in the Fertile Crescent of the Middle East. The great civilization of Sumer, which invented writing 5,000 years ago for us all, owes its downfall partly to soil salinity. That region of the Mesopotamian floodplain of the Tigris and Euphrates rivers once supported 17 to 25 million people; today this once-fertile river delta has been transformed into a desert. To most experts, salt heads the list of probable reasons for this disaster. There are some archaeologists who suspect that the same thing happened to the Anasazi over at Chaco Canyon in New Mexico, which for several centuries a thousand years ago was the focus of Anasazi culture in this area.

Farmers in the American West don't use modern irrigation technology now, so long as the taxpayers are willing to build more dams. In fact, American irrigation practices are not all that different from Pakistan's. But irrigation has been vastly improved by the drip irrigation systems developed primarily in Israel, where farming the desert is a matter of national pride. And, because the Israelis make do with far less water, I wonder how many of the Southwest's demands for more agricultural water—the American taxpayers are currently footing the $1,300 million bill for the Central Arizona Project and it's got $2,300 million to go—are simply a consequence of a wasteful technology.

Consider those giant garden-sprinklers and the large percentage of the water they emit that simply evaporates before landing on those round plots. Even the evaporation from the less wasteful flood-the-fields techniques is significant. In Israel, they cover all irrigation canals to limit evaporation. They run pipes down each row, dripping the water out of little holes directly into the soil where it is needed, wetting the soil only to root depth and no more. Moisture meters buried in the ground measure how dry the soil actually is. Microcomputers, hooked up to those moisture sensors, control the water valves so that they deliver only the amount of water actually needed to maintain the proper amount of moisture in the soil. This avoids waste through overwatering. The water is also delivered in the middle of the night, to further minimize evaporative loss. And the Israelis erect clear plastic tents over some fields to trap the humid air, slowing further evaporation. If one flies over Israel's northern valley, the Hula, there seem to be giant ponds dotting the landscape—but they're actually fields covered by giant plastic canopies.

And since the Israelis don't soak the depths of the soil with water, they run less long-term risk of ruining their soil via salt wicking up through it. It would be tragic if the rest of the world finally switched to drip irrigation only after its soil was ruined, its civilizations already scarred by starving mobs.

In the Southwest, building more dams is simply another way of subsidizing the farmer. To allow the farmer to get along without installing efficient technology, we raise taxes rather than food prices. This approach, in the minds of many thoughtful people, is appallingly wasteful and, most importantly, a luxury we cannot afford because of what it does to the land. It not only drowns beautiful canyons beneath new lakes, but risks crippling the soil that the lake water irrigates—by allowing the salt of the earth to rise to the surface.

Busy as beavers have been our dam-builders. A few years ago, Dave Brower came along on this float trip. He stood atop one of the tailing piles below a test hole here at Mile 39 to recount how the battle over this particular dam was fought by the boatmen and the Sierra Club. And temporarily won—but reminded everyone that fighting the dam-builders was a continuing battle, that we only have to fail once to lose permanently.

In a lighter vein, Brower compared dam-builders to beavers: they simply can't stand the sight of running water. Actually, that is quite a good characterization of beaver behavior if you substitute "sound" for "sight." It was once thought that beavers were terribly intelligent agricultural engineers, executing a preconceived plan to flood lowlands so as to raise more trees to eat. Instead, it seems that beavers have a strong instinct to shove mud and sticks toward the sound of running water. In fact, someone who wanted to investigate this took a loudspeaker, placed it up on a dry riverbank, and played a tape of a burbling brook. The beavers plastered the hi-fi speaker, not the river, with mud and sticks. When the tape was turned off, they stopped, presumably feeling some sense of accomplishment.

So beaver dams are built (and repeatedly repaired) thanks to this primitive instinct, stopping noise. Who would have ever predicted that dams and irrigation would emerge as a result of an animal's liking peace and quiet? One wonders if human dam-builders are operating on a similarly unreasoning principle, a blind expediency that we can no longer afford; our irrigation practices hardly seem to be the product of the insightful, reflective intelligence on which we humans pride ourselves.

Earplugs might slow down the beavers, but it will take more to stop the dam-builders. Eternal vigilance is also the price of wilderness.

Writing Exercise

Choose a piece of technology that has influenced your life. By analyzing how it has affected you (psychologically, socially, intellectually, etc.), decide whether you are a Technocratic Booster or Technology Obstructionist. Write a letter to a friend describing your feelings and urging your friend to take action regarding the piece of technology—e.g., buy it (it will change your life!), organize a protest in your area against accquiring it, put your decision off until the technology improves, etc.

APPLYING SCIENTIFIC CONCEPTS TO HUMAN LIFE

Scientific theories and technological heuristics are often applied to human life, with good and not-so-good results. We hear people say, echoing Herbert Spencer (a British philosopher and contemporary of Darwin), that only the fittest survive, but this argument is often used to rationalize oppression. The concept of stress and equilibrium from engineering is often applied to human psychology. And scientific and technical idioms, loosely applied, have found a secure place in business administration: we hear of flow charts, optimum utilization, critical mass in decision making, evolutionary modes in the stock market, and others.

In the following article, Daniel Menaker, a writer for the *New Yorker* magazine, speculates on how entropy—described in the second law of thermodynamics—may affect the human condition.

Notes and Comment

Daniel Menaker

A friend writes: I read, and try to understand, as much as I can about new discoveries and theories in astronomy, cosmology, particle physics, and higher mathematics. I read a fair amount and understand only a little, but there is something in those articles in the *Times* and *Science News* and *Scientific American*—something compelling in the various terminologies ("dark matter," "flavor," "parquet deformation," "lepton," "pancake theory," "naked beauty," "gauge boson") and something grand in the concepts under discussion—that draws me in every time. Once, quite a while ago, I read an article about entropy, which is the process of degradation of all matter and energy in the universe into a final state of motionless uniformity. When entropy has finished its work, in other words, the universe will be so much still, evenly distributed chaff—or that's how I apprehended it, at any rate. I'm sure that in the great mind of nature entropy is as neutral and indifferent as gravity or continental drift, but it seems to me one of those purely empirical patterns which the mind of man simply cannot help endowing with moral and emotional significance, however vaingloriously or foolishly. In the nineteenth century, people used the Darwinian law of the survival of the fittest to justify a whole panoply of bizarre and usually cruel social and political ideas. Heisenberg's uncertainty principle threw philosophers and theologians into an electronlike spin, because it was thought to undermine in a fundamental and irremediable way precisely what they were looking for—order. (It was even reported that the discovery of the uncertainty principle caused some neurotics to fear that all the atoms of oxygen in their room or Central Park or Europe, or wherever they happened to be, would simultaneously jump somewhere else, thus asphyxiating them. That's not the same thing as imposing human values on scientific information, of course, but it's weird.) Entropy, for its part, might seem to furnish good grounds for everything from not shining one's shoes or not voting to falling into utter melancholic despair: if everything is ultimately going to run down anyway, what's the point of winding up anything for what will be no more than a trice in comparison with the endless aeons of gelid stasis that lie in wait for the cosmos? And entropy also tempts the mind as a titanic gloomy metaphor for the decay and deterioration that all of us witness every day—the word "degradation," which is part of the dictionary's definition of entropy, is not meant to carry any moral value, but it would be almost impossible to close one's ears tightly enough to its connotations to keep them out. Roses fade, tires become bald, candles gutter, porches sag, neighborhoods decline, political ideals lose their steam, and one's own energy inevitably ebbs—as the hours go by at play, as the days undermine friendship, as the months eat away at enthusiasm for work, as the years erode plans and dreams. People who habitually dwell on such thoughts are emotionally disturbed—clinically depressed—yet the principle of entropy does seem to provide such a cast of mind with something that at least *looks like* a rational basis.

From Menaker, D. 29 October 1984. Notes and comment. *New Yorker,* 35–36. Reprinted by permission.

All this is going through my mind right now because of an experience that began for me a few weeks ago—an experience of a kind that nearly everyone has sometime, but one that I suspect always seems uniquely difficult. My aunt—my father's sister—after years of proverbially fierce independence as a widow in her house near Claverack and in her apartment in New York, suddenly and inexplicably took a very bad turn. Well, I suppose her age—ninety-two—does make it explicable. She has no children and I'm her only nephew, and my mother and father are a bit too far along in their own journeys to do more than lend moral and financial support—though that is a lot—so it fell to me to go up to the country and get my aunt out of the hospital, close her house, bring her back to New York, and install her in her apartment with round-the-clock homemaker-companions. For, as I was told by friends who live nearby in the country, in the space of a week she had grown physically feeble and mentally confused, and even though there was no reason for her to stay in the hospital, where she had gone for tests (inconclusive), she could no longer be alone. Before I left home, through the hospital upstate I hired a nurse's aide to help for the two days it would take to arrange things for the winter up there. Then I rented a car and drove up the Taconic Parkway and did what needed to be done. It involved such tasks as taking down screens and putting up storm windows, finding a notary who could witness my aunt's granting of power of attorney to me, cleaning out the refrigerator, arranging for help in New York by telephone, and seeing to it that a leak in the roof was mended—and also trying to comfort my aunt, who thought that the hospital was a prison and believed that she was about to go on a trip to Paris which would prove to be perilous. She could barely get about, her eyes had lost their keen interest in the world around her, and her once sharp—even caustic—wit had disappeared.

This was the first time I had ever had to try to take over in such a situation, and though it isn't exactly the most cheerful sort of occasion for pride I will say that I did the best I could. It wasn't perfect, and things were still a bit haywire—the power of attorney may be questionable, because of my aunt's mental incapacity, and one of the people who stay with her in New York may not work out—but as I started and have continued to attempt to sort things out and organize them I have found resources in myself that I didn't know I had; it's almost as though I had developed some new faculty or capacity, as a kind of compensation for, or answer to, distant early warnings of a slow decline in other powers and abilities. It has been like a river finding a new channel, or a new river springing up. And it has been mirrored on the outside by what has seemed to me an immediate strengthening and deepening of the ties that already existed among my friends and family, and the creation of new ones. People have come forward to help with expenses and in other ways. When I got my aunt to her house from the hospital, two young men, the sons of her nearest neighbors, at once appeared and, without being asked, took the porch furniture off the porch and put it in the barn, where they knew it belonged for the winter. I couldn't have done that job myself. And a cousin I hadn't heard from in years called me and offered a thousand dollars a year toward the help my aunt will need if she is to be able to remain outside a nursing home.

I may be taking intellectual refuge from painful feelings in speculating about whether this sad episode embodies larger patterns, but, surely, trying to find a broad context, or even just a metaphor, for one's personal experiences is or ought to be a universal human goal—it accounts for our taxonomical name. Anyway, I can't help doing it. I can't help seeing my aunt's decline as a sudden and grievous symbol of the principle of entropy, and the initiation—in me and in others, alone and together—of ways of coping with this crisis as a confounding of the principle: which may be nothing more than a fancy description of hope.

Writing Exercise

Choose a prominent scientific law or theory or an axiom from engineering and apply it, as Menaker has done, to the human condition.

EVALUATING SCIENTIFIC ACHIEVEMENT

Scientists and engineers are constantly evaluating and re-evaluating the generative theories in their own disciplines, trying to separate the unverified from the verified and thus to isolate what may come to be thought of as scientific truth.

The following selections cover the fields of primatology and engineering. Stephen Jay Gould, a professor at Harvard University, praises the work of Jane Goodall and assesses its importance. (Recall Goodall's work, *The Chimpanzees of Gombe,* excerpted in Chapter 5.) And Henry Petroski argues that engineers can only learn what designs will succeed if they understand those that have failed and, consequently, that the study of design failure should be part of every engineer's curriculum.

Animals and Us

Stephen Jay Gould

A traditional scientist might argue that if better understanding requires a radical departure from these human impositions of manners, bridles, and morals, then noninterfering study of natural populations should receive our highest priority. For no species could such a goal be more important or more difficult than for chimpanzees.

We have known for a long time that the African apes (chimps and gorillas) are our closest genealogical relatives among primates. Only recently has biochemical data been converging upon a surprising conclusion within this general knowledge: of the three species (chimp, gorilla, and human) the closest relatives by recency of common ancestry are not chimp and gorilla, as always assumed, but chimp and human. Thus the striking features held in common by chimp and gorilla, knuckle-walking in particular, are not unique specializations derived after the split from humans, but retained ancestral traits that must also have been present in the common forebear of chimps and humans.

The best estimate (based on genetic similarities) for the chimp-human split is only six to eight million years ago. Chimps, in other words, are evolutionarily much closer to us than we had ever dared to imagine. . . . Therefore, accepting the evolutionist's basic premise that genealogical closeness is the key to similarity of the most meaningful kind in biology, chimps vastly surpass all other species in their promise as sources of insight, based on methods of comparative zoology, about ourselves and our history. Since chimps are confined to dwindling areas of native vegetation in Africa, it goes without saying that we had better move fast (and with extensive resources) if we wish to benefit from this source of knowledge in its natural environment. If we blow this one too, we will never get another chance. And there can be only one closest species.

In this light, I believe that Jane Goodall is one of the intellectual heroes of this century. More than twenty-five years ago, when she was in her twenties, by sheer gumption, and virtually alone, she began her studies of a chimpanzee community at Gombe, in Tanzania. She has persisted ever since, and continues still, by struggle and courage—and she has amassed in this quarter century our first adequate and comprehensive account of the natural behavior of our closest animal relative. What in all natural history could possibly be of more interest and importance to us? This large and well-illustrated book (never forget that all primates are primarily visual animals) is a long progress report of work that must continue.

From Gould, S. J. 25 June 1987. Animals and us. *New York Review of Books,* 20–24.

Natural historians insist that their work can only proceed beyond the anecdotal by continuous, repeated close observation—for there are no essences, there is no such thing as "the chimpanzee." You can't bring a few into a laboratory, make some measurements, calculate an average, and find out, thereby, what chimpness is. There are no shortcuts. Individuality does more than matter; it is of the essence. You must learn to recognize individual chimps and follow them for years, recording their peculiarities, their differences, and their interactions. It may seem quaint to some people who fail to grasp the power of natural history that this great work of science largely tells stories about individual creatures with funny names like Jomeo, Passion, and David Greybeard. When you understand why nature's complexity can only be unraveled this way, why individuality matters so crucially, then you are in a position to understand what the sciences of history are all about. I treasure this book most of all for its quiet and unobtrusive proof, by iterated example rather than theoretical bombast, that close observation of individual differences can be as powerful a method in science as the quantification of predictable behavior in a zillion identical atoms (we need both styles in their proper slots of our multifarious world).

Close, long-continued, and maximally noninterfering observation has paid its greatest dividend in thoroughly revising the clinical or romantic views of chimpanzees that prevailed previously. Against the clinical view, Goodall has taught us how the primary features of chimp society at any time are not direct consequences of first principles or measures of simple quantities (size, number of aggressive encounters), but irreducible and unique features of individual personalities and their complex interactions. The position of the dominant, or alpha, male, for example, has not generally been held by the biggest or strongest at Gombe, but by chimps who work hard at bettering their social positions and who combine inventiveness with careful and calculating attention to social interactions. (Mike, for example, bluffed his way from low rank in 1963 to alpha in 1964 by learning to hit empty kerosene cans together while charging his superiors. Goblin, the current incumbent, is decidedly below average in size, but a master of social manipulation in forging and breaking alliances at propitious times.)

Against the romantic view of chimps as peaceful inhabitants of the abundant, primeval forest, Goodall's attention to detail has reformulated our view toward something closer to Hobbes's epithet for the lives of their closest relatives—nasty, brutish, and short. Not that chimps are brutal killers or nasty schemers; rather, they show such a wide range of behaviors in such flexible situations, from an abundance of enormous affection and kindness, to cannibalism of infants and murder with intent to kill. As in human society, the effects of occasional brutality outweigh a thousand acts of unrecorded kindness in setting the basic events of social history. Of sixty-six recorded deaths, more than half resulted from disease, but 20 percent occurred as the result of injury in fights or falls.

Continued close observation has also affirmed another important principle of historicity and individuality. The current state of a chimpanzee society cannot be assessed by accumulating the ordinary, predictable events of daily life (if so, the experimentalist's procedure of occasional random sampling might suffice). Major features are consequences of rare large-scale events (the analogues of war, famine, and pestilence in our lives)—true historical particulars that can only be appreciated by watching, not predicted from theory. Goodall teaches us, for example, how much about group size, composition, and dynamics of the Gombe chimpanzees can be explained by three particular events: a polio outbreak (probably transmitted from humans) in 1966 leading to six deaths and six cripplings; the split of her community into two groups during 1970–1972, and the later annihilation of one by males of the other; and the killing and cannibalism of up to ten infants by one female (named, of all things, Passion) and her adult daughter, following their "discovery" of this food source in 1974. They ganged up on other females and seized their babies, leading to the death of all but one infant in the Gombe community during the four-year period of their depredations.

The study of chimps places us in a double bind with respect to our need for reliable knowledge about their social lives: our affection, arising from an acute perception of biological affinity, draws us inexorably to these animals, yet our need for distancing to break the bonds of bias and expectation has never been greater. We cannot, as their review holds, see animals in any other than human terms; yet, in some immediate sense, we must struggle for maximal distance.

This struggle, waged consciously, courageously, and above all humanely, by Goodall, is an underlying and coordinating theme of this volume. She admits and defends her proper involvement (for objectivity is not an erasure of emotions, but a firm recognition of their inevitable role and presence). She writes: "I readily admit to a high level of emotional involvement with individual chimpanzees—without which, I suspect, the research would have come to an end many years ago." She loved David Greybeard, the wise old male who comforted her and taught her so much during her greenhorn years—as she hated the cannibal Passion who nearly extinguished an entire generation. (Can we ever begin to appreciate the power of her temptation to intervene, and the absolute need, so respected by Goodall, to let events run their course, while recording the consumption of infants in all possible observational detail?)

Goodall began by cultivating a closeness that she had to break when she realized that such relationships were not only distorting the behavior of individual chimpanzees, but also threatening to alter the character of entire communities. I felt her emotional pain and appreciated what she has done all the more (for science can be one hell of a harsh taskmaster) in reading:

> *In the early years of the research I actively encouraged social contact—play or grooming—with six different chimpanzees. For me personally, these contacts were a major breakthrough; they meant that I had won the trust of creatures who initially had fled when they saw me in the distance. However, once it became evident that the research would continue into the future, it was necessary to discourage contacts of this sort.*

(Goodall had, perhaps significantly, altered the history and community structure of Gombe chimpanzees during the early years of her research by feeding bananas essentially ad libitum to chimps who visited her camp. One can well imagine the impact of introducing such a consistent and abundant resource in a world marked by unpredictability and seasonality of foodstuffs. This magnet brought many chimps into camp, and into levels of social contact that would probably not have been achieved otherwise. Goodall recognized the problem and has, ever since, greatly restricted the number and timing of bananas supplied in camp. But the introduction of this resource, and its later restriction, may have been a major influence in first establishing an enlarged community and then precipitating the split that led to the eventual annihilation of one subgroup by the other.)

I do not recount this episode to criticize. Goodall's decisions were correct, and all of us who pursue science under difficult conditions in the bush know that purist research designs are pipe dreams, and that pragmatism must often rule. (You can't just match off into dense foliage and find chimps; you must first make contact and build trust in order to win acceptance and establish the possibility of following in the wild.) I mention this crucial incident only to point out once again how subtle and complex our necessary interaction with such a flexible and intelligent creature must be.

Yet the most instructive examples lie not in overt incidents, but in more general uses of language, or the way questions were posed. As an obvious case, Goodall did not recognize for many years how much females participate in hunts for meat (though males are more active) because "until 1976 females were only rarely the subject of full-day or consecutive-day follows: the new information on hunting has largely been acquired by rectifying that bias."

Speaking of conventionalities imposed by human perceptions, I was shaken by occasional statements in the worst tradition of the chain of being. "It is evident that chimpanzees have made considerable progress along the road to humanlike love and compassion." Or, comparing her chimps with their more sociable cousin, *Pan paniscus,* the pygmy chimp: "It sounds like a utopian society—and viewed against this, it would seem that the Gombe chimpanzees have a long way to go." But the Gombe chimps are not on any road, and the metaphor can only produce a biased itinerary.

More subtly, I found again and again that both the scheme of research and the design of this volume follow our perception of interest, and I wondered over and over, page after page, what we might be missing that is important to chimpanzees. Thus chapters on behaviors similar to actions that either trouble us when expressed by humans, or are conventionally judged crucial to the evolution of our own intelligence, are given prominence in length and word (hunting, territoriality, and the analogues of warfare), while calmer and less obtrusive behaviors without clear human analogues (the extraordinary attention given to mutual grooming, in particular) are not as intensely pondered. Thus Goodall presents a long speculation on the role of warfare in precipitating the evolution of intelligence, but offers no similar pride of place to childhood play, or to the complex relationships and mutual interchange of grooming for that matter.

What are we missing by parsing the behavior of chimpanzees into the conventional categories recognized largely from our own behavior? What other taxonomies might revolutionize our view—for taxonomies are theories of order? What are we missing because we must place all we see into slots of our usual taxonomy? Why do chimps eat a walnutshell full of termite clay every day? Must we view such an action as an ecological adaptation (good source of concentrated soil nutrients, Goodall cogently suggests)—or might it represent something that does not lie within any of our categories, so that we miss its meaning entirely or record the wrong thing? What does the prison of our language do to the possibilities of interpretation? What would a taxonomy based on things *not* done look like? How can we comprehend the soul of a creature whose watchword is flexibility with a language that parcels actions into discrete categories given definite names? And if this be a problem for chimps, how can we hope to understand ourselves?

Animals are, and must be, our mirror. And I know no better reason for a continued struggle to understand them, and a persistent drive to define relationships of respect, than this necessary wedlock. But what are we missing because we have made ourselves, for reasons of blind vanity and hubris, the measure of all things? Gunnar Myrdal, whose recent death we all lament, grappled with this question in his masterpiece on human intolerance:

> *There must still be . . . countless errors . . . that no living man can yet detect, because of the fog within which our type of Western culture envelops us. Cultural influences have set up the assumptions about the mind, the body, and the universe with which we begin; pose the questions we ask; influence the facts we seek; determine the interpretation we give these facts; and direct our reaction to these interpretations and conclusions.*

This, we must understand, is not merely "An American Dilemma," but the central problem of all our natural knowledge.

The Limits of Design

Henry Petroski

Daedalus, whose mythical wing-making has earned him the title of first aeronautical engineer, is said to have cursed his skill when he spied the wings of his son Icarus floating on the sea. But Icarus had to share some of the blame for what may well have been the first structural failure in the history of air travel. He had been warned by his father, the designer of wax and feather wings, that he should not fly too high with the untested new invention.

Instead of wandering about the labyrinth dreaming of the perfect and indestructible wing, Daedalus accepted the compromise of a design that could be fouled by the water or melted by the sun. Everyone and everything has its limitations and its breaking point, but that does not mean that we and all our designs are total failures. Just as people are not expected to push themselves too hard or to overextend themselves, so machines and structures are not expected by their designers to be pushed too hard or to be overloaded. Daedalus foresaw, as all engineers must foresee, the ways in which his structure could fail, for it is only by recognizing the possible ways of failure that a successful structure can be designed to resist the forces that might tear part from part. Had Icarus used the wings within the proper altitude, they would to this day be declared a success and the first instruments of successful, albeit mythical, manned flight.

What Daedalus did was presumably the best he could with the technology and resources available to him. He had limited materials, so wax and feathers had to do. Neither did he and Icarus have limitless time, for the hungry Minotaur was also in the labyrinth. The tragedy of the myth of the first flight is that the flight might have worked—within the myth at least—had the repeated cautions of Daedalus been heeded by his son.

Even today, when engineers make wings of metal for airplanes that carry myriad strangers as well as their own sons, the designers still must caution the pilots and maintenance men about the proper use of the wings. This is done through operating and maintenance manuals, which provide the proper limitations and procedures, but as accidents attributed to pilot error and improper maintenance procedures testify, there is still a bit of Icarus about the airport.

The object of engineering design is to obviate failure, but the truly fail-proof design is chimerical. The ways in which a structure or machine can fail are many, and their effects range from blemishes to catastrophes. The designer David Pye has written of the compromise of design:

> *The requirements for design conflict and cannot be reconciled. All designs for devices are in some degree failures, either because they flout one or another of the requirements or because they are compromises, and compromise implies a degree of failure.*
>
> *Failure is inherent in all useful design not only because all requirements of economy derive from insatiable wishes, but more immediately because certain quite specific conflicts are inevitable once requirements for economy are admitted; and conflicts even among the requirements of use are not unknown.*
>
> *It follows that all designs for use are arbitrary. The designer or his client has to choose in what degree and where there shall be failure. Thus the shape of all designed things is the product of arbitrary choice. If you vary the terms of your compromise—say, more speed, more heat, less safety, more discomfort, lower first cost—then you vary the shape of the thing designed. It is quite impossible for any design to be "the logical outcome of the requirements" simply because, the requirements being in conflict, their logical outcome is an impossibility.*

This is all simply to say that not even engineers can have their cake and eat it too, though they, being human, may sometimes try. To a student of the history of technology, the immobilization of New York's fleet of Grumman Flexible buses because of the occurrence of cracked steel frames was but another chapter in the Iron-Age-old story of man against manufacturing. Ever since the first ironworker—call him John Smith—tried to extract ore from the geological chaos of our terrestrial sphere and fashion it into tools, the integrity of his products has been subject to the seemingly mercurial qualities of the metal.

Swords, among the earliest iron objects, were wrought in a variety of ways to produce a keen edge and a tough blade that would not fail a warrior in battle. Yet many an ancient sword must have broken at a most unpropitious time, and the fallen warrior's comrades in arms would certainly have wanted to know why, and whether their swords would fail next. John Smith, if he wanted to stay in the sword-making business, had to apply some of his profit and experience toward research and development to come up with more reliable swords.

These were no doubt easier promised than delivered, and one can easily speculate that the paucity of flawless swords raised those few apparently indestructible ones to legend. Ironically, the vastly improved reliability of modern weapons probably owes more to John Smith's understanding of why a common blade cracked or snapped in two than to his forging of an Excalibur.

The same holds true in peaceful applications of iron and steel, and the cracks in New York's buses pointed to weaknesses of design that may have inconvenienced commuters but that would be corrected in future designs to build a sounder bus frame that could take the potholes in any city's streets. Yet the nagging question arises, Does this iterative process of design by failure ever end? Will there be a day when the designers will be able to say with assurance and finality, This is a flawless design? Yes, the process can converge on a design as reliable as is reasonable; but, no, it can never be guaranteed to produce a perfectly flawless product. Design involves assumptions about the future of the object designed, and the more that future resembles the past the more accurate the assumptions are likely to be. But designed objects themselves change the future into which they will age.

It follows that departures from traditional designs are more likely than not to hold surprises. Good design minimizes the effects of surprises by anticipating troublesome details and by overdesigning for an extra measure of safety. While John Smith himself may have produced an Excalibur only by accident, the experience of failures that he could pass on to his descendants has enabled them eventually to make Excalibur a brand name and put a virtually flawless sword in every kitchen warrior's hand. This was possible because of the historical time scale over which the perfection of the sword and knife evolved. With little change in form or function from century to century, the Smiths could concentrate on the all-important metallurgy of the blade.

In contrast, it is barely two hundred years since the first iron bridge at Coalbrookdale was erected at the beginning of the Industrial Revolution. And fewer than five generations separate the introduction of the railroad train in England and the Grumman Flexible bus in New York. Unfortunately, merely knowing the history of technology does not absolve one from repeating it. While there are similarities between a nineteenth-century railway car and a late twentieth-century bus, the dissimilarities dominate. The design of a new generation mass-transit vehicle to negotiate metropolitan traffic while satisfying guidelines for energy efficiency, accommodating the handicapped, and meeting a volume of other federal specifications is a formidable task. Though there is never any excuse for a faulty design, even under the worst constraints, there should be room for understanding. For no manufacturer wants his design to fail or endanger life. Not only is it morally wrong, it is also bad for business.

The successful transportation of men to the moon and back has demonstrated that lack of experience alone does not necessarily condemn a design to failure. It is rather the combination of inexperience, distracted by overly restrictive requirements, coupled with the pressures of deadlines, and aggravated by concerns for profit margins that initiates the cracking up of bus frames and their designers. John Smith no longer works alone. He must leave his forge regularly to go over the account books with John Doe, the supplier of his capital, and to be briefed on revised regulations by John Law, the granter of his license.

It is not only in the high technological business of building mass-transit buses that our accelerating socio-economic system breaks down. Computer models that predict the behavior of the economy have come increasingly to be relied upon to justify major economic decisions, and yet these models are not necessarily any more infallible than the ones that predict the fatigue life of a bus frame. Thus the same tools that apparently free us from the tedium of analyzing the wheel condemn us to reinvent it. We have come to be a society that is so quick to change that we have lost the benefits of one of mankind's greatest tools—experience. We are redesigning the commonest of vehicles as routinely as we are restructuring the fundamental economy. Changes are being made so radically that the relevance of lessons learned from earlier generations is not recognized. It is as if we are beating our swords into ploughshares so frenetically and carelessly as to produce blades that might fracture on striking the first pebble in their path.

Sir Alfred Pugsley, a pioneer in the study of metal fatigue in military aircraft and an articulate spokesman for the cause of structural safety, wrote: "A profession that never has accidents is unlikely to be serving its country efficiently." He was merely putting in crass terms what is a constant goal of structural engineering design in a rapidly changing society: to build safe structures more economically. The limits of structures are not always so conclusively demonstrated on their maiden flight as were the wings that fell from Icarus' arms. When an airplane flies for thousands of hours without incident, it is of course no proof that its success is a result of excessive strength that translates to more weight that in turn translates to permanent excess baggage. However, knowing that they have allowed for uncertainties in the strength of materials and uncertainties in loads on the wings and uncertainties in stress calculations with factors of safety, the engineers naturally wonder how much unnecessary weight is contained in the structure of the airplane. Thus when designing the next generation of the airplane, engineers are hard pressed to respond to questions as to why such a large factor of safety must be maintained. After all, in the intervening years they have come to understand the materials better. They have measured the loads during successful test flights of the airplane design. And they have acquired computers that enable them to make more and presumably better, more accurate stress calculations than ever before. If they do not build a lighter and more economical new plane their competitors will, and the first designers will have failed to build on their experience.

This is not unlike the way cathedrals and bridges and buses and virtually every engineering structure has evolved. With each success comes the question from society, from taxpayers, from government, from boards of directors, from engineers themselves as to how much larger, how much lighter, and how much more economically the next structure can be made. And it is not only the structural factor of safety that is skimped upon. Workmanship and style can go the way of strengh. But the phenomenon is not new to the computer age. Henry Adams, forever seeing the degeneration of society, could write in *Mont-Saint-Michel and Chartres,* "The great cathedrals after 1200 show economy, and sometimes worse. The world grew cheap, as worlds must."

What happens, of course, is that success ultimately leads to failure: aesthetic failure, functional failure, and structural failure. The first can take away the zest for life, the second the quality of life, and the third life itself. Structural failure usually reverses the trend toward less and less safe structures of the kind that failed, however, either through the abandonment of that line of structures or through their being strengthened or used more conservatively. There is always pressure for relevant building codes, factors of safety, and engineering practice to be made more conservative after a failure, and in this way failures lead to new successes. The process would appear to be cyclic.

The lessons of history are clear for the structural engineer. Innovation need not be doomed to failure, for there are Iron Bridge, the Eads Bridge, and the Brooklyn Bridge—all standing as monuments to cautious exploration with new materials. There is the legacy of the Crystal Palace and there is the Empire State Building reminding us that innovative and rapid construction need not be inferior. And there is the flag on the moon to remind us that what has never been done before may be done now with the use of computers. Yet the same history tells us also that materials in new environments can lead to cracked and embrittled nuclear power plants. And it tells us that fast-track construction can leave convention-goers without convention centers. And it tells us that the computer does not necessarily make a better bus. Are the lessons of history clear for the structural engineer?

Yes, at least one lesson is clear, that innovation involves risk. Some innovations are successful and their engineers are heroes. Some innovations are failures and their engineers are goats. Yet failure in innovation should be no more opprobrious to the engineer who has prepared himself as well as he could for his attempt to build a longer bridge than to a pole vaulter who fails to make a record vault after practicing his event and using his pole to its capacity. It is the engineer who has tried to do what he is not prepared to do, or who has made the same mistake that has led to failure before, who is acting irresponsibly. He is letting down his profession as surely as the pole vaulter who refuses to practice and who uses the same kind of pole that had broken under his lighter opponent is letting down his team. The well-prepared engineer can and does build beyond experience without hubris as surely as the well-trained pole vaulter goes after a new record.

Writing Exercise

Following the directions of your instructor for audience and length, write a summary of the most recent developments in one field or subfield of science and technology, and speculate on the most promising areas for further research. This assignment might fulfill the needs of a short research paper.

KEEPING TECHNOLOGY TO THE HUMAN SCALE

The selections that follow are a spinoff from those in the first section of this chapter, where we discussed the boosters and the obstructionists. These center on a recent debate in science and technology, namely, Should technological mechanisms be kept within the range of the human scale, especially in rural regions and Third World countries?

E. F. Schumacher brought this argument into the public arena in the 1970s with his book *Small is Beautiful,* from which his excerpt comes. Wendell Berry offers a sympathetic perspective. And Samuel Florman eloquently refutes Schumacher's premises from technical, political, social, aesthetic, and moral perspectives.

Technology with a Human Face

E. F. Schumacher

The modern world has been shaped by its metaphysics, which has shaped its education, which in turn has brought forth its science and technology. So, without going back to metaphysics and education, we can say that the modern world has been shaped by technology. It tumbles from crisis to crisis; on all sides there are prophecies of disaster and, indeed, visible signs of breakdown.

If that which has been shaped by technology, and continues to be so shaped, looks sick, it might be wise to have a look at technology itself. If technology is felt to be becoming more and more inhuman, we might do well to consider whether it is possible to have something better—a technology with a human face.

Strange to say, technology, although of course the product of man, tends to develop by its own laws and principles, and these are very different from those of human nature or of living nature in general. Nature always, so to speak, knows where and when to stop. Greater even than the mystery of natural growth is the mystery of the natural cessation of growth. There is measure in all natural things—in their size, speed, or violence. As a result, the system of nature, of which man is a part, tends to be self-balancing, self-adjusting, self-cleansing. Not so with technology, or perhaps I should say: not so with man dominated by technology and

---

From Schumacher, E. F. Technology with a human face. In *Small is beautiful: Economics as if people mattered,* 146–51. New York: Harper & Row.

specialisation. Technology recognises no self-limiting principle—in terms, for instance, of size, speed, or violence. It therefore does not possess the virtues of being self-balancing, self-adjusting, and self-cleansing. In the subtle system of nature, technology, and in particular the super-technology of the modern world, acts like a foreign body, and there are now numerous signs of rejection.

Suddenly, if not altogether surprisingly, the modern world, shaped by modern technology, finds itself involved in three crises simultaneously. First, human nature revolts against inhuman technological, organisational, and political patterns, which it experiences as suffocating and debilitating; second, the living environment which supports human life aches and groans and gives signs of partial breakdown; and, third, it is clear to anyone fully knowledgeable in the subject matter that the inroads being made into the world's non-renewable resources, particularly those of fossil fuels, are such that serious bottlenecks and virtual exhaustion loom ahead in the quite foreseeable future.

Any one of these three crises or illnesses can turn out to be deadly. I do not know which of the three is the most likely to be the direct cause of collapse. What is quite clear is that a way of life that bases itself on materialism, *i.e.* on permanent, limitless expansionism in a finite environment, cannot last long, and that its life expectation is the shorter the more successfully it pursues its expansionist objectives.

If we ask where the tempestuous developments of world industry during the last quarter-century have taken us, the answer is somewhat discouraging. Everywhere the problems seem to be growing faster than the solutions. This seems to apply to the rich countries just as much as to the poor. There is nothing in the experience of the last twenty-five years to suggest that modern technology, as we know it, can really help us to alleviate world poverty, not to mention the problem of unemployment which already reaches levels like thirty per cent in many so-called developing countries, and now threatens to become endemic also in many of the rich countries. In any case, the apparent yet illusory successes of the last twenty-five years cannot be repeated: the threefold crisis of which I have spoken will see to that. So we had better face the question of technology—what does it do and what should it do? Can we develop a technology which really helps us to solve our problems—a technology with a human face?

The primary task of technology, it would seem, is to lighten the burden of work man has to carry in order to stay alive and develop his potential. It is easy enough to see that technology fulfils this purpose when we watch any particular piece of machinery at work—a computer, for instance, can do in seconds what it would take clerks or even mathematicians a very long time, if they can do it at all. It is more difficult to convince oneself of the truth of this simple proposition when one looks at whole societies. When I first began to travel the world, visiting rich and poor countries alike, I was tempted to formulate the first law of economics as follows: "The amount of real leisure a society enjoys tends to be in inverse proportion to the amount of labour-saving machinery it employs." It might be a good idea for the professors of economics to put this proposition into their examination papers and ask their pupils to discuss it. However that may be, the evidence is very strong indeed. If you go from easy-going England to, say, Germany or the United States, you find that people there live under much more strain than here. And if you move to a country like Burma, which is very near to the bottom of the league table of industrial progress, you find that people have an enormous amount of leisure really to enjoy themselves. Of course, as there is so much less labour-saving machinery to help them, they "accomplish" much less than we do; but that is a different point. The fact remains that the burden of living rests much more lightly on their shoulders than on ours.

The question of what technology actually does for us is therefore worthy of investigation. It obviously greatly reduces some kinds of work while it increases other kinds. The type of work which modern technology is most successful in reducing or even eliminating is skilful, productive work of human hands, in touch with real materials of one kind or another. In an advanced industrial society, such work has become exceedingly rare, and to make a decent living by doing such work has become virtually impossible. A great part of the modern neurosis may be due to this very fact; for the human being, defined by Thomas Aquinas as a being with brains and hands, enjoys nothing more than to be creatively, usefully, productively engaged with both his hands and his brains. Today, a person has to be wealthy to be able to enjoy this simple thing, this very great luxury: he has to be able to afford space and good tools; he has to be lucky enough to find a good teacher and plenty of free time to learn and practise. He really has to be rich enough not to need a job; for the number of jobs that would be satisfactory in these respects is very small indeed.

The extent to which modern technology has taken over the work of human hands may be illustrated as follows. We may ask how much of "total social time"—that is to say, the time all of us have together, twenty-four hours a day each—is actually engaged in real production. Rather less than one-half of the total population of this country is, as they say, gainfully occupied, and about one-third of these are actual producers in agriculture, mining, construction, and industry. I do mean *actual producers,* not people who tell other people what to do, or account for the past, or plan for the future, or distribute what other people have produced. In other words, rather less than one-sixth of the total population is engaged in actual production; on average, each of them supports five others beside himself, of which two are gainfully employed on things other than real production and three are not gainfully employed. Now, a fully employed person, allowing for holidays, sickness, and other absence, spends about one-fifth of his total time on his job. It follows that the proportion of "total social time" spent on actual production—in the narrow sense in which I am using the term—is, roughly, one-fifth of one-third of one-half, *i.e.* 3½ per cent. The other 96½ per cent of "total social time" is spent in other ways, including sleeping, eating, watching television, doing jobs that are not *directly* productive, or just killing time more or less humanely.

Although this bit of figuring work need not be taken too literally, it quite adequately serves to show what technology has enabled us to do: namely, to reduce the amount of time actually spent on production in its most elementary sense to such a tiny percentage of total social time that it pales into insignificance, that it carries no real weight, let alone prestige. When you look at industrial society in this way, you cannot be surprised to find that prestige is carried by those who help fill the other 96½ per cent of total social time, primarily the entertainers but also the executors of Parkinson's Law. In fact, one might put the following proposition to students of sociology: "The prestige carried by people in modern industrial society varies in inverse proportion to their closeness to actual production."

There is a further reason for this. The process of confining productive time to 3½ per cent of total social time has had the inevitable effect of taking all normal human pleasure and satisfaction out of the time spent on this work. Virtually all real production has been turned into an inhuman chore which does not enrich a man but empties him. "From the factory," it has been said, "dead matter goes out improved, whereas men there are corrupted and degraded."

We may say, therefore, that modern technology has deprived man of the kind of work that he enjoys most, creative, useful work with hands and brains, and given him plenty of work of a fragmented kind, most of which he does not enjoy at all. It has multiplied the number of people who are exceedingly busy doing kinds of work which, if it is productive at all, is so only in an indirect or "roundabout" way, and much of which would not be necessary at all if technology were rather less modern. Karl Marx appears to have foreseen much of this when he wrote: "They want production to be limited to useful things, but they forget that the production of too many useful things results in too many useless people," to which we might add: particularly when the processes of production are joyless and boring. All this confirms our suspicion that modern technology, the way it has developed, is developing, and promises further to develop, is showing an increasingly inhuman face, and that we might do well to take stock and reconsider our goals.

Taking stock, we can say that we possess a vast accumulation of new knowledge, splendid scientific techniques to increase it further, and immense experience in its application. All this is truth of a kind. This truthful knowledge, as such, does *not* commit us to a technology of giantism, supersonic speed, violence, and the destruction of human work-enjoyment. The use we have made of our knowledge is only one of its possible uses and, as is now becoming ever more apparent, often an unwise and destructive use.

A Good Scythe

Wendell Berry

When we moved to our little farm in the Kentucky River Valley in 1965, we came with a lot of assumptions that we have abandoned or changed in response to the demands of place and time. We assumed, for example, that there would be good motor-powered solutions for all of our practical problems.

One of the biggest problems from the beginning was that our place was mostly on a hillside and included a good deal of ground near the house and along the road that was too steep to mow with a lawn mower. Also, we were using some electric fence, which needed to be mowed out once or twice a year.

When I saw that Sears Roebuck sold a "power scythe," it seemed the ideal solution, and I bought one. I don't remember what I paid for it, but it was expensive, considering the relatively small amount of work I needed it for. It consisted of a one-cylinder gasoline engine mounted on a frame with a handlebar, a long metal tube enclosing a flexible drive shaft, and a rotary blade. To use it, you hung it from your shoulder by a web strap, and swept the whirling blade over the ground at the desired height.

It did a fairly good job of mowing, cutting the grass and weeds off clean and close to the ground. An added advantage was that it readily whacked off small bushes and tree sprouts. But this solution to the mowing problem involved a whole package of new problems:

1. The power scythe was heavy.
2. It was clumsy to use, and it got clumsier as the ground got steeper and rougher. The tool that was supposed to solve the problem of steep ground worked best on level ground.
3. It was dangerous. As long as the scythe was attached to you by the shoulder strap, you weren't likely to fall onto that naked blade. But it *was* a naked blade, and it did create a constant threat of flying rock chips, pieces of glass, etc.
4. It enveloped you in noise, and in the smudge and stench of exhaust fumes.
5. In rank growth, the blade tended to choke—in which case you had to kill the engine in a hurry or it would twist the drive shaft in two.
6. Like a lot of small gas engines not regularly used, this one was temperamental and undependable. And dependence on an engine that won't run is a plague and a curse.

When I review my own history, I am always amazed at how slow I have been to see the obvious. I don't remember how long I used that "labor-saving" power scythe before I finally donated it to help enlighten one of my friends—but it was too long. Nor do I remember all the stages of my own enlightenment.

The turning point, anyhow, was the day when Harlan Hubbard showed me an old-fashioned, human-powered scythe that was clearly the best that I had ever seen. It was light, comfortable to hold and handle. The blade was very sharp, angled and curved precisely to the path of its stroke. There was an intelligence and refinement in its design that made it a pleasure to handle and look at and think about. I asked where I could get one, and Harlan gave me an address: The Marugg Company, Tracy City, Tennessee 37387.

I wrote for a price list and promptly received a sheet exhibiting the stock in trade of the Marugg Company: grass scythes, bush scythes, snaths, sickles, hoes, stock bells, carrying yokes,

---

From Berry, W. 1981. A good scythe. In *The gift of good land: Further essays cultural and agricultural,* 171–75. San Francisco: North Point Press.

whetstones, and the hammers and anvils used in beating out the "dangle" cutting edge that is an essential feature of grass scythes.

In due time I became the owner of a grass scythe, hammer and anvil, and whetstone. Learning to use the hammer and anvil properly (the Marugg Company provides a sheet of instructions) takes some effort and some considering. And so does learning to use the scythe. It is essential to hold the point so that it won't dig into the ground, for instance; and you must learn to swing so that you slice rather than hack.

Once these fundamentals are mastered, the Marugg grass scythe proves itself an excellent tool. It is the most satisfying hand tool that I have ever used. In tough grass it cuts a little less uniformly than the power scythe. In all other ways, in my opinion it is a better tool:

1. It is light.
2. It handles gracefully and comfortably even on steep ground.
3. It is far less dangerous than the power scythe.
4. It is quiet and makes no fumes.
5. It is much more adaptable to conditions than the power scythe: in ranker growth, narrow the cut and shorten the stroke.
6. It always starts—provided the user will start. Aside from reasonable skill and care in use, there are no maintenance problems.
7. It requires no fuel or oil. It runs on what you ate for breakfast.
8. It is at least as fast as the power scythe. Where the cutting is either light or extra heavy, it can be appreciably faster.
9. It is far cheaper than the power scythe, both to buy and to use.

Since I bought my power scythe, a new version has come on the market, using a short length of nylon string in place of the metal blade. It is undoubtedly safer. But I believe the other drawbacks remain. Though I have not used one of these, I have observed them in use, and they appear to me to be slower than the metal-bladed power scythe, and less effective on large-stemmed plants.

I have noticed two further differences between the power scythe and the Marugg scythe that are not so practical as those listed above, but which I think are just as significant. The first is that I never took the least pleasure in using the power scythe, whereas in using the Marugg scythe, whatever the weather and however difficult the cutting, I always work with the pleasure that one invariably gets from using a good tool. And because it is not motor driven and is quiet and odorless, the Marugg scythe also allows the pleasure of awareness of what is going on around you as you work.

The other difference is between kinds of weariness. Using the Marugg scythe causes the simple bodily weariness that comes with exertion. This is a kind of weariness that, when not extreme, can in itself be one of the pleasures of work. The power scythe, on the other hand, adds to the weariness of exertion the unpleasant and destructive weariness of strain. This is partly because, in addition to carrying and handling it, your attention is necessarily clenched to it; if you are to use it effectively and safely, you *must* not look away. And partly it is because the power scythe, like all motor-driven tools, imposes patterns of endurance that are alien to the body. As long as the motor is running there is a pressure to keep going. You don't stop to consider or rest or look around. You keep on until the motor stops or the job is finished or you have some kind of trouble. (This explains why the tractor soon evolved headlights, and farmers began to do daywork at night.)

These differences have come to have, for me, the force of a parable. Once you have mastered the Marugg scythe, what an absurd thing it makes of the power scythe! What possible sense can there be in carrying a heavy weight on your shoulder in order to reduce by a very little the use of your arms? Or to use quite a lot of money as a substitute for a little skill?

The power scythe—and it is far from being an isolated or unusual example—is *not* a labor saver or a shortcut. It is a labor maker (you have to work to pay for it as well as to use it) and a long cut. Apologists for such expensive technological solutions love to say that "you can't turn back the clock." But when it makes perfect sense to do so—as when the clock is wrong—of *course* you can!

Small Is Dubious

Samuel C. Florman

While reading the newspapers the morning after President Carter's energy address to the nation in April 1979, I was struck by a statement attributed to Mr. Carter's pollster and adviser, Patrick Caddell: "The idea that big is bad and that there is something good to smallness is something that the country has come to accept much more today than it did ten years ago. This has been one of the biggest changes in America over the past decade."

Since the nation had just been exhorted to embark on the most herculean technological, economic, and political enterprises, the reference to smallness seemed to me to be singularly inapt. Waste is to be deplored, of course, as is inefficiency. But bigness? I had not realized that the small-is-beautiful philosophy had reached the White House.

Of all the ideas that have emanated from the opponents of technology, none has been more confusing and potentially dangerous than "small is beautiful." The early literature of antitechnology, from Jacques Ellul's *The Technological Society* (1964) through Theodore Roszak's *Where the Wasteland Ends* (1972), for all its faults, was relatively harmless. It advocated a passive resistance to high technology, or at most a "withdrawal" by sensitive souls from the prevailing culture. The publication in 1973 of E. F. Schumacher's book *Small Is Beautiful* heralded a more aggressive mood. Schumacher and his disciples are not content to brood. Their doctrine of smallness advocates a headlong retreat into the past, or worse, a heedless rush into a future that will not work. This new movement is all the more to be feared because it has such a seductive rallying cry.

The initial success of Schumacher's book was in some respects deserved. It could not help but strike a responsive chord in anyone who was familiar with many abortive attempts to transfer high technology to the developing nations. Undoubtedly some foolish, insensitive things were done, and these deserved to be criticized. (On the other hand, some of the "small" technologies introduced into nations such as India have also been fiascos—for example, wind-powered water pumps installed in areas where there are long seasons of windlessness, bio-gas generators that fail to work in the cool of winter, and specially subsidized village soap industries that collapse in competition with factory-produced brands.)

It is not my purpose, however, to argue on behalf of any particular master plan for India or any other Third-World nation. Nor do I suggest that every society model itself after the United States. If Schumacher had restricted himself to the topic of technology transfer from developed to undeveloped lands, his contribution would have been interesting and responsible, even if not altogether convincing. Unfortunately, with convoluted missionary zeal he transmuted his approval of primitive village life into a critique of Western industrial society.

When Schumacher's book arrived on the scene in 1973, the United States was reeling under the effects of an oil embargo, an environmental crisis, the just-ended Vietnam War, and Watergate. Our problems were formidable, and our large institutions seemed to be showing signs of strain. Somehow the phrase "small-is-beautiful" had a soothing ring. It sounded like an idea whose time had come. In fact, it is an idea that does not bear scrutiny, and if it takes hold in our thinking, it has the potential for doing much harm.

To an engineer in the United States, the debate about whether technologies should in principle be large or small, hard or soft, high or low, is almost incomprehensible. Even when the word *appropriate* is used, the arguments seem absurd. Of course, everyone agrees that technologies should be "appropriate"—but to define "appropriate" as "small" or even "intermediate," as so many people are doing, is to beg the question. When engineers are confronted with a problem of design, in which the objective is to satisfy certain human needs or

---

desires, often the solution that appears to be most effective, economical, elegant, satisfactory, suitable, fit, proper, *appropriate*—call it what you will—is a technology that is large in scale.

Take, for example, the use of water. One of the oldest technologies—indeed a technology that is associated with the beginnings of civilization—is the control and distribution of water, the *sharing* of water by large numbers of people. Obviously our society could not survive without large-scale study and allocation of our water resources, without reservoirs, dams, water treatment plants, pumping stations, canals, and aqueducts. It is not possible for each small group of people to drill a well or divert a stream. Engineers have not arbitrarily decided to make water distribution a large technology; their engineering solutions have been inherent in the very scheme of things. Other technologies are also by their nature "large"—railroads, highways, airports—most of the undertakings that fall into the category of what we call public works.

This does not mean that engineers are committed only to large technologies, nor that they are mesmerized by massive projects. A tiny hearing aid or a pacemaker or a small transistor radio is just as exciting an engineering achievement as a bridge or a dam or a Boeing 747. Engineers are constantly trying to evaluate the appropriate scale for their designs. Most households have their own small refrigerators making their own small quantity of ice. It would be no great trick to make the ice at a central location and distribute it to each home by vacuum tube, or even to fly ice in from the Arctic and drop it at each doorstep by helicopter. However, we seek the system that seems to "make sense," including economic sense.

In the design of tools and machines, as in the design of systems, engineers are also pragmatic. A gargantuan printing press seems right for a weekly magazine with a circulation of several million people; a small press or copying machine is appropriate for a local stationer; and for artisans who do etchings or woodcuts, there are beautifully designed and crafted hand tools. A thousand similar examples could be given, but there is no need to belabor the obvious.

Often we find that large technologies and small are intertwined. Consider the backpack—small technology, the very essence of the counterculture life-style—yet this item is made of aluminum and nylon, products of two very large energy-intensive, mass-production technologies. Any small technology that uses metal—say, the bicycle—is dependent upon the large technologies of mining and metallurgy.

A concern for "appropriate" technology would seem to call for both large and small solutions in an ever-changing creative flux. The world around us demonstrates a need for both large and small—whales and plankton, oceans and pools, redwood trees and wildflowers, huge migrating flocks and tiny isolated organisms. By demonstrating flexibility and variety, we fit naturally into the biosphere. Why, then, has an impassioned debate arisen over the question of small technologies versus large? Why do we find hostility directed against so many of the things that engineers build and deem worthy; dams, highways, power networks, large systems of all sorts?

Certainly the small-is-beautiful concept cannot be argued from a purely technical point of view. Our large systems have been, and are, technically successful. They are not perfect; they are merely superior to other systems presently available.

It is Schumacher's contention that a nation of small communities, populated by self-reliant craftsmen, is not only socially desirable, but also economically efficient. He said this in his writings, and also in a series of lectures, one of which I heard him deliver shortly before his untimely death in 1977. This idea sounds fine in a symposium on the human condition, and it was enthusiastically received by his audience; but the concept overlooks the enormous practical and psychological difficulties that stand in its way. Anyone who walks through a large city can see that the life-support systems for millions of people, many of whom are socially and educationally disadvantaged, cannot derive from neighborhood gardens, community bakeries, and roof-top solar panels. Even the modest government initiative in urban homesteading, in which abandoned homes are turned over to families at no cost, foundered because only a select few people have the technical knowledge and business acumen needed to restore an abandoned house to livable condition.

We could, of course, try to reduce our dependence upon high technology by doing away with large cities altogether, and move everybody "back to the earth." Such an experiment in self-reliance has been tried in our time by the Cambodian Communists, and what ensued was one of the great calamities of human history.

Neither Schumacher nor his followers have seriously advocated such apocalyptic designs. They would be willing to let us keep our large cities if only we could dispense with the enormous technological and social systems upon which these cities depend.

A prime target of their critcism has been the nationwide grid of electrical power. Amory B. Lovins, a British physicist, became an instant celebrity when he dealt with this subject in an article entitled "Energy Strategy: The Road Not Taken?" which appeared in the October 1976 issue of *Foreign Affairs.* The article, which argued forcefully for the development of small, or "soft," energy technologies, was extensively quoted in the international press, entered into the Congressional Record, and discussed in *Business Week. Foreign Affairs* received more reprint requests for this article than for any other they had published.

The existing "hard" system for distributing electrical power has deficiencies that are obvious enough. Electricity is created in huge central plants by boiling water to run generators. Whether the heat that boils the water is furnished by oil, coal, gas, atomic power, or even by a solar device, a great deal of energy is wasted in the process; an additional amount is lost in transmission over long lines. By the time the electricity arrives in our home or factory and is put to use, about two-thirds of the original energy has been dissipated. Also, the existence of the infrastructure of the power industry itself—tens of thousands of workers occupying enormous office complexes—costs the system more energy, and costs the consumer more money.

The alternative proposed by Lovins is the creation of small, efficient energy-creating installations in the buildings where the energy is used or, at most, at the medium scale of urban neighborhoods and rural villages. Direct solar plants are the preferred system, although Lovins also mentions small mass-produced diesel generators, wind-driven generators, and several other technologies still in the development stage.

Whatever the advantages of this concept (and Lovins's technical calculations have been challenged by a host of experts), its widescale adoption would obviously entail the manufacture, transport, and installation of millions of new mechanisms. This cannot but be a monumental industrial undertaking requiring enormous outlays of capital and energy. Then these mechanisms would have to be maintained. We all resent the electric and phone companies but, when service is interrupted, a competent crew arrives on the scene to set things right. It is easy to say that the solar collectors or windmills in our homes will be serviced by our independent neighborhood mechanic, but this is a prospect that must chill the blood of anyone who has ever had to have a car repaired or tried to get a plumber in an emergency.

The critics of high technology are fond of pointing out that large systems are "vulnerable." When there is a blackout, it is apt to be widespread. When there is an earthquake or a hurricane—or a war or sabotage—the potential for extended damage is great. This is true; but the vulnerability of large systems to large failures is more than offset, I believe, not only by the benefits of civilization being worth the risks, but by the self-healing capability of large communities. Cities that have been half-leveled by earthquakes or floods are miraculously regenerated in astonishingly brief periods of time, except that *miraculously* is the wrong word to use. It is large organizations, not miracles, that make it possible to tend to the injured, care for the homeless, and rebuild the metropolis. In primitive communities sufferers from storm, earthquake, or avalanche are left to scrape at rubble with little more than their bare hands, unless, of course, help is sent from the big cities. Also when struck by plague, drought, or famine, the residents of poor, rural areas simply suffer and die—unless, again, rescue missions are mounted by the much-maligned centralized bureaucracies.

According to Schumacher, Lovins, et al., large bureaucracies are uneconomical. If, for example, we could do away with the administrations of the large utilities companies, we would realize great savings. This is the age-old dream of "eliminating the middleman." But in the real world it appears that the middleman does perform a useful function. How else can we explain the failure of the cooperative buying movement, which is based on the idea that people can band together to eliminate distribution costs? The shortcomings of large organizations are universally recognized, and *bureaucratic* has long been a synonym for *inefficient.*

But, like it or not, large organizations with apparently superfluous administrative layers seem to work better than small ones. Chain stores are still in business, while mom and pop stores continue to fail. Local power companies, especially, are a vanishing breed. Decisions made in the marketplace do not tell us everything, but they do tell us a lot more than the fantasies of futurists.

This is not to say that the situation cannot change. If a handy gadget becomes available that will heat a house economically using wind, water, sunlight, or moonlight, most of us will rush out to buy it. On the other hand, if the technological breakthroughs come in the power-plant field—perhaps nuclear fusion or large-scale conversion of sunlight to electricity—then I, for one, will be pleased to continue my contractual arrangements with the electric company.

Such an open-minded approach has no appeal to Lovins. Quoting Robert Frost on two roads diverging in a wood, he asserts that we must select one way or the other, because we cannot travel both. The analogy is absurd, since we are a pluralistic society of almost a quarter billion people, not a solitary poet, and it has been our habit to take every road in sight. Will it be wasteful to build power plants that may soon be obsolete? I think not. If a plant is used for an interim period while other technologies are developed, it will have served its purpose. When billions of dollars are spent each year on constantly obsolescing weapons that we hope we will never have to use, it does not seem extravagant to ask for some contingency planning for our vital energy systems.

Our resources are limited, of course, and we want to allocate them sensibly. At this time it is not clear whether the most promising energy technologies are "hard" or "soft" or, as is most likely, some combination of both. The "soft" technologies are not being ignored. Research and development funds are being granted to a multitude of experimental projects. At the same time, we are working on improvements to our conventional systems. What else could a responsible society do? We must assume that the technologies that prevail will be those that prove to be most cost-effective and least hazardous. Lovins claims that political pressures are responsible for government support for hard technologies, particularly nuclear. There may be some truth to this, but such pressures have a way of cancelling each other out. A new product attracts sophisticated investors, and before long there is a new lobbyist's office in Washington. The struggle for markets and profits creates a jungle in which the fittest technologies are likely to survive.

Technological efficiency, however, is not a standard by which the small-is-beautiful advocates are willing to abide. Lovins makes this clear when he states that even if nuclear power were clean, safe, and economic, "it would still be unattractive because of the political implications of the kind of energy economy it would lock us into." As for making electricity from huge solar collectors in the desert or from exploiting temperature differences in the oceans or from solar energy collected by satellites in outer space—these also will not do, according to Lovins, "for they are ingenious high-technology ways to supply energy in a form and at a scale inappropriate to end-use needs." Finally, he admits straight out that the most important questions of energy strategy "are not mainly technical or economic but rather social and ethical."

So the technological issue is found to be a diversion, not at all the heart of the matter. We can show a thousand ways in which large technologies are efficient and economical, and it will have availed us nothing. The *political* consequences of bigness, it would appear, are what we have to fear. A centralized energy system, Lovins tells us, is "less compatible with social diversity and personal freedom of choice" than the small, more pluralistic approach he favors.

But is it not paranoiac to speak of losing our political freedom by purchasing electrical power from federally supervised utility companies? Is there any evidence to show that large technological systems lead to political despotism? More than 200 years ago Americans agreed to give up their individual militias and entrust the national defense to a national army. Was that a mistake? Has it led to a police state? Of course not. Diversity and freedom in the United States are protected and encouraged by strong central institutions.

I grew up on a diet of Western movies, and I cannot forget the small communities in which nice families were constantly being harassed by bad guys. Help invariably arrived in the form of the U. S. marshal, who came from the "big" society to protect people in the "small" society from the predations of their neighbors. In my adult life I have found that the situation has not changed much. Exploitation thrives in small towns and in small businesses. The personal freedoms that we hold so dear are achieved through big government, big business, big labor unions, big political parties, and big volunteer organizations. We want as much independence as we can get, of course; but independence flourishes best within the protective framework of a great national democracy. This is the beautiful paradox of America.

When large organizations challenge our well-being, as indeed they do—monopolistic corporations, corrupt labor unions, and so forth—our protection comes not from petty insurrections, but from that biggest of all organizations, the federal government. And when big government itself is at fault, the remedy can only be shake-ups and more sensible procedures, not elimination of the bureaucracy that is a crucial element of our democracy. Surely we would not be freer citizens if we made our own electricity or pumped our own water, much less if we had to buy either from our friendly neighborhood utility. Small-is-beautiful makes no more sense politically than it does technologically.

Writing Exercise

Is there any rational way to reconcile Florman and Schumacher's perspectives? Are there grains of truth in both arguments? If you were moderating a panel discussion between these two formidable opponents, what cases in the history of science and technology would you have them discuss? Write an outline for a debate between these two positions that you will present at a professional colloquium.

WRITING AS A HUMANIZING ENDEAVOR

The writing you are called upon to do as a professional is often not a humanizing kind of writing. You may have to write for a deadline and for a manager who may not completely understand your work, and most probably you will have to fit your writing into a busy schedule.

Conversely, if the right time and opportunity present themselves, writing can be both relaxing and a way of bringing together many of the disparate events that you are exposed to. That is, the act and process of writing can be a way of making sense out of our normally chaotic life experiences.

In the following selection, Richard Selzer, a medical doctor, reveals how writing has transformed his life as a surgeon.

The Exact Location of the Soul

Richard Selzer

Someone asked me why a surgeon would write. Why, when the shelves are already too full? They sag under the deadweight of books. To add a single adverb is to risk exceeding the strength of the boards. A surgeon should abstain. A surgeon, whose fingers are more at home in the steamy gullies of the body than they are tapping the dry keys of a typewriter. A surgeon,

---

From Selzer, R. 1976. The exact location of the soul. In *Mortal lessons: Notes on the art of surgery,* 15–23. New York: Simon & Schuster.

who feels the slow slide of intestines against the back of his hand and is no more alarmed than were a family of snakes taking their comfort from such an indolent rubbing. A surgeon, who palms the human heart as though it were some captured bird.

Why should he write? Is it vanity that urges him? There is glory enough in the knife. Is it for money? One can make too much money. No. It is to search for some meaning in the ritual of surgery, which is at once murderous, painful, healing, and full of love. It is a devilish hard thing to transmit—to find, even. Perhaps if one were to cut out a heart, a lobe of the liver, a single convolution of the brain, and paste it to a page, it would speak with more eloquence than all the words of Balzac. Such a piece would need no literary style, no mass of erudition or history, but in its very shape and feel would tell all the frailty and strength, the despair and nobility of man. What? Publish a heart? A little piece of bone? Preposterous. Still I fear that is what it may require to reveal the truth that lies hidden in the body. Not all the undressings of Rabelais, Chekhov, or even William Carlos Williams have wrested it free, although God knows each one of those doctors made a heroic assault upon it.

I have come to believe that it is the flesh alone that counts. The rest is that with which we distract ourselves when we are not hungry or cold, in pain or ecstasy. In the recesses of the body I search for the philosophers' stone. I know it is there, hidden in the deepest, dampest cul-de-sac. It awaits discovery. To find it would be like the harnessing of fire. It would illuminate the world. Such a quest is not without pain. Who can gaze on so much misery and feel no hurt? Emerson has written that the poet is the only true doctor. I believe him, for the poet, lacking the impediment of speech with which the rest of us are afflicted, gazes, records, diagnoses, and prophesies.

I invited a young diabetic woman to the operating room to amputate her leg. She could not see the great shaggy black ulcer upon her foot and ankle that threatened to encroach upon the rest of her body, for she was blind as well. There upon her foot was a Mississippi Delta brimming with corruption, sending its raw tributaries down between her toes. Gone were all the little web spaces that when fresh and whole are such a delight to loving men. She could not see her wound, but she could feel it. There is no pain like that of the bloodless limb turned rotten and festering. There is neither unguent nor anodyne to kill such a pain yet leave intact the body.

For over a year I trimmed away the putrid flesh, cleansed, anointed, and dressed the foot, staving off, delaying. Three times each week, in her darkness, she sat upon my table, rocking back and forth, holding her extended leg by the thigh, gripping it as though it were a rocket that must be steadied lest it explode and scatter her toes about the room. And I would cut away a bit here, a bit there, of the swollen blue leather that was her tissue.

At last we gave up, she and I. We could no longer run ahead of the gangrene. We had not the legs for it. There must be an amputation in order that she might live—and I as well. It was to heal us both that I must take up knife and saw, and cut the leg off. And when I could feel it drop from her body to the table, see the blessed *space* appear between her and that leg, I too would be well.

Now it is the day of the operation. I stand by while the anesthetist administers the drugs, watch as the tense familiar body relaxes into narcosis. I turn then to uncover the leg. There, upon her kneecap, she has drawn, blindly, upside down for me to see, a face; just a circle with two ears, two eyes, a nose, and a smiling upturned mouth. Under it she has printed SMILE, DOCTOR. Minutes later I listen to the sound of the saw, until a little crack at the end tells me it is done.

So, I have learned that man is not ugly, but that he is Beauty itself. There is no other his equal. Are we not all dying, none faster or more slowly than any other? I have become receptive to the possibilities of love (for it is love, this thing that happens in the operating room), and each day I wait, trembling in the busy air. Perhaps today it will come. Perhaps today I will find it, take part in it, this love that blooms in the stoniest desert.

All through literature the doctor is portrayed as a figure of fun. Shaw was splenetic about him; Molière delighted in pricking his pompous medicine men, and well they deserved it.

The doctor is ripe for caricature. But I believe that the truly great writing about doctors has not yet been done. I think it must be done *by* a doctor, one who is through with the love affair with his technique, who recognizes that he has played Narcissus, raining kisses on a mirror, and who now, out of the impacted masses of his guilt, has expanded into self-doubt, and finally into the high state of wonderment. Perhaps he will be a nonbeliever who, after a lifetime of grand gestures and mighty deeds, comes upon the knowledge that he has done no more than meddle in the lives of his fellows, and that he has done at least as much harm as good. Yet he may continue to pretend, at least, that there is nothing to fear, that death will not come, so long as people depend on his authority. Later, after his patients have left, he may closet himself in his darkened office, sweating and afraid.

There is a story by Unamuno in which a priest, living in a small Spanish village, is adored by all the people for his piety, kindness, and the majesty with which he celebrates the Mass each Sunday. To them he is already a saint. It is a foregone conclusion, and they speak of him as Saint Immanuel. He helps them with their plowing and planting, tends them when they are sick, confesses them, comforts them in death, and every Sunday, in his rich, thrilling voice, transports them to paradise with his chanting. The fact is that Don Immanuel is not so much a saint as a martyr. Long ago his own faith left him. He is an atheist, a good man doomed to suffer the life of a hypocrite, pretending to a faith he does not have. As he raises the chalice of wine, his hands tremble, and a cold sweat pours from him. He cannot stop for he knows that the people need this of him, that their need is greater than his sacrifice. Still . . . still . . . could it be that Don Immanuel's whole life is a kind of prayer, a paean to God?

A writing doctor would treat men and women with equal reverence, for what is the "liberation" of either sex to him who knows the diagrams, the inner geographies of each? I love the solid heft of men as much as I adore the heated capaciousness of women—women in whose penetralia is found the repository of existence. I would have them glory in that. Women are physics and chemistry. They are matter. It is their bodies that tell of the frailty of men. Men have not their cellular, enzymatic wisdom. Man is albuminoid, proteinaceous, laked pearl; woman is yolky, ovoid, rich. Both are exuberant bloody growths. I would use the defects and deformities of each for my sacred purpose of writing, for I know that it is the marred and scarred and faulty that are subject to grace. I would seek the soul in the facts of animal economy and profligacy. Yes, it is the exact location of the soul that I am after. The smell of it is in my nostrils. I have caught glimpses of it in the body diseased. If only I could tell it. Is there no mathematical equation that can guide me? So much pain and pus equals so much truth? It is elusive as the whippoorwill that one hears calling incessantly from out the night window, but which, nesting as it does low in the brush, no one sees. No one but the poet, for he sees what no one else can. He was born with the eye for it.

Once I thought I had it: Ten o'clock one night, the end room off a long corridor in a college infirmary, my last patient of the day, degree of exhaustion suitable for the appearance of a vision, some manifestation. The patient is a young man recently returned from Guatemala, from the excavation of Mayan ruins. His left upper arm wears a gauze dressing which, when removed, reveals a clean punched-out hole the size of a dime. The tissues about the opening are swollen and tense. A thin brownish fluid lips the edge, and now and then a lazy drop of the overflow spills down the arm. An abscess, inadequately drained. I will enlarge the opening to allow better egress of the pus. Nurse, will you get me a scalpel and some . . . ?

What happens next is enough to lay Francis Drake avomit in his cabin. No explorer ever stared in wilder surmise than I into that crater from which there now emerges a narrow gray head whose sole distinguishing feature is a pair of black pincers. The head sits atop a longish flexible neck arching now this way, now that, testing the air. Alternately it folds back upon itself, then advances in new boldness. And all the while, with dreadful rhythmicity, the unspeakable pincers open and close. Abscess? Pus? Never. Here is the lair of a beast at whose malignant purpose I could but guess. A Mayan devil, I think, that would soon burst free to fly about the room, with horrid blanket-wings and iridescent scales, raking, pinching, injecting God knows what acid juice. And even now the irony does not escape me, the irony of my patient as excavator excavated.

With all the ritual deliberation of a high priest I advance a surgical clamp toward the hole. The surgeon's heart is become a bat hanging upside down from his rib cage. The rim achieved—now thrust—and the ratchets of the clamp close upon the empty air. The devil has retracted. Evil mocking laughter bangs back and forth in the brain. More stealth. Lying in wait. One must skulk. Minutes pass, perhaps an hour. . . . A faint disturbance in the lake, and once again the thing upraises, farther and farther, hovering. Acrouch, strung, the surgeon is one with his instrument; there is no longer any boundary beween its metal and his flesh. They are joined in a single perfect tool of extirpation. It is just for this that he was born. Now—thrust—and clamp—and *yes.* Got him!

Transmitted to the fingers comes the wild thrashing of the creature. Pinned and wriggling, he is mine. I hear the dry brittle scream of the dragon, and a hatred seizes me, but such a detestation as would make of Iago a drooling sucktit. It is the demented hatred of the victor for the vanquished, the warden for his prisoner. It is the hatred of fear. Within the jaws of my hemostat is the whole of the evil of the world, the dark concentrate itself, and I shall kill it. For mankind. And, in so doing, will open the way into a thousand years of perfect peace. Here is Surgeon as Savior indeed.

Tight grip now . . . steady, relentless pull. How it scrabbles to keep its tentacle-hold. With an abrupt moist plop the extraction is complete. There, writhing in the teeth of the clamp, is a dirty gray body, the size and shape of an English walnut. He is hung everywhere with tiny black hooklets. Quickly . . . into the specimen jar of saline . . . the lid screwed tight. Crazily he swims round and round, wiping his slimy head against the glass, then slowly sinks to the bottom, the mass of hooks in frantic agonal wave.

"You are going to be all right," I say to my patient. "We are *all* going to be all right from now on."

The next day I take the jar to the medical school. "That's the larva of the botfly," says a pathologist. "The fly usually bites a cow and deposits its eggs beneath the skin. There, the egg develops into the larval form which, when ready, burrows its way to the outside through the hide and falls to the ground. In time it matures into a full-grown botfly. This one happened to bite a man. It was about to come out on its own, and, of course, it would have died."

The words *imposter, sorehead, servant of Satan* spring to my lips. But now he has been joined by other scientists. They nod in agreement. I gaze from one gray eminence to another, and know the mallet-blow of glory pulverized. I tried to save the world, but it didn't work out.

No, it is not the surgeon who is God's darling. He is the victim of vanity. It is the poet who heals with his words, stanches the flow of blood, stills the rattling breath, applies poultice to the scalded flesh.

Did you ask me why a surgeon writes? I think it is because I wish to be a doctor.

Writing Exercise

1. Following Selzer's model, describe a writing experience you have had that has been particularly positive, either because you undertook it on your own initiative or because it helped you to make sense of a confusing set of circumstances. Your essay should be about four pages long, with the audience being a group of your professional peers.

Part Three

Composing a Technical Report or Scientific Paper

10

Doing Library Research

You will often need to gather material for your writing from books and journal articles in the library. Many students feel overwhelmed by such a task, mainly because they don't know how to start their search or how to keep accumulating material.

HOW TO BEGIN

The first step is to take advantage of the resources all around you: your librarians, your professors, and your friends.

Your Librarians

If you are unfamiliar with your college library, take a library tour, which will show you how the resources are arranged (what's where), how to use the card catalog, how to use the microfilm and microfiche readers, how to use the online system for finding important journals and books, and how to ask librarians for help. The librarians are actually there to help you; the more you ask, the more you'll find out.

If you are unfamiliar with how libraries are arranged by subject, get a list called the Library of Congress (LC) Subject Headings, which will show you how the LC system groups knowledge.

Your Friends and Professors

When you are conducting any search for material, seek help from your friends and your professors. Your friends can often give you valuable shortcuts and tips about using the library: which microfilm reader is the easiest one to use, from a mechanical standpoint; which librarian will be the most friendly; which books or articles he or she has used recently that were really helpful; how to get a book back if it's checked out, etc. That is, your friends can give you the low down on how the system can be made to work for you.

Finally, your professors intimately know the field in which you'll be doing research. They can tell you the two or three best new books in the area and the names of the professional journals that will most likely contain articles on your topic.

CONDUCTING THE BASIC SEARCH

After you get information and suggestions from these three sources, you go the library, notepad and pen in hand. By following this set of steps—a kind of algorithm for research—you can accumulate a list of sources.

Read Encyclopedias and Dictionaries

Familiarize yourself with your subject by reading entries about it in encyclopedias and dictionaries. The following reference books will often have lists of important books at the end of the article. Copy these down and try to use them.

Encyclopaedia britannica (15th ed., 32 vols., 1985). Check the section with shorter entries (the Micropaedia) and that for longer entries (the Macropaedia.).

McGraw-Hill encyclopedia of science and technology. A multivolume source for all fields of technology and science.

International encyclopedia of the social sciences (1968).

International encyclopedia of statistics (1978). Explains statistical concepts and procedures.

McGraw-Hill modern scientists and engineers. Articles about twentieth-century engineers and scientists.

Dictionary of scientific biography. Lengthy biographies of the great scientists.

Using the Card Catalog to Find Books

Go to the card catalog and check for books under the general subject heading of your topic. If you are unsure of the subject heading, ask a librarian for the LC Subject Headings list, the same ones according to which your card catalog is arranged. (If your library is still converting from the Dewey Decimal System, ask your librarian for help.)

You can double-check your list from the card catalog by looking in the subject volumes of *Books in print,* which, as its name says, lists all books currently in print. If you find a book there that is not in your card catalog, you can order it on interlibrary loan (ask your librarian for assistance).

Once you have a list of books and their call (access) numbers, pull three or four of them off the shelves. See if they are relevant for your topic. If they are, read the bibliography of each and record the names and publication information of the works that you feel will be most useful for your project. This will provide you with a growing list of primary references, both books and articles. You can then go to some of the other books and articles from your original search and repeat the process. In a short time, you will have an extensive record. By concentrating on the most recently published works, you will have an up-to-date reference list.

Using Periodical Indexes to Find Articles

Repeat the same process of accumulation for journal articles. Journal articles, though, are harder to collect than books. You have to find the relevant article in a subject-specific or a field-specific index, and finding the right index often requires some work. Here's what to do.

Applied Science and Technology Index

The first index to use is the *Applied science and technology index* (or *ASTI*). This covers over 200 journals in scientific and technical fields and is organized alphabetically by subject. Beginning with the most recent year, look up your subject and see if there are any relevant articles, Copy down the citations, and go to the library shelves to find the article.

Readers' Guide to Periodical Literature

If you need to find articles on science and technology that are written for the nontechnical audience, use the *Readers' guide to periodical literature.* Like *ASTI,* it is organized by subject.

Two Other Indexes

For specialized subjects in engineering, use the *Engineering index monthly and author index.* It includes abstracts. For the computer field, use *Computer literature index.*

When you find useful articles, be sure to *expand your search* by reviewing the Works Cited list at the end of the article. The entries should represent the best and most recent scholarship and research. If you get new titles from several current articles, you will probably have almost all the sources you need.

Using Indexes for Scholarly Journals

ASTI does not reference some of the purely academic or scholarly journals: those that contain articles, but no professional advertising. If your professor wishes you to use articles from scholarly journals, you must take one more step.

You need to find an index that covers scholarly journals and that lists their articles under subject categories. Such an index will be like *ASTI,* but it will be limited to one subject. To find such an index, use Sheehy's *Guide to Reference Books.* Find your subject in Sheehy, and then look under the subheading "Indexes" for the index specific to your field. Locate this index in the card catalog, find the book, and use it as you did *ASTI*.

Ulrich's International Periodical Directory

To supplement your search, use *Ulrich's international periodical directory* (pronounced "ol-ricks"), which is arranged by subject and lists the names of all the journals published in that area. You can copy down the names of journals, find them on the library shelves, and thumb through them, looking for a relevant article. Or you can consult that journal's annual index, usually bound with the journal, for your subject.

Following these steps should give you a substantial list.

Outside of books and journal articles, you might want to consult three additional sources of information: newspaper indexes, government documents, and computerized databases.

Using Newspaper Indexes

Use the *New York Times index* to find articles on science and technology for the general audience. The *New York Times* publishes a science and technology section in its Tuesday issue.

You will most probably locate interesting citations on a microfilm reader in your library. To understand exactly how to use the citation (how to read the page number, for instance), you may want to look in the front of the volume you are reading for the guide for understanding citations.

For articles concerning technology and business, consult the *Wall Street Journal index*.

Finding Government Documents

There are thousands of government documents published each year on a wide variety of subjects. Many of them report scientific and technical research being carried out by government agencies or by subcontractors.

Getting copies of these documents is not particularly easy, unless you live near a government document center. But the work involved in locating such documents often pays off because they offer state-of-the-art information. These publications are usually not listed in the card catalog or in subject indexes or abstracting services.

The first step is to consult the *Monthly catalog of United States government publications*. Indexed by author, subject, and title, it covers documents from 1970 to the present. When you find a citation that looks relevant to your research, copy down its access number, which uses the classification system of the Superintendent of Documents (i.e., it differs from the Dewey Decimal system and the Library of Congress system). With the access number, find out where government documents are shelved in your library; they will be in a special section and filed by access number. If your library doesn't have the document, ask the librarians how you can get it.

Using Computerized Databases

Within our lifetime, most research will be conducted on computer databases: electronic depositories of information. More databases become available each month. To remain current about what's available, you can check the *Directory of online databases* (New York: Cuadra/Elsevier), and look under the subject heading.

Your librarian can tell you which databases your library has available. One of the most popular is Dialog Information Services and its subdivision, Knowledge-Index, located in Palo Alto, California. Among its more than 250 databases are Agricola (agriculture), Biosis (biology), Heilbron (chemistry), Microcomputer Index, Engineering Literature Index, Food Science and Technology Abstracts, the National Technical Information Service Database (NTIS: government publications), Mathsci, Medline (for medical literature), International Pharmaceutical Abstracts, a sports medicine database (sport), and the Academic American Encyclopedia.

You can search these databases by subject, as well as by author. In addition to getting citations for your research, you can often have the abstracts printed out on your local computer network. It is also possible to order a copy of an article before logging off the database.

While database searching is a powerful research tool, it can be expensive ($20–250 per hour depending on the database used). However, if you prepare for your search by knowing the most efficient key words to use, you can often limit an average search to $20–30. Your college or academic department may help to defray the cost. As these databases become more numerous, connect charges will probably decrease.

OTHER USEFUL SOURCES

References

Books in print

The cumulative book index: List of books in the English language

Herner, B. 1980. *A brief guide to sources of scientific and technical information.* Washington, D.C.: Information Resources Press.

Malinowsky, H. R. 1980. *Science and engineering literature: A guide to current reference sources.* 3d ed. Littleton, Col.: Libraries Unlimited.

The national union catalog

Science books and films

Abstracts (annuals)

Biological abstracts

Chemical abstracts

Dissertation abstracts international

Food science and technology abstracts

Forestry abstracts

Geo abstracts

Information science abstracts

International abstracts of biological sciences

Petroleum abstracts

Physics abstracts

Pollution abstracts

Psychological abstracts

Science abstracts

Sociological abstracts

Statistical Abstract of the United States

World textile abstracts

Indexes (annuals)

Biological and agricultural index

Businesss periodicals index

Computer literature index

The engineering index

General science index

Science citation index

Social sciences index

Dictionaries and Handbooks

Chambers dictionary of science and technology (1978)

CRC handbook of chemistry and physics. 1986. Edited by R. C. Weast. Boca Raton, Fl.: CRC Press.

Emiliani, C. 1987. *Dictionary of the physical sciences: Physics, chemistry, geology, and cosmology.* London: Oxford University Press.

Handbook of physical constants. 1969. Rev. ed. by Sidney B. Clark. New York: Geological Society of America.

McGraw-Hill dictionary of scientific and technical terms (1984)

Encyclopedias

Encyclopedia of the biological sciences (1970)

Encyclopedia of chemical technology (1978–1984)

Encyclopedia of computer sciences and technology (1983)

The encyclopedic dictionary of physics
Encyclopedia of physics (1981)
Prentice-Hall encyclopedia of mathematics (1982)

Computer Retrieval

OCLC (Online Computer Library Center)

11

Documentation Format in Science and Technology

When you refer, in your writing, to the published work of other authors, or when you quote directly from them, you need to provide your reader with the reference for the material used. By citing your sources, you enable your readers to pursue their interests through these sources, and you establish a historical context for your own work.

Failure to cite references constitutes *plagiarism*—the act of aggrandizing your self by stealing the work of others. Most plagiarism is detected, and stiff penalties usually ensue.

DISCIPLINE-SPECIFIC FORMATS

Many professional disciplines—engineering, chemistry, medicine, biology, etc.—have developed their own special way for documenting sources and for listing the elements in any bibliographical citation. Professional organizations for these disciplines have developed specific style manuals. These include the *APA style guide* (the American Psychological Association, for social sciences), the *ACS style guide* (American Chemical Society), the *CBE style guide* (Council of Biology Editors, for natural sciences), the *GPO style manual* (Government Printing Office, for government publications), and the *MLA handbook* (Modern Language Association, for the humanities).

THE AUTHOR-DATE SYSTEM

When you write professionally, your discipline—or the journal or publishing house you are writing for—will dictate the specific format for your documentation. For your academic writing in undergraduate and graduate school, it is easiest just to follow one standard system for documentation.

The most developed style manual is the *Chicago Manual of Style* (13th ed. Chicago: University of Chicago Press, 1982), which recommends the *author-date* system for textual references in scientific or technical papers. Instead of using the cumbersome system of footnotes linked to a list at the end of the paper, you simply insert the last name of the author and the year of publication within parentheses immediately following the material that needs to be documented, as in this example:

> This type of analysis became popular after the discovery of X-rays (Rogers 1984).

Then, in a Reference List at the end of the paper, you list the book alphabetically by the author's last name. If there is more than one work by the same author, arrange the entries chronologically, with the most recent appearing first.

FORMS FOR THE AUTHOR-DATE REFERENCE IN THE TEXT

1. Page number, other specific division, or equation needed because of the specificity of the reference:

    (Rogers 1984, 75)

    (Rogers 1984, 76–87)

    (Rogers 1984, sec. 4)

    (Rogers 1984, eq. [35])

2. Both volume and page number needed:

    (Rogers 1984, 4:135)

3. Two authors:

    (Rogers and Abrams 1985)

4. Three authors:

    (Rogers, Abrams, and Wilkins 1986)

5. More than three authors:

    (Rogers et al. 1980) ["et al." is an abbreviation for the Latin "et alii," meaning "and others"; it does not need to be underlined]

    (Rogers and others 1980)

6. Two authors with the same last name:

    (Rogers and Rogers 1978)

7. Author published more than one work in the same year, and both are used in the report:

    (Rogers 1984a)

    The second source would then be

    (Rogers 1984b).

    In the Reference List use the date followed by the lowercase letter.
    Or a shortened version of the title can be used:

    (Rogers, *Aorta Reconstruction,* 1983)

8. Institute, group, or organization as author:

    (The Hemlock Society 1987)

9. Two or more citations for the same statement:

    (Rogers 1984; Milner 1985)

10. Ideally, the reference should be placed just before a mark of punctuation, often before a period ending the sentence. If such placement is awkward or unworkable, insert the reference at the most logical place in the sentence.

THE REFERENCE LIST

At the end of your book, paper, or report, list alphabetically the books, articles, and other sources that you have referred to in the parenthetical citations. Every citation includes the following elements, in the order listed.

For a Book

1. The author or the authoring body or the editor(s).
2. The year of publication.
3. The title and full subtitle, underlined. For titles, capitalize the first word, all proper nouns, and the first word in a subtitle following a colon. This method is called sentence punctuation. If appropriate, include here the title of the series or the number of volumes of a multivolume work.
4. The edition, if appropriate.
5. The city of publication, a colon, and the name of the publishing house or agency.

For an Article in a Periodical

1. The author, the authoring body, or the editor(s).
2. The month and year of publication; or day, month, and year.
3. The title and full subtitle of the article, not underlined. For titles, capitalize the first word, all proper nouns, and the first word in a subtitle following a colon.
4. The title of the journal, underlined or in italics, with each word capitalized.
5. The volume number, a colon, and the inclusive page numbers.

For all entries, use hanging indentation: indent five spaces all lines after the first. Initials may be used instead of an author's first name. Names of journals may be abbreviated according to the conventions of the individual discipline. If a publication has no author, use the title as the author.

Examples of Citations for Other Types of Documents

Book

Calvin, W. H. 1986. *The river that flows uphill: A journey from the big bang to the big brain.* San Francisco: Sierra Club Books.

Article with Two Authors

Gould, J. L., and C. G. Gould. May 1981. The instinct to learn. *Science* 2(4): 44–50.

The second author's name is not inverted and is preceded by a comma. The "2(4)" means "volume 2, number 4."

Edited Volume

Harris, J. E., and E. F. Wente, eds. 1980. *An X-ray atlas of the royal mummies.* Chicago: University of Chicago Press.

Later Edition

Le Gros Clark, W. E. 1978. *The fossil evidence for human evolution.* 3d ed. Edited by B. Campbell. Chicago: University of Chicago Press.

Symposium Paper Published in Conference Proceedings

Mulik, P. R. 1981. High-temperature removal of alkali and particulates in pressurized gasification systems. Paper presented at the American Society of Mechanical Engineers Conference, March 9–12, 1981, Houston, Tex. ASME Paper No. 81-GT-67.

Public Document

U.S. Environmental Protection Agency. Office of Research and Development. 1979. *Energy alternatives and the environment.* EPA-600/9-80-009. (The acronym and set of numbers are an access code for the document, should you want to order and read it.)

Book Issued by an Organization

Monger, T. G., and J. H. McMullan. August 1983. *Investigations of enhanced oil recovery through use of carbon dioxide.* DOE/BC/10344-8. NTIS/DE83015656. 83p. ("DOE" is the Department of Energy; "NTIS" is the National Technical Information Service.)

If a publication has no author, use the title as the author, or the issuing agency.

Dissertation

Mann, E. A. 1968. The paleodemography of Australopithecus. Ph.D. diss., University of California, Berkeley.

Personal Communication, Interview, or Lecture

Fisher, S. Professor of Physics, Quantum University. Letter to author, 16 February 1987.

Fisher, S. Interview with author, 16 February 1987.

Stiles, N. Physics 233 Lecture, 22 March 1986.

Newspaper Article

Browne, M. W. 1 December 1987. Laser devised to examine cell interior. *New York Times,* 17, 20.

Unpublished Lab Report, Log, Etc.

Markham, N. E. 4 April 1987. Robotics: Imitating the human thumb. Submitted to Dr. R. Roper, Computer Engineering 255, West Virginia University.

Questionnaire

Follow-up study on lay-offs at Seneca Glass, Inc. July 1985. Data derived from questionnaire administered to 65 laid-off glass blowers in Morgantown, W.V.

Information Derived from a Database Service

Benum, P., et al. 1977. Porous ceramics as a bone substitute in the medial condyle of the tibia: An experimental study in sheep. Long-term observations. *Acta Orthop Scand* 48(2):150–57. DIALOG, INDEX MEDICUS, item 1293575 77198575.

12

Revision, Readability, and Style

REVISION STRATEGIES

A professional document in science and technology must meet professional standards of readability. It must have

Significant content;

Correct format;

No errors in punctuation, grammar, and spelling; and

Clear organization and style.

After you write your first complete draft—a process that usually requires a series of partial drafts, designed to develop and clarify your thesis—you must turn your attention to revision. The revision process involves turning writer-based prose into reader-based prose (recall Chapter 3), turning the draft you write to develop your ideas into a document designed for someone else to read. Converting writer-based prose to reader-based prose is hard work mainly because authors are usually so close to what they write that they have trouble imagining how someone else could not immediately understand and agree with their thesis.

The revision process, therefore, demands that you distance yourself from what you have written so you can see your draft as if for the first time, as your reader will.

Proofreading for Common Errors

The first step in the distancing process is to proofread for errors in punctuation, grammar, and spelling. If you write at a computer and have a spelling checker program, you can use that software to check your spelling. If you don't have the spell function, get a ruler and read each line with the ruler under it in order to find spelling errors. Using the ruler isolates the line for your eye, making it easier to spot errors. Professional proofreaders, working without computer spelling checkers, often use the ruler method and read the page from bottom to top and each line from right to left. They do this in order to interrupt the mind's perceptual expectations of what is coming next, thus enabling themselves to pick out errors they might otherwise miss.

Second, check your draft for errors in grammar and punctuation. If you need a refresher on these matters, see Appendix C.

Third, read each sentence out loud to make sure that all the words are there and that all the sentences make sense.

Reader-based prose demands no errors in punctuation, grammar, or spelling. You do not want to cause confusion from mistakes that you could have easily corrected.

Checking the Document Flow

Some of us like serendipitous trips, where there is no daily itinerary and where unexpected side trips and unplanned diversions replace planned events. But most professional writing mirrors the ideal of a trip planned in advance, with the route clearly laid out, changes in direction and route number marked in advance to allow a smooth transition, and a description of the destination that indicates why the trip is worthwhile.

Writer-based prose often flows like a meandering stream, coming back on itself and even making no sure forward progress. But reader-based prose tends to flow straight to its mark, with eddies carefully charted and then subordinated to the overall plan.

Though these metaphors of the trip and the stream are a bit simplistic, they do capture the feeling of a document whose *organizational flow* is purposive and progressive. A successful professional document in science and technology uses *highlighting techniques* to reveal its organization easily to the reader.

Highlighting Techniques

Using effective highlighting techniques dramatically improves the readability of your document. Organizational cues and graphical elements are the most effective highlighting techniques. Two of the basic organizational cues for highlighting are labelling organizational sections and using in-text guideposts. (Graphic elements are discussed in Chapter 14.)

Labelling Organizational Sections

For much of technical and scientific writing, the organizational formats are preset. A journal article, for instance, will almost always include an abstract, an introduction, a materials and methods section, a discussion, a conclusion, and a reference section. When using such formats, you use the labels for each part as a heading, usually centering them on the page and underlining or bold-facing them for emphasis. Headings give your readers the sense that they share with you a common organizational expectation.

The table of contents, which is an outline of the piece, can be arranged with either Roman or Arabic numbers. The Roman method uses letters for subheadings; the Arabic uses the decimal method.

Roman

I.
  A.
    1.
    2.
  B.
    1.
    2.
II.

Arabic

1.
  1.1
  1.2
    1.2.1
    1.2.2
2.
  2.1

To organize your pages effectively, center first-level headings (e.g., 1.1 above) and bold face or capitalize them. Then, for contrast, put second-level headings (e.g., 1.2.1) flush left and underline them. This simple strategy allows your readers to keep track of what section they are in. And it gives each page some extra white space, which increases readability. Each section, if it is to be subdivided, should have at least two subsections.

Using In-Text Guideposts

Typographically emphasized headings will guide readers from one large section to the next, but readers also have to be guided from sentence to sentence and from paragraph to paragraph, as if they were following a string laid out through a maze. There are four in-text guideposts that improve readability:

Guidepost Sentences

Often called "topic sentences," these sentences state a thesis idea that you intend to develop in depth. While it is conventional to think that these occur at the beginning of a paragraph, they just as often occur in the middle or at the end, depending on how the paragraph is organized. If you think of a paragraph as being composed of a thesis idea (T), a restriction or narrowing of that idea (R), and an illustration (I), a paragraph can be arranged as TRI, TIR, RTI, RIT, ITR, and IRT.

Though paragraphs can be organized flexibly they usually contain a main guidepost sentence. If you are writing a paragraph that contains only an illustration (I) of a principle, be sure that the next paragraph contains a thesis sentence.

Using Transitional Words and Phrases

Guidepost sentences are the magnetic cores that hold paragraphs in a stable orbit. Transitional words and phrases allow you to develop or link ideas without your readers feeling that they have been cut loose from a familiar mooring. Some common transitional words and phrases are *therefore, consequently, as you can see, Illustration 6 shows that, as was proved in Section 3, on the other hand,* and so forth. These words and phrases flesh out for your readers the relationships between and among the elements of your document.

There is, however, a danger of using them too much. In a headlong rush to use transition words, students sometimes use them as a biologist might use pins to secure the parts of a skeleton. Don't pepper your prose with adverbs of transition; too many of them give the impression that you are using them to cover up weak links in your thinking process. Use them judiciously.

Using Periodic Summaries

Every four or five paragraphs, or every couple of pages, give your reader a chance to collect his or her breath. Give a brief summary (one or two sentences) of what you have just said and that hints at where you are about to go. These summaries are like museum guides. They remind you what you were just supposed to have seen and appreciated and they prepare you for what you are to see next.

As with transitional words, don't be heavy handed with these summaries. A light touch will ensure readability more than will a hammer blow.

Using Echoic Words and Phrases

Many style manuals advise writers to avoid needless repetition. In following such advice, however, you should not delete the repetition of key words and phrases, especially if they are unfamiliar to your reader. In fact, letting such words echo (repeat) throughout your document can give it a kind of resonance.

Learning theorists believe that learning is based on judicious repetition: you learn something (A), and then you learn more by adding to it (A+B). Gradually you build up a structure of knowledge, organized under hierarchical groupings (general classifications of things). By repeating unfamiliar words and concepts, you enable your readers to become familiar with them and thereby to become observer-experts by the time they have finished reading your document.

STRENGTHENING YOUR STYLE

After you have revised your document for organization, transitions, and highlighting, you can turn your energies to strengthening the style. Most style guides in science and technology recommend a strong, assertive, positive, action-based style. They prefer action to stasis. In grammatical terms, they prefer

1. Strong verbs ("possesses") instead of weak verbs ("has");
2. Strong verbs ("recommend") instead of dead, inactive nouns ("recommendation");

3. Active voice instead of passive voice;
4. Fewer negative constructions ("not").

As with all recommendations, you don't want to overdo these. Writing only with strong verbs, all in the active voice, and with no negative constructions can result in a style that feels like an aging boxer warming up on a punching bag: there's some power there, but there's no balance or subtlety.

The following subsections review some common problems in style and provide some examples of student prose for practice in revising.

Active and Passive Voice

In the active voice, the subject of the sentence performs the action described by the verb:

> The Ecology Action Committee decided to oppose the Army Corps of Engineers' plan to raise the level of Deep Creek Lake.

In the passive voice, a person, group, or force other than the subject is performing the action described by the verb:

> The decision was made to oppose the plan of raising the level of Deep Creek Lake.

Current stylistic conventions dictate that authors should use the active voice in most forms of writing. When writing technical reports, however, the passive voice is acceptable in the materials and methods section and in descriptions of trial runs, field tests, etc. Authors may use the passive voice in the abstract, if the journal or authorizing agency requests it. An example of the passive voice in an abstract follows:

> The mine-safety investigations were conducted in strict accordance with the standards of the U.S. Bureau of Mines. All phases of mine operation for all shifts were monitored, with the results being forwarded by the investigation team to the Bureau. The investigators determined that two shafts were insufficiently lighted and that the small-loading-car brake mechanisms had not been regularly serviced. The mine owners were cited for a total of 17 violations.

The passive voice is appropriate in these contexts because it gives greater emphasis to the experiment rather than to the experimenters. In science and technology, it does not usually matter who performed the experiment. How it was performed and the ability of other researchers to duplicate it (and the alleged results) are the key issues.

Although it is an asset in reporting experiments, the passive voice weakens your prose in most other contexts. In the example about Deep Creek Lake, the use of the passive voice obscures the identity of the group that made the decision and makes it appear that the decision arose somehow all by itself.

The most inexcusable use of the passive voice occurs in bureaucratic prose, in which committees and other groups describe their activities in the passive voice in order to conceal those responsible for making decisions. In the following example, culpable and negligent parties escape censure, or even identity:

It was decided to use DDT to spray the crops, despite the possibility that adverse climatic conditions might allow the chemical to spread beyond its intended range.

A particularly ungraceful extension of this principle occurs when an abstract noun that refers to a bureaucratic process functions as the subject for a passive verb:

The decision to cancel the program and redirect the use of the funds was made before the end of the grant period.

Those whose grant-in-aid was cancelled could validly ask, How was the decision made? Who made it? When was it made? Is there any redress?

Those who use the "bureaucratic passive" style may think that it sounds official and impressive. But anyone sensitive to language will see through this ploy. The impression of authority in prose comes from a sure tone and a fair handling of the details presented.

As a general rule, not more than 20 percent of all the verbs in your document should be in the passive voice.

Stacked Nouns and Adjective Strings

Emerging technologies often enrich the language by creating new words. Where would we be, for instance, without *state-of-the-art* accessories, *digital* watches, *field-specific* vocabularies, and *plate tectonics?* But amidst the rush to formalize a new technology by creating new vocabulary, certain unfortunate forms tend to arise. Though such forms usually have a short life span if left alone, professional writers should not help their interim survival.

Stacked Nouns

One feature of such vocabularies is stacked nouns—a long string of nouns whose meaning is unclear, unless you know the technology first-hand. Stacked nouns look like this:

the water restraint pressure mechanism

Although this group of words sounds formal and impressively technical, the reader has to work too hard to figure out its meaning. We have a perfectly good word—valve—for discussing the regulation of water pressure. The meaning can be made clear by using this word and then injecting a present participle ("regulating") to enliven the idea:

the valve regulating water pressure

Adjective Strings

Further problems arise when stacked nouns are mixed with strings of adjectives:

The goal of the project is to provide in-house computer-based systems analysis expertise.

What is wrong with this kind of stacking?

In traditional grammar and punctuation, two or more adjectives that modify a noun are marked off by commas, in order to indicate that the adjectives modify the noun and not each other. "The big, red, delicious apple" identifies an apple that has three characteristics: it is big and red and delicious. If we punctuated this series of words as "the big red delicious apple," the meaning would be confused. The phrase could mean "the Big Red, delicious apple" or "the big, Red Delicious apple" or "some specific, big apple that is also red and delicious," etc.

Likewise, referring to the original example, does "in-house" modify "computer-based," meaning that the computers are in-house? Or does "computer-based" go with "systems analysis" or with "expertise"? And, what does "in-house" modify? In short, we don't know what the string means. Its clear, denotative meaning has been lost.

The string could be improved by adding one comma:

> in-house, computer-based systems analysis expertise

Now we know that "in-house" does not principally modify "computer-based." The comma makes it refer to "expertise." The next step is deciding what the rest means. We could ask the author, but having to do this undermines the power of prose to stand on its own and to explain itself.

If we had to revise this phrase, we would have to consider three options, each with a different meaning:

1. In-house expertise dealing with computer-based systems analysis
2. In-house, computer-based expertise dealing with systems analysis
3. Expertise dealing with in-house, computer-based systems analysis.

By leaving out traditional punctuation and by trying to condense information more tightly than normal English grammar allows, the author of this string has obscured his or her meaning.

Maybe the sentence means:

> The goal of the project is to use our own people for computer-based systems analysis.

Or, if this rendering is too informal, maybe this:

> The goal of the project is to provide in-house expertise for computer-based systems analysis.

Improving these kind of strings usually involves adding the words that make explicit the implied connections.

Here is another example and a revision:

> Confusing: Hot gas desulfurization utilizing zinc ferrite sorbents has high potential for gasifier combined-cycle and molten carbonate fuel cell types of application.

> Clear: Hot-gas desulfurization using zinc ferrite sorbents has a high potential for application in gasifier, combined-cycle fuel cells and in molten carbonate fuel cells.

Strings of Prepositional Phrases

Don't string more than three prepositional phrases along in a row.

Incorrect: The probes were chosen on the basis of their relative resistance to physical deterioration at high temperatures in corrosive atmospheres.

Corrected: The probes were chosen because of their relative resistance to physical deterioration, both at high temperatures and in corrosive atmospheres.

Sentence Problems

Imprecise Words Beginning a Sentence

Beginning sentences with "there is," "this is," and "it is" usually weakens your style.

Incorrect: The carburetor is designed to feed gas and air into the engine at a fixed rate. This allows the car's engine to run.

Corrected: By feeding the gas and air at a fixed rate, the carburetor allows the car's engine to run.

Words Repeated Unnecessarily

Incorrect: One of the most common household tools is the screwdriver. The screwdriver has two basic forms: the regular, flat-head screwdriver and the Phillips screwdriver.

The repetitive use of "screwdriver" adds no new information and thus bores the reader. The sentences sound like jack-hammer prose.

Corrected: The screwdriver, a common household tool, comes in two types: flat-head and Phillips.

Too Many State-of-Being Verbs, Lack of Variety in Sentence Structure, and Needless Repetition

State-of-being verbs function as equal signs (e.g., *is, has, to be,* etc.), lack of variety in sentence structure usually means that most sentences have the simple subject-verb-object arrangement and lack introductory clauses or informative interrupters (such as appositives), and needless repetition occurs when terms or concepts are repeated, with no new information being added. These three faults often occur together, as in the following description of a golf-ball retriever:

The steel-ring cup functions as a golf-ball scoop. The cup is made of thick, steel wiring which is actually one long piece of steel wire spun into a circular spiral. The cup is formed by decreasing the diameter of the wire circle with each successive loop. The cup resembles a spring with a nonuniform diameter. The bottom of the cup has a small diameter loop and the rim of the cup has a large diameter loop.

By changing some of the state-of-being verbs to action verbs, by varying the style, and by cutting out redundancy, the passage can be shortened and tightened:

> Made of .5″ steel wire spun into a spiral, the cup resembles a spring with a nonuniform diameter.

Pompus Diction

As much as possible, try not to sound like an inept politician trying to obscure an issue by using circumlocutions. That is, come to the point, and say it simply.

> Incorrect: It is interesting to note that some in the population suffered mortal consequences from the lead compound in the flour.
>
> Corrected: Some people died after eating the flour that contained a lead compound.

Revising for Clarity

Long, diffuse sentences should be broken down into manageable units to increase their ease of readability. And internal confusions should be clarified.

> Incorrect: In keeping with these principles, this work can be summarized as follows: the redesign of the electronic circuitry for use of a single transmitting/receiving acoustic transducer necessary for nonintrusive monitoring, and design, fabrication, and testing or sensor probes using metal and ceramic materials chosen on the basis of their relative resistance to physical deterioration at high temperatures in corrosive atmospheres.

What's wrong here?

1. The sentence is 61 words long, too long for normal ease of readability. Sentences in technical writing should average 15–25 words in length.
2. The commas obscure the two parts, especially those between "monitoring" and "testing."
3. "For use of" is non-idiomatic.
4. "Using" creates confusion. What is actually doing this using?
5. An excessively long prepositional string at the end taxes the reader's normal retention span.

Here's a revision, with the sections clearly highlighted and with the other errors corrected.

> In keeping with these principles, this work has two goals:
>
> 1. To redesign the electronic circuitry used in the type of single transmitting/receiving, acoustic transducers that are required for nonintrusive monitoring; and
> 2. To design, fabricate, and test a variety of sensor probes, using metal and ceramic materials chosen because of their relative resistance to physical deterioration both at high temperatures and in corrosive atmospheres.

Sexist Language

Don't give the impression that you believe the world contains only one sex. When you are not talking about a specific, named person, use "he or she" or make the subject (*and* verb) plural.

Incorrect: If a young engineer wants more financial security, he should invest in an independent retirement plan.

Corrected: If a young engineer wants more financial security, he or she should invest in an independent retirement plan.

Corrected: If young engineers want more financial security, they should invest in an independent retirement plan.

Prose to Revise

Here are some examples of scientific and technical prose written by undergraduates. Revise them according to the basic rules of grammar and punctuation (see Appendix C for a review) and the guidelines given in this chapter. Above all, strive for simplicity and economy of expression.

1. Instructions About Stringing a Tennis Racket

The tension clamp is much more complicated than the hand clamp, this is what grips the string without damaging it by using brass bearings to give a firm yet soft grip. Before inserting the string in the tension clamp hand tighten the string somewhat so there is no slack. This holds the string intact as you lower the tension bar, and after clamping the string with the hand clamp will automatically release it as you raise the arm to its original position.

2. Drill Strings

The drill string is a network of connected pipes extended into an oil, gas or water well. Its purpose is to allow the drill bit to reach the bottom of the hole in order to achieve penetration. There are two types of drilling methods. The first is cable tool drilling. This is where drilling is accomplished by the pounding action of a steel bit that is alternately raised by a steel cable and allowed to fall delivering sharp, successive blows to the bottom of the hole. This principle is the same as that employed in drilling through concrete with an air hammer, or in driving a nail through a board. Since in this drilling case the cable acts primarily as the drill string, there is no changing of drill string. It involves just a simple matter of letting out more cable. The second type of drill method is called rotary tool drilling. In the rotary method, the hole is drilled by a rotating bit to which a downward force is applied. The bit is fastened to and rotated by a drill string. This string of pipe is composed of high quality drill pipe and drill collars with new sections or joints being added as deeper penetration occurs.

3. A Description of a Door

The construction of the door is that of a wooden frame covered by two thin sheets of wood paneling. About one-third the height from the bottom of the door is the locking mechanism. When the knob of the locking mechanism is turned the bolt of the mechanism slides in the door perpendicular to the door's thickness allowing the door to move. When the knob is released, the bolt slides out of the door. If the door is in the same plane as the wall when the knob is released, the door is locked, closing the entrance to the room or building. The door moves using hinges located opposite the locking mechanism. The hinges are made up of two metal plates with small cylindrical holes. There is spacing between the cylinders. The cylinders of the two plates mesh together to form one long cylinder. A metal rod is put in the cylinder to keep the two plates together yet it lets them pivot about the rod. One metal plate is connected to the door and the other is connected to the wall. The door has three hinges all connected to the same side of the door.

4. A Filter Press

The filter press consists of a cylindrical cell with a top cap and a bottom cap. The cell is about six inches tall and about four inches across the top. On the bottom of the base cap is a small filtrate tube which allows the liquid to leave the cell. Inside the base cap is a rubber gasket with a screen on top. A piece of filter paper is located on top of the screen and another rubber gasket rests on top of this. After the base cap is assembled, the cell is placed in the cap and filled with mud. The top cap is then assembled by placing a rubber gasket in it. This top cap is then placed on the cell. This construction is referred to as the mud cup.

5. Small Dams

A dam is a barrier constructed to prevent the flow of water or loose solid material. More commonly, a dam is a barrier built across a watercourse. Dams are generally classified according to size and more importantly according to the amount of adverse impact to life and property which would result due to the failure of the dam. If failure of the dam would cause loss of life and/or public or private property, very stringent design standards are imposed by regulatory authorities on the dam. For example, the USDA Soil Conservation Service description of a Class (C) dam is

"Dams located where failure may cause loss of life, serious damage to houses, industrial and commercial buildings, important public utilities, main highways, or railroads." In the SCS classification system, Class (C) of (A) through (C) requires the most stringent design standards.

The service of small dams are invaluable to the farmer who is concerned with agricultural water usage, the environmental engineer who is concerned with the control of sediment and the conservationist who is concerned with flood control and land usage. When designed and constructed in accordance with accepted engineering practice, these structures are a dependable and safe tool which can be used to enhance man's ability to control his environment.

6. Diagnostic-Related Groups in the Medical Profession: The Effects of DRG's on Quality of Care For Home Health Care Clients

With the coming of Diagnosis Related Groups to the health care industry so came the largest change in the medical professions history. Health care in America will never be the same.

According to congressional reports federal expenditures for health care had risen 22.2% from 1969 through 1982. In 1984 expenditures exceeded 1 billion dollars for health care. As a result, the federal government implemented diagnosis related group (DRG)-based payments for Medicare. DRGs brought about a major shift in the way hospitals were reimbursed for care of Medicare patients. DRGs spelled the movement away from retrospective payment to that of prospective payment. Under the DRG program, hospitals are paid fixed amounts for a given diagnosis and not for services rendered once the patient has been discharged.

To date, research has dominated in the areas of the effects of DRGs on hospitals, effects of DRGs on patients ability to pay and how they have effected the home health care industry as a whole.

The changes that have occurred since the adoption of the Diagnosis Related Groups are significant. Home health care was not as sophisticated technically at the beginning of this century as it is today.

In the past health insurance agencies favored care in costly institutions. There were no incentives for hospitals or physicians to look beyond the hospital walls for less costly alternatives. Physicians misunderstood home care. There is still an unsettled debate about home cares ability to reduce health care spending. There was evidence that adding more health coverage to existing benefit programs, both Medicare and

Medicaid, would do nothing more than add to the total costs of health care.

The home care industry has historically been composed of non-profit agencies, both private (visiting nurses associations) and public (public health department), which have provided supportive services in the homes of patients. Among the supportive services were skilled nursing care; home health aide services, speech, physical, and occupational therapy; and medical/social services. Most home health care agencies did not employ specially trained nurses to take care of complex cases that required advanced measures.

Because of the incredible burden that DRG's have placed on hospitals, they have been forced to change many of the behaviors that were once the norm. Today there is an increase in the total number of admissions, hospitals are more selective in the types of admissions, there is reduced or eliminated amenities (for example, patient admission kits and televisions), a decrease in the number of services provided and a decrease in the length-of-stay.

Of all these changes, those that will have the greatest impact on the home health system include changes in admissions policy and discharge policy.

If you are interested in more recommendations about style, read what the professionals say later in this chapter, and look at two books, both classics in their field:

Strunk, W. Jr. 1979. *The elements of style,* 3d ed. New York: Macmillan.

Williams, J. M. 1981. *Style: Ten lessons in clarity and grace.* Glenview, Ill.: Scott, Foresman.

THE FOG INDEX: A PRACTICAL READABILITY SCALE

Since the late 1940s, reading specialists have created several objective measurements of readability. The two most widely used scales are the Flesch Reading Ease Scale (Flesch 1948) and the Gunning Fog Index (Gunning 1968). The Fog Index, the simpler of the two, measures the grade level of a piece of writing (i.e., its assumed readability) by manipulating two factors: the average sentence length and the percentage of hard words.

Before we describe this index, however, a caveat is in order. Most current research in readability concludes that, whatever "readability" really is, it cannot be contained within the narrow limits of sentence length and long words. Short, choppy sentences are often harder to follow than a smooth, longer sentence with adequate transitions. Long words ("democracy") may often be easier to understand than short ones ("watt"). Familiarity with the subject matter and the ease of following organizational patterns (by prose signposts) may also be more important than more easily quantifiable properties. And passive constructions, negative constructions, long noun strings, and nominalizations interfere more with readability than do long words by themselves (e.g., see Klare 1977; Redish and Selzer 1985; and Selzer 1983).

But readability formulas can be computerized and are becoming popular in business and industry. The wordprocessing network you use in your professional job may include a readability software program that can be used to analyze what you write. And your manager may require you to revise your reports, letters, and memos until they fall within the guidelines of such programs. Therefore, it is important that you know how to apply a readability formula, and what its strengths and weaknesses are.

Though current readability formulas are simplistic, the three examples of prose that follow do show that the Fog Index has limited applicability—it can be used to give a general estimate of reading difficulty. The first example comes from a professional journal, whose articles are read by specialists; the second comes from *Scientific American,* whose articles are written for a cross-disciplinary group of scientists and engineers; the third is from Victor Weisskopf, writing about science for high school students. The Fog Indexes—18.45 (for the abstract); 15.608 (for the description); 11.75 and 9.97, respectively—show that the three documents are written at different grade levels, 18 being graduate level and beyond, 15 being junior year in college, 11 being eleventh grade, and 9 being ninth grade. For reference, the *New York Times* has an average Fog Index of 11–12, *Time* magazine about 11. Professional prose should never, or almost never, exceed the upper limit of 18.

To calculate the Fog Index, use the following formula:

Reading Level (Grade) = .4(ASL + %HW),

where "ASL" is Average Sentence Length and "%HW" is the Percentage of Hard Words. It is usually best to use one or more passages of at least 100 words each. To get the average sentence length, divide the number of words by the number of sentences. Be sure to treat independent clauses as separate sentences. For instance, "in school we read, we learned, and we wasted some time" is three sentences.

To get the percentage of hard words, divide the number of hard words by the total number of words. A hard word is a word of 3 syllables or more, but NOT a word that is (1) a combination of short words (bookkeeper); (2) a verb form that becomes polysyllabic by adding *-ed, -es, or -ing;* or (3) a proper name.

Such indexes are often included in editing software along with spelling checkers and other style guides. As computers become more sophisticated, their ability to identify stylistic problems will increase. Using these programs efficiently requires a nonsimplistic approach: no computer program or readability scale can give a definitive analysis of the effectiveness of a piece of writing. The only reliable standard is the reaction of sophisticated readers and users.

Finally, these computer programs and readability formulas should be used only during the revision process to spot excesses and eccentricities and to give you confidence that your prose is free from simple errors.

Fog Index Examples

Two-Phase Flow in Random Network Models of Porous Media

J. Koplik and T. J. Lasseter

Abstract

To explore how the microscopic geometry of a pore space affects the macroscopic characteristics of fluid flow in porous media, we have used approximate solutions of the Navier-Stokes equations to calculate the flow of the two fluids in random networks. The model pore space consists of an array of pores of variable radius connected to a random number of nearest neighbors by throats of variable length and radius. The various size and connectedness distributions may be arbitrarily assigned, as are the wetting characteristics of the two fluids in the pore space. The fluids are assumed to be incompressible, immiscible, Newtonian, and of equal viscosity.

Fog Index = 18.45

103 words, 4 sentences, 21 hard words

ASL = 25.75

%HW = 20.38

Physical Principles

In this section, we describe the flow calculation in physical terms, postponing a detailed discussion of the numerical procedure to the next section.

We begin by modeling the irregular, convoluted, and random pore space of a realistic porous material by a more tractable network, as shown in Figure 1. The figure and all subsequent discussions in this paper refer to 2D systems, but the same principles would apply in 3D (although we do not necessarily expect the results to be the same). We have circular pores of variable radius centered on a regular lattice, connected randomly by straight channels (throats) of variable radius.

Fog Index = 15.608

102 words, 5 sentences, 19 hard words

ASL = 20.4

%HW = 18.62

Brownian Motion

B. H. Lavenda

Topic Note [not figured in FI]

Observing the random course of a particle suspended in a fluid led to the first accurate measurement of the mass of the atom. Brownian motion now serves as a mathematical model for random processes.

It sometimes happens that a drop of water is trapped in a chunk of igneous rock as the rock cools from its melt. In the early 19th century the Scottish botanist Robert Brown discov-

---

From Koplik, J., and T. J. Lasseter. February 1985. Two-phase flow in random network models of porous media. *Society for Petroleum Engineers,* 89–90.

From Lavenda, B. H. February 1985. Brownian motion. *Scientific American,* 70.

ered such a drop in a piece of quartz. The water, Brown reasoned, must have been inaccessible for many millions of years to spores or pollen carried by the wind and rain. He focused a microscope on the drop of water. Suspended in the water were scores of tiny particles, ceaselessly oscillating with a completely irregular motion. The motion was familiar to Brown: he had previously happened to observe such oscillations during his studies of pollen grains in water. The new experiment, however, ruled out the explanation he had put forward earlier, namely that "vitality is retained by (the 'molecules' of a plant) long after the (plant's) death." Brown rightly concluded that the agitation of the particles trapped inside the quartz must be a physical phenomenon rather than a biological one, but he could not be more specific.

Fog Index = 11.75

171 words, 10 sentences, 21 hard words

ASL = 17.1

%HW = 12.28

(The following paragraph is repeated from Chapter 2.)

Gravity on Earth and in the Sky

V. F. Weisskopf

Gravity is a well-known phenomenon here on Earth. All things around us, large or small, are attracted by the earth—they fall downward when they are not held up by some support. The attraction of every piece of matter by the earth is the best-known example of a force in nature. Still, tremendous effort and centuries of thinking were needed before mankind recognized that the motion of the moon around the earth and of the planets around the sun is based upon the same force. It was long thought that the laws governing heavenly bodies were different from those that held on Earth.

Fog Index = 9.97

103 words, 6 sentences, 8 hard words

ASL = 17.17

%HW = 7.76

Anomalous Example

The Fog Index for the following example is 5.816 (ASL 7.17; %HW 7.37). As you can see, the passage is hardly at the fifth or sixth grade level. The Fog Index is unreliable for prose containing words (e.g., "set, bounded, infinite, point of accumulation, interval, closed, point, nest") that are symbols for abstract ideas, as in mathematics.

From Weisskopf, Victor F. 1979. Gravity on Earth and in the sky. In *Knowledge and wonder: The natural world as man knows it,* 2d ed. Cambridge: MIT Press.

Theorem:

Let S be a set. Let it be bounded. Let it be infinite. Then there is at least one point of accumulation of S.

Proof:

S lies in a closed interval. Call it I_1. Divide I_1 into two parts. Each point in S lies in one part or the other. Therefore, at least one of the parts contains an infinite number of points of S. Call this part I_2. Divide I_2 into two parts. At least one of the parts contains an infinite number of points of S. Call this part I_3. Keep it up. You get a nest of closed intervals I_n. There is one point common to all the intervals of the nest. This point is an accumulation point of S.

Reference List

Flesch, R. 1948. A new readability yardstick. *Journal of Applied Psychology* 32:221–33.
Gunning, R. 1968. *The technique of clear writing.* Rev. ed., 38. New York: McGraw-Hill.
Klare, G. R. Second Quarter 1977. Readable technical writing: Some observations. *Technical Communication* 24, 2.
Redish, J. C., and J. Selzer. Fourth Quarter 1985. The place of readability formulas in technical communication. *Technical Communication* 32(4):46–52.
Selzer, J. 1983. "What constitutes a "readable" technical style?" In *New essays in technical and scientific communication,* edited by Paul V. Anderson, R. John Brockman, and Carolyn R. Miller, 71–89. Framingdale, N.Y.: Baywood.

THE PROFESSIONALS DISCUSS REVISION, READABILITY, AND STYLE

Professional writers and rhetoricians have written extensively about revision, readability, and style. Here follows a short anthology that will enrich your understanding of the writing and revision process.

Donald M. Murray, a journalist and a professor at the University of New Hampshire, has won the Pulitzer Prize for editorial writing. In the following essay he discusses the importance of revision and stresses that writing is a craft that can be practiced and learned.

The Maker's Eye: Revising Your Own Manuscripts

Donald M. Murray

When the beginning writer completes his first draft, he usually reads it through to correct typographical errors and considers the job of writing done. When the professional writer

---

From Taylor, A. E., and W. R. Mann. 1972. *Advanced calculus.* 2d ed, 483. Lexington, Mass.: Xerox College Publishing.

From Murray, D. M. 1973. The maker's eye: Revising your own manuscripts. In *The writer,* 133–37. New York: International Creative Management. Reprinted by permission of International Creative Management, Inc., New York.

completes his first draft, he usually feels he is at the start of the writing process. Now that he has a draft he can begin writing.

That difference in attitude is the difference between amateur and professional, inexperience and experience, journeyman and craftsman. Peter F. Drucker, the prolific business writer, for example, calls his first draft "the zero draft"—after that he can start counting. Most productive writers share the feeling that the first draft—and most of those which follow—is an opportunity to discover what they have to say and how they can best say it.

Detachment and Caring

To produce a progression of drafts, each of which says more and says it better, the writer has to develop a special reading skill. In school we are taught to read what is on the page. We try to comprehend what the author has said, what he meant, and what are the implications of his words.

The writer of such drafts must be his own best enemy. He must accept the criticism of others and be suspicious of it; he must accept the praise of others and be even more suspicious of it. He cannot depend on others. He must detach himself from his own page so that he can apply both his caring and his craft to his own work.

Detachment is not easy. Science fiction writer Ray Bradbury supposedly puts each manuscript away for a year and then rereads it as a stranger. Not many writers can afford the time to do this. We must read when our judgment may be at its worst, when we are close to the euphoric moment of creation. The writer "should be critical of everything that seems to him most delightlful in his style," advises novelist Nancy Hale. "He should excise what he most admires, because he wouldn't thus admire it if he weren't . . . in a sense protecting it from criticism."

The writer must learn to protect himself from his own ego, when it takes the form of uncritical pride or uncritical self-destruction. As poet John Ciardi points out, ". . . the last act of the writing must be to become one's own reader. It is, I suppose, a schizophrenic process, to begin passionately and to end critically, to begin hot and to end cold; and, more important, to be passion-hot and critic-cold at the same time."

Just as dangerous as the protective writer is the despairing one, who thinks everything he does is terrible, dreadful, awful. If he is to publish, he must save what is effective on his page while he cuts away what doesn't work. The writer must hear and respect his own voice.

Remember how each craftsman you have seen—the carpenter eyeing the level of a shelf, the mechanic listening to the motor—takes the instinctive step back. This is what the writer has to do when he reads his own work. "The writer must survey his work critically, coolly, and as though he were a stranger to it," says children's book writer Eleanor Estes. "He must be willing to prune, expertly and hard-heartedly. At the end of each revision, a manuscript may look like a battered old hive, worked over, torn apart, pinned together, added to, deleted from, words changed and words changed back. Yet the book must maintain its original freshness and spontaneity."

It is far easier for most beginning writers to understand the need for rereading and rewriting than it is to understand how to go about it. The publishing writer doesn't necessarily break down the various stages of rewriting and editing, he just goes ahead and does it. One of our most prolific fiction writers, Anthony Burgess, says, "I might revise a page twenty times." Short story and children's writer Roald Dahl states, "By the time I'm nearing the end of a story, the first part will have been reread and altered and corrected at least 150 times. . . . Good writing is essentially rewriting. I am positive of this."

There is nothing virtuous in the rewriting process. It is simply an essential condition of life for most writers. There are writers who do very little rewriting, mostly because they have the capacity and experience to create and review a large number of invisible drafts in their minds before they get to the page. And many writers perform all of the tasks of revision simultaneously, page by page, rather than draft by draft. But it is still possible to break down the process of rereading one's own work into the sequence most published writers follow and which the beginning writer should follow as he studies his own page.

Seven Elements

Many writers at first just scan their manuscript, reading as quickly as possible for problems of subject and form. In this way, they stand back from the more technical details of language so they can spot any weaknesses in content or in organization. When the writer reads his manuscript, he is usually looking for seven elements.

The first is *subject*. Do you have anything to say? If you are lucky, you will find that indeed you do have something to say, perhaps a little more than you expected. If the subject is not clear, or if it is not yet limited or defined enough for you to handle, don't go on. What you have to say is always more important than how you say it.

The next point to check is *audience*. It is true that you should write primarily for yourself, in the sense that you should be true to yourself. But the aim of writing is communication, not just self-expression. You should, in reading your piece, ask yourself if there is an audience for what you have written, if anyone will need or enjoy what you have to say.

Form should then be considered after audience. Form, or genre, is the vehicle which will carry what you have to say to your audience, and it should grow out of your subject. If you have a character, your subject may grow into a short story, a magazine profile, a novel, a biography, or a play. It depends on what you have to say and to whom you wish to say it. When you reread your own manuscript, you must ask yourself if the form is suitable, if it works, and if it will carry your meaning to your reader.

Once you have the appropriate form, look at the *structure*, the order of what you have to say. Every good piece of writing is built on a solid framework of logic or argument or narrative or motivation; it is a line which runs through the entire piece of writing and holds it together. If you read your own manuscript and cannot spot this essential thread, stop writing until you have found something to hold your writing together.

The manuscript which has order must also have *development*. Each part of it must be built in a way that will prepare the reader for the next part. Description, documentation, action, dialogue, metaphor—these and many other devices flesh out the skeleton so that the reader will be able to understand what is written. How much development? That's like asking how much lipstick or how much garlic. It depends on the girl or on the casserole. This is the question that the writer will be answering as he reads his piece of writing through from beginning to end, and answering it will lead him to the sixth element.

The writer must be sure of his *dimensions*. This means that there should be something more than structure and development, that there should be a pleasing proportion between all of the parts. You cannot decide on a dimension without seeing all of the parts of writing together. You have to examine each section of the writing in its relationship to all of the other sections.

Finally, the writer has to listen for *tone*. Any piece of writing is held together by that invisible force, the writer's voice. Tone is his style, tone is all that is on the page and off the page, tone is grace, wit, anger—the spirit which drives a piece of writing forward. Look back to those manuscripts you most admire, and you will discover that there is a coherent tone, an authoritative voice holding the whole thing together.

Potentialities and Alternatives

When the writer feels that he has a draft which has subject, audience, form, structure, development, dimension, and tone, then he is ready to begin the careful process of line-by-line editing. Each line, each word has to be right. As Paul Gallico has said, ". . . every successful writer is primarily a good editor."

Now the writer reads his own copy with infinite care. He often reads aloud, calling on his ear's experience with language. Does this sound right—or this? He reads and listens and revises, back and forth from eye to page to ear to page. I find I must do this careful editing at short runs, fifteen or twenty minutes, or I become too kind with myself.

Slowly the writer moves from word to word, looking through the word to see the subject. Good writing is, in a sense, invisible. It should enable the reader to see the subject, not the writer. Every word should be true—true to what the writer has to say. And each word must be precise in its relation to the words which have gone before and the words which will follow.

This sounds tedious, but it isn't. Making something right is immensely satisfying, and the writer who once was lost in a swamp of potentialities now has the chance to work with the most technical skills of language. And even in the process of the most careful editing, there is the joy of language. Words have double meanings, even triple and quadruple meanings. Each word has its own tone, its opportunity for connotation and denotation and nuance. And when you connect words, there is always the chance of the suddden insight, the unexpected clarification.

The maker's eye moves back and forth from word to phrase to sentence to paragraph to sentence to phrase to word. He looks at his sentences for variety and balance in form and structure, and at the interior of the paragraph for coherence, unity and emphasis. He plays with figurative language, decides to repeat or not, to create a parallelism for emphasis. He works over his copy until he achieves a manuscript which appears effortless to the reader.

I learned something about this process when I first wore bifocals. I thought that when I was editing I was working line by line. But I discovered that I had to order reading (or, in my case, editing) glasses, even though the bottom section of my bifocals have a greater expanse of glass than ordinary glasses. While I am editing, my eyes are unconsciously flicking back and forth across the whole page, or back to another page, or forward to another page. The limited bifocal view through the lower half of my glasses is not enough. Each line must be seen in its relationship to every other line.

When does this process end? Most writers agree with the great Russian novelist Tolstoy, who said, "I scarcely ever reread my published writings, but if by chance I come across a page, it always strikes me: all this must be rewritten; that is how I should have written it."

The maker's eye is never satisfied, for he knows that each word in his copy is tentative. Writing, to the writer, is alive, something that is full of potential and alternative, something which can grow beyond its own dream. The writer reads to discover what he has said—and then to say it better.

A piece of writing is never finished. It is delivered to a deadline, torn out of the typewriter on demand, sent off with a sense of frustration and incompleteness. Just as the writer knows he must stop avoiding writing and write, he also knows he must send his copy off to be published, although it is not quite right yet—if only he had another couple of days, just another run at it, perhaps. . . .

In this selection, Robert Graves, the author of *I, Claudius,* and Alan Hodge examine the prose style of Sir James Jeans (a British scientist), write a revision (a "fair copy") of one of Sir James's paragraphs, and add a comment.

Examinations and Fair Copies: Sir James Jeans (From *The Stars in Their Courses,* 1931)

Robert Graves and Alan Hodge

Text

[Scientists can weigh stars by calculating the amount of gravitational pull that components of the same stellar system exert on one another.]

The results are interesting. Our[20] *sun proves*[8a] *to be of about average weight, or perhaps somewhat over.*[0] *Taken as a whole,*[23] *the stars shew only a small range*

From Graves, R., and A. Hodge. 1947. Examinations and fair copies: Sir James Jeans. In *The reader over your shoulder: A handbook for writers of English prose.* 2d ed., 222–27. New York: Vintage Books. Copyright © 1943 and renewed 1971 by Robert Graves and Alan Hodge. Reprinted by permission of Random House, Inc.

in weight[12/H1]*; if we compare*[14a] *the sun to a man of average weight, most*[10] *of the weights of the stars lie*[14b] *between those of a boy*[8b] *and a heavy man. Yet a few exceptional stars have quite exceptional weights.*[12/16] *A colony of four stars,*[H2] *27 Canis Majoris, is believed to have a total weight*[8c] *nearly 1,000 times that of the sun,*[H3] *although this is not certain. An ordinary binary system,*[H2] *Plaskett's star, is believed, this time with fair certainty,*[8d] *to have a total weight of more than 140 suns.*[H3] *But such great weights are very exceptional.*[12] *It is very rare*[3] *to find a star with*[14c] *ten times the weight of the sun,*[H4] *and no star yet found*[14d] *has as little as a tenth of the sun's weight.*[H4] *Thus on the whole the stars shew only a very moderate range in weight.*[H1/12]

Examination

3. What?

It is very rare to find a star with ten times the weight of the sun, . . .

"Very rare" is hardly admissible as a scientific expression except when applied to occurrences of particular birds, flowers, diseases and other organic phenomena, of which a census is impracticable. Here the reader deserves to be told whether one in fifty, or one in five hundred, or one in five thousand, of the starweights recorded at reputable observatories is ten times as much as that of the sun.

8. Inappropriate word or phrase

(a) Our sun proves to be of about average weight, or perhaps somewhat over.

If Sir James really has scientific proof of the stars' comparative weight, then there is no need for a "perhaps"; but if he doubts the accuracy of the calculation, then the results *suggest* rather than *prove* to him that the sun is a little heavier than the average star.

(b) . . . If we compare the sun to a man of average weight, most of the weights of the stars lie between those of a boy and a heavy man.

A "boy" is a term implying sexual immaturity, not weight. Some boys weigh more than an average man; some are infants and weigh only a few pounds. Here "boy" must be qualified by some phrase suggesting a narrower range of weight than from five pounds to one hundred and fifty.

(c) . . . 27 Canis Majoris, is believed to have a total weight . . .

An astronomer should avoid the word "believe" wherever the question is one of reckoning rather than faith.

(d) . . . Plaskett's star, is believed, this time with fair certainty . . .

An astronomer should also avoid such illogical colloquialisms. A thing is either certain, or it is uncertain. It would be better here to say that the second reckoning is less widely disputed than the first.

10. Misplaced word

. . . most of the weights of the stars . . .

It should be "the weights of most of the stars."

12. Duplication

Taken as a whole the stars shew only a small range in weight; . . .
Yet a few exceptional stars have quite exceptional weights. . . .
But such great weights are very exceptional. . . .
Thus on the whole the stars shew only a very moderate range in weight.

Sir James should have remembered that the Bellman in "The Hunting of the Snark" was considered eccentric because he said things merely three times over, if he wanted them to be true.

14. Material omission

(a) . . . if we compare the sun to a man of average weight . . .

The intended comparison is not of the sun to a man, but of the weight of the sun to that of a man of average size.

(b) . . . most of the weights of stars lie between . . .

A logical link has been omitted: "it will be found that, proportionately, . . ."

(c) It is very rare to find a star with ten times the weight of the sun, and no star yet found has as little as a tenth of the sun's weight.

If it was necessary for the sake of clarity to put "as little" into the second half of the sentence, then "as much" should have been put into the first half, before "ten times the weight of the sun," to show that "ten times or more" is meant, not merely "ten times."

(d) . . . and no star yet found has as little as a tenth of the sun's weight.

Many stars have been found but not yet examined: "and weighed" should therefore have been inserted after "found."

16. Undeveloped theme

Yet a few exceptional stars have quite exceptional weights.

Unless the stars are exceptional in other ways also, which should be briefly summarized, there is no point in classifying them "exceptional."

20. Irrelevancy

Our sun proves to be of about average weight. . . .

"Our" is used perhaps to remind the reader that besides the Solar system there are others that consist of sun and planets; and that "our" sun is not, as has for centuries been assumed, the Lord of the Visible Universe. But this "our" is a dangerous irrelevancy in a passage containing the phrase "a weight of more than 140 suns"—for these 140 then seem to be suns selected from 140 systems, rather than "our sun" multiplied by 140.

23. Logical weakness

Taken as a whole, the stars shew only a small range in weight . . .

He probably means "most stars show only . . ." But "taken as whole," they show a very wide range: for "the whole" includes the exceptionally heavy ones, which are two or three thousand times heavier than the exceptionally light ones.

H. Elegant variation

(1) . . . only a small range in weight . . .
. . . a very moderate range in weight . . .
(2) . . . a colony of four stars . . .
. . . an ordinary binary system . . .

(The second phrase means a colony of two stars.)

(3) . . . a total weight nearly 1,000 times that of the sun . . .
. . . a total weight of more than 140 suns . . .
(4) . . . ten times the weight of the sun . . .
. . . as little as a tenth of the sun's weight . . .

This variation, pardonable in a copy of ornate Latin verses, seems out of place in a scientific exposition.

O. Second thoughts

Our sun proves to be of about average weight, or perhaps somewhat over.

It confuses the reader to be told something, and then to have this qualified with a contradiction. What is meant is: "of perhaps a little more than average weight."

Fair Copy

[Scientists can weigh stars by calculating the amount of gravitational pull that components of the same stellar system exert on one another.]

The results are interesting. They suggest that the sun is a star of little more than average weight. On the whole, stars have only a small range in weight. If the weight of a medium-sized man were to stand for that of the sun, then the weights of most other stars would, in proportion, be found to lie between those of a large man and a ten-year-old boy. Of all the stars that have been weighed, only one in every [so many] has proved to be as much as ten times heavier than the sun; none has yet proved to be as little as ten times lighter. Among the few exceptionally heavy stars are the four components of the system 27 Canis Majoris: *they have been reckoned as being, together, nearly 1,000 times heavier than the sun. The accuracy of this figure is disputed by astronomers; but at least they generally agree that the two components of a system named "Plaskett's Star" are, together, over 140 times heavier than the sun.*

Comment

Sir James Jeans has set himself the task of translating the theories of physicists from mathematical formulae into ordinary English. A late-Victorian education seems to have taught him to shun a bald style; and his experience at Cambridge and Princeton to have taught him that the best way to make students in the lecture-room remember things is to repeat himself constantly. But such repetition is unnecessary in writing, if the points are clearly made in the first place: a reader can turn back the pages and refresh his memory whenever he wishes. And when repetition is disguised by constant variation of language, with the object of making the passage seem less tedious, the reader becomes confused. He is not quite sure whether the second and third repetitive phrases are the exact equivalent of the first or whether there is meant to be a subtle difference between them.

It is remarkable that nearly all scientists, at the point where they turn from mathematical or chemical language to English, seem to feel relieved of any further obligation to precise terminology. The sentence "It is very rare to find a star with ten times the weight of the sun, and no star yet found has as little as a tenth of the sun's weight" would, if the words were translated into a mathematical formula, be found lacking in three necessary elements, and to have an inexplicable variation of symbols in the elements given. A more scientific presentation of the sentence is: "Of all the stars that have been weighed, only one in every [so many] has proved to be as much as ten times heavier than the sun; and none has yet proved to be as little as ten times lighter." If the second part of the sentence is thus phrased in the same style as the first, they both become easier to understand and remember.

William Zinsser has written for the *New York Herald Tribune, Life,* and the *New York Times.* The author of 10 books, he has taught writing at Yale University and is executive editor of the Book-of-the-Month Club. Here Zinsser praises the virtue of simplicity in writing style and discusses the nature of scientific and technical writing.

Simplicity

William Zinsser

Clutter is the disease of American writing. We are a society strangling in unnecessary words, circular constructions, pompous frills and meaningless jargon.

Who can understand the viscous language of everyday American commerce and enterprise: the business letter, the interoffice memo, the corporation report, the notice from the bank explaining its latest "simplified" statement? What member of an insurance or medical plan can decipher the brochure that tells him what his costs and benefits are? What father or mother can put together a child's toy—on Christmas Eve or any other eve—from the instructions on the box? Our national tendency is to inflate and thereby sound important. The airline pilot who wakes us to announce that he is presently anticipating experiencing considerable weather wouldn't dream of saying that there's a storm ahead and it may get bumpy. The sentence is too simple—there must be something wrong with it.

But the secret of good writing is to strip every sentence to its cleanest components. Every word that serves no function, every long word that could be a short word, every adverb which carries the same meaning that is already in the verb, every passive construction that leaves the reader unsure of who is doing what—these are the thousand and one adulterants that weaken the strength of a sentence. And they usually occur, ironically, in proportion to education and rank.

During the late 1960s the president of a major university wrote a letter to mollify the alumni after a spell of campus unrest. "You are probably aware," he began, "that we have been experiencing very considerable potentially explosive expressions of dissatisfaction on issues only partially related." He meant that the students had been hassling them about different things. I was far more upset by the president's English than by the students' potentially explosive expressions of dissatisfaction. I would have preferred the presidential approach taken by Franklin D. Roosevelt when he tried to convert into English his own government's memos, such as this blackout order of 1942:

> *Such preparations shall be made as will completely obscure all Federal buildings and non-Federal buildings occupied by the Federal government during an air raid for any period of time from visibility by reason of internal or external illumination.*

"Tell them," Roosevelt said, "that in buildings where they have to keep the work going to put something across the windows."

Simplify, simplify. Thoreau said it, as we are so often reminded, and no American writer more consistently practiced what he preached. Open *Walden* to any page and you will find a man saying in a plain and orderly way what is on his mind:

> *I love to be alone. I never found the companion that was so companionable as solitude. We are for the most part more lonely when we go abroad among men than when we stay in our chambers. A man thinking or working is always alone, let him be where he will. Solitude is not measured by the miles of space that intervene between a man and his fellows. The really diligent student in one of the crowded hives of Cambridge College is as solitary as a dervish in the desert.*

How can the rest of us achieve such enviable freedom from clutter? The answer is to clear our heads of clutter. Clear thinking becomes clear writing: one can't exist without the other. It is impossible for a muddy thinker to write good English. He may get away with it for a paragraph or two, but soon the reader will be lost, and there is no sin so grave, for he will not easily be lured back.

Who is this elusive creature the reader? He is a person with an attention span of about twenty seconds. He is assailed on every side by forces competing for his time: by newspapers

---

Zinsser, W. 1980. Simplicity. In *On writing well: An informal guide to writing nonfiction,* 7–13. New York: Harper & Row.

FIGURE 12-1

5 --

is too dumb or too lazy to keep pace with the ~~writer's~~ train of thought. My sympathies are ~~entirely~~ with him. ~~He's not so dumb.~~ If the reader is lost, it is generally because the writer ~~of the article~~ has not been careful enough to keep him on the ~~proper~~ path.

This carelessness can take any number of ~~different~~ forms. Perhaps a sentence is so excessively ~~long and~~ cluttered that the reader, hacking his way through ~~all~~ the verbiage, simply doesn't know what it ~~the writer~~ means. Perhaps a sentence has been so shoddily constructed that the reader could read it in any of several ~~two or three different~~ ways. ~~He thinks he knows what the writer is trying to say, but he's not sure.~~ Perhaps the writer has switched pronouns in mid-sentence, or ~~perhaps he~~ has switched tenses, so the reader loses track of who is talking ~~to whom,~~ or ~~exactly~~ when the action took place. Perhaps Sentence B is not a logical sequel to Sentence A -- the writer, in whose head the connection is ~~perfectly~~ clear, has not bothered to provide ~~given enough thought to providing~~ the missing link. Perhaps the writer has used an important word incorrectly by not taking the trouble to look it up. ~~and make sure.~~ He may think that "sanguine" and "sanguinary" mean the same thing, but ~~I can assure you that~~ the difference is a bloody big one. ~~to the reader.~~ The reader ~~He~~ can only ~~try to~~ infer ~~what~~ (speaking of big differences) what the writer is trying to imply.

Faced with these ~~such a variety of~~ obstacles, the reader is at first a remarkably tenacious bird. He ~~tends to~~ blames himself— ~~He~~ he obviously missed something, ~~he thinks,~~ and he goes back over the mystifying sentence, or over the whole paragraph,

and magazines, by television and radio and stereo, by his wife and children and pets, by his house and his yard and all the gadgets that he has bought to keep them spruce, and by that most potent of competitors, sleep. The man snoozing in his chair with an unfinished magazine open on his lap is a man who was being given too much unnecessary trouble by the writer.

It won't do to say that the snoozing reader is too dumb or too lazy to keep pace with the train of thought. My sympathies are with him. If the reader is lost, it is generally because the writer has not been careful enough to keep him on the path.

This carelessness can take any number of forms. Perhaps a sentence is so excessively cluttered that the reader, hacking his way through the verbiage, simply doesn't know what it means. Perhaps a sentence has been so shoddily constructed that the reader could read it in any of several ways. Perhaps the writer has switched pronouns in mid-sentence, or has switched tenses, so the reader loses track of who is talking or when the action took place. Perhaps Sentence B is not a logical sequel to Sentence A—the writer, in whose head the connection

6 --

piecing it out like an ancient rune, making guesses and moving on. But he won't do this for long. ~~He will soon run out of patience.~~ The writer is making him work too hard, ~~-- harder than he should have to work --~~ and the reader will look for ~~a writer~~ one who is better at his craft.

The writer must therefore constantly ask himself: What am I trying to say? ~~in this sentence?~~ Surprisingly often, he doesn't know. ~~And~~ Then he must look at what he has ~~just~~ written and ask: Have I said it? Is it clear to someone ~~who is coming upon~~ encountering the subject for the first time? If it's not, ~~clear,~~ it is because some fuzz has worked its way into the machinery. The clear writer is a person ~~who is~~ clear-headed enough to see this stuff for what it is: fuzz.

I don't mean ~~to suggest~~ that some people are born clear-headed and are therefore natural writers, whereas ~~other people~~ others are naturally fuzzy and will ~~therefore~~ never write well. Thinking clearly is ~~an entirely~~ a conscious act that the writer must ~~keep forcing~~ force upon himself, just as if he were ~~starting out~~ embarking on any other ~~kind of~~ project that ~~calls for~~ requires logic: adding up a laundry list or doing an algebra problem. ~~or playing chess.~~ Good writing doesn't ~~just~~ come naturally, though most people obviously think ~~it's as easy as walking.~~ it does. The professional

is clear, has not bothered to provide the missing link. Perhaps the writer has used an important word incorrectly by not taking the trouble to look it up. He may think that "sanguine" and "sanguinary" mean the same thing, but the difference is a bloody big one. The reader can only infer (speaking of big differences) what the writer is trying to imply.

Faced with these obstacles, the reader is at first a remarkably tenacious bird. He blames himself—he obviously missed something, and he goes back over the mystifying sentence, or over the whole paragraph, piecing it out like an ancient rune, making guesses and moving on. But he won't do this for long. The writer is making him work too hard, and the reader will look for one who is better at his craft.

The writer must therefore constantly ask himself: What am I trying to say? Surprisingly often, he doesn't know. Then he must look at what he has written and ask: Have I said it? Is it clear to someone encountering the subject for the first time? If it's not, it is because some fuzz has worked its way into the machinery. The clear writer is a person clear-headed enough to see this stuff for what it is: fuzz. [See Figure 12–1.]

I don't mean that some people are born clear-headed and are therefore natural writers, whereas others are naturally fuzzy and will never write well. Thinking clearly is a conscious act that the writer must force upon himself, just as if he were embarking on any other project that requires logic: adding up a laundry list or doing an algebra problem. Good writing doesn't come naturally, though most people obviously think it does. The professional writer is forever being bearded by strangers who say that they'd like to "try a little writing sometime" when they retire from their real profession. Good writing takes self-discipline and, very often, self-knowledge.

Many writers, for instance, can't stand to throw anything away. Their sentences are littered with words that mean essentially the same thing and with phrases which make a point that is implicit in what they have already said. When students give me these littered sentences I beg them to select from the surfeit of words the few that most precisely fit what they want to say. Choose one, I plead, from among the three almost identical adjectives. Get rid of the unnecessary adverbs. Eliminate "in a funny sort of way" and other such qualifiers—they do no useful work.

The students look stricken—I am taking all their wonderful words away. I am only taking their superfluous words away, leaving what is organic and strong.

"But," one of my worst offenders confessed, "I never can get rid of anything—you should see my room." (I didn't take him up on the offer.) "I have two lamps where I only need one, but I can't decide which one I like better, so I keep them both." He went on to enumerate his duplicated or unnecessary objects, and over the weeks ahead I went on throwing away his duplicated and unnecessary words. By the end of the term—a term that he found acutely painful—his sentences were clean.

"I've had to change my whole approach to writing," he told me. "Now I have to *think* before I start every sentence and I have to *think* about every word." The very idea amazed him. Whether his room also looked better I never found out.

Writing is hard work. A clear sentence is no accident. Very few sentences come out right the first time, or the third. Keep thinking and rewriting until you say what you want to say.

13

Computers and Wordprocessing in Technical Writing

Computers with wordprocessing software have transformed both the teaching of writing and its practice. Most colleges and universities now make computers available to their students for a variety of purposes, such as writing computer programs, number crunching, wordprocessing, creation of graphics, and data filing and retrieval.

As a professional in the fields of science and technology, you will use computers constantly. You will probably be using either individual, stand-alone personal computers (PCs) or a terminal hooked up to a computer network run by mainframes (large computer banks). Using a stand-alone PC gives you the advantage of customizing your programs and files to your own needs. Being part of a computer network usually enables you to do larger number-crunching runs and to send your documents by electronic mail to other users.

For scientific and technical writing, computerized wordprocessing offers four advantages over writing by hand:

1. Watching what you write (type in) come up on your computer screen enables you to see your writing more objectively and more clearly. You see it immediately in the form it will assume in hard copy (on paper). Thus, you are able to spot trivial errors, scan sentences, and check for paragraph organization.

2. Since everything you write is electronically stored, the revision process is streamlined. No longer do you have to consider *typing over the whole paper* as the price of making important revisions. Inserting and deleting material (words, sentences, paragraphs, and whole pages) is as easy as hitting a few keystrokes, and you can then print out a fresh copy. Veteran writers testify that this ease of revision is the main advantage of the wordprocessor.

3. Sophisticated software gives you the ability to create graphics and visuals and integrate them with the paper you are writing. With such software you can create a complete scientific or technical document, without having to use hand-drawn graphics, and insert them in spaces you leave in the text.

4. Other sophisticated software offers capability in proofreading and stylistic revision. You can have a spelling checker electronically proofread your document for spelling errors. It will cue you for each incorrect spelling and allow you to make corrections. Other software will highlight errors in style and suggest corrections.

Many software programs can perform these tasks. For wordprocessing, there are simple programs and more difficult ones. For instance, PFS:Write is a wordprocessing program that you can learn to use in less than 30 minutes. It is menu-driven, which means that prompts appear on the screen telling you what it is possible to do. A simple program like this is sufficient for producing most college-level reports.

Somewhat more sophisticated programs are available from American publishers, such as HBJ Writer by Harcourt Brace Jovanovich. HBJ Writer is a wordprocessor complete with a spelling checker. In addition, it offers prewriting aids (freewriting, nutshelling, and planning) and reviewing and revising aids. This is a multipurpose piece of software, taking you from the first ideas to first draft, to revision, to finished draft.

TEXTRA, from the publishing house of W. W. Norton & Company, offers wordprocessing and a spelling checker. Easy to learn, it features an on-screen listing of function keys and other commands. It also offers an on-line electronic handbook, covering brainstorming techniques, style, grammar, punctuation, mechanics, and documentation (formats for bibliographical entries, following the MLA style). When you have a question, for instance, about commas, the software will display comma rules in a boxed insert on the screen.

At the more complicated end of these programs are WordStar and WordPerfect, which are command-driven programs. To use them, you refer to a set of commands, supplied on a handy command sheet that you keep at your terminal. Though more complex to use, these programs give you much greater ability to move and format text.

After you create a text and save it, you can check for spelling mistakes by using a spelling checker. These software programs contain a dictionary of usually 50,000 to 80,000 words, and you can add to this dictionary words that you commonly use that are specific to your discipline. (I wrote this textbook on a computer and used my spelling checker *[The Word]* after finishing each chapter and each set of revisions. In the process, I added about 250 words to the update dictionary.)

Other software can analyze the readability of a document, isolate grammatical problems, and identify wordy or inappropriate phrasing. Two of these, IBM's Epistle and AT&T's Writer's Workbench, require extensive memory.

Others can be used on conventional double-sided, disk-drive computers. Of these, two are especially interesting, because they are both inexpensive and effective. PC-Style analyzes your document and gives you the number of words, sentences, and words per sentence. It also calculates the percentage of long words, personal words, and action verbs and provides the number of syllables per word as well as the readability level. In addition, it ranks the readability, personal tone, and sense of action versus stasis (action verbs versus helping verbs and intransitive verbs). You can add words to the directories (of long words, personal words, and action verbs) in order to evaluate your document more precisely. PC-Style is available from ButtonWare and works on an MS-DOS computer.

RightWriter uses pattern recognition techniques to analyze your document. More complex than PC-Style, RightWriter displays your text on the screen and gives interlinear comments, analysis, and suggestions for revision on such problems as redundancy, ambiguity, wordiness, negatives, colloquial words, jargon, misspellings, misused words, and use of hard words.

It isolates adjectives and nominalizations that could be turned into verbs, thus increasing the strength of the writing. And it makes recommendations about improving these features by using over 3,000 rules and a 45,000-word dictionary. You then decide how to revise. Finally, it gives a readability index and a strength index and finds errors in grammar and punctuation.

You can understand how corporate managers might want all employees to apply such software to their writing before those drafts circulate outside the employees' own unit. As computers advance and as our understanding of effective style becomes more sophisticated, these software programs will gain in popularity, and other programs will take their place or will be developed.

RightWriter works with MS-DOS computers and with PFS:Write, WordStar, WordPerfect, and many other wordprocessing packages. RightWriter can be ordered from DecisionWare/RightSoft.

14

Graphics

What is to be sought in designs for the display of information is the clear portrayal of complexity. Not the complication of the simple; rather the task of the designer is to give visual access to the subtle and the difficult—that is, the revelation of the complex. (Tufte 1983, 191)

Graphic elements—drawings, photographs, schematics, tables of data, graphs, and charts—are the life blood of technical writing. Just think of all the graphic elements that have caught your attention in the selections in this textbook. The eye naturally gravitates toward visual representations. It enjoys them and learns from them.

If well-conceived and effectively rendered, an illustration can indeed be worth a thousand words. Scientists and engineers use graphics to visualize some object or concept in a more complete or emphatic way than writing alone would allow. Think, for instance, of Watson and Crick's schematic of the DNA molecule. (See Figure 5-3, page 70.) Their dramatic rendering uses the idea of a ship's staircase to reveal the complex interrelationships of bases in the molecule. This graphic had a dramatic effect on the field of biology and led to a revised method of picturing and imagining molecular structure. Consider how difficult it would be to gain the same effect through words alone. D'Arcy Thompson's drawing of the nautilus with a geometric and mathematical overlay presents his innovative idea with startling clarity and effectiveness.

Engineers and scientists use scribbles, incomplete drawings, attempted sketches, and so forth to enrich and stimulate their thinking process. Many develop engineering designs by going through many stages of visual renderings. Leonardo DaVinci's notebooks and Thomas Edison's logs and notebooks dramatically show this process of thinking via graphics: if you can see the design or the idea, you can also see its advantages and flaws and can revise from there. In more work-a-day terms, getting a graph or table just right often allows you to see your data clearly, whereas a muddled or confusing graphic often implies a failure of conceptualization.

It is also important to realize that many engineers and scientists report that they read professional publications not in the linear order in which they appear but in the order that is most important to them:

Abstract

Graphics, especially tables, graphs, and summaries of data

Conclusion

Introduction

Body

That is, they get an overview first (in the abstract), then look at the graphics to check out the data, then read the conclusion, and then, if time allows, read the rest. Scientists and engineers often do not read a technical document with the same leisure and linear attention with which they would read a novel. Graphics, then, are not just "visual aids." They are an organic part of any technical or scientific document.

ACTUAL PRACTICE

As an undergraduate, you will be inserting graphics into your document from three sources: (1) photocopied material; (2) engineering-type line drawings of your own, drawn according to principles you have learned in drafting classes; and (3) graphic elements that you draw yourself or that you generate using computer graphics software. (Since this computer capability is expensive and many universities and colleges cannot afford it for their students, you will not necessarily have this option. Therefore, computer graphics will not be discussed further here. You should understand, however, that some of the best contemporary graphics in science and technology are being produced by such computer methods.)

As a practicing scientist or engineer, you will also have the added advantage of a graphics or illustration department in your company or firm. Such a group will produce graphics for you, based on the ideas and sketches you give them.

For purposes of undergraduate technical writing, you will want to choose graphics of high quality for your text, and you will want to know how to use them most effectively. Some principles follow.

QUALITY IN GRAPHICS

Graphics should be integrated into the text, placed effectively on the page, labelled clearly, and based on simple, clear designs.

Integrate It Into the Text

Refer to the graphic in your text as "Figure 1" or "Table 1." Make this reference part of the grammar of the sentence:

> Cost estimates for repairing the bridge were beyond our budget (Figure 1).

Place It Effectively on the Page

Place the graphic in the text so it is as close as possible to the line or section that it illustrates. Leave space on the typed page so you can tape in the graphic later.

Label It Clearly

Label each graphic with a title that identifies the object and describes the concept. If the illustration is a graph, label the axes. Add descriptive language if appropriate.

Make Simple, Clear Designs

The graphic element should show what it is designed to show—neither more nor less. Fancy graphics usually conceal a confused or overly simple idea.

For more information on these basic principles, see Arthur T. Turnbull and Russell N. Baird's *The Graphics of Communication,* especially pages 247 to 296.

If you photocopy a graphic and insert it in your document, always cite the source. Standard academic regulations against plagiarism apply to graphics. It is customary to type the source of your graphic in its bottom right-hand corner.

Be sure to avoid these common faults in using graphic elements:

1. Don't use pictures or drawings of common objects as seen at normal size. For instance, don't include a conventional picture of a bicycle pump just because you are describing it. You can assume that your reader has seen such a pump. Instead, have a cut-away drawing of the interior, which most people haven't seen. Then describe the features that can be seen in the cut-away.

2. Don't add useless decoration or extra elements just because you think they look attractive or professional. Understated simplicity is always best.

Excellent advice on using graphic elements based on statistical information can be found in Edward Tufte's *The visual display of quantitative information.* Tufte gives the following summary about the principles of graphic excellence:

> Graphic excellence is the well-designed presentation of interesting data—a matter of *substance,* of *statistics,* or *design.*
>
> Graphic excellence consists of complex ideas communicated with clarity, precision, and efficiency.
>
> Graphic excellence is that which gives to the viewer the greatest number of ideas in the shortest time with the least ink in the smallest space.
>
> Graphic excellence is nearly always multivariate.
>
> And graphic excellence requires the telling the truth about the data. (1983, 51)

EXAMPLES OF GRAPHIC ELEMENTS

Graphic elements can be pictorial or quantitative. Pictorial graphics include drawings, photographs, and schematics. Quantitative graphics include tables, graphs, and charts.

Pictorial Graphics

The purpose of using drawings or photographs in a technical document is to enable readers to visualize something that they cannot easily see. Examples are unlimited. Think of, for instance,

An electron microscope picture of a macrophage (a white cell involved in fighting invasive disease);

A drawing of the docking of two spacecraft;

A time-stop photograph of a drop of milk entering a bowl of milk, or a bullet passing through a series of balloons: the kind of stop-photography pioneered by Dr. Edgerton of M.I.T.;

A cut-away showing the inside of the internal combustion engine; or

An X-ray or magnetic resonance image of some part of the human body.

The purpose of such photographs or drawings is to render visible something that is usually not seen, or not able to be seen, with the naked eye. Such pictorial graphic elements reveal a hitherto unseen world and inspire our curiosity. They are the tokens of wonder that display the successes of science and technology.

There are many examples of pictorial graphics in this book. Among them are

The hanger-rod–box-beam connection of the walkway of the Hyatt-Regency, showing the original and revised plans. The drawings dramatically highlight the difference of the two designs.

Robert Hooke's drawing of a flea. It inspires amazement from its viewers.

Benjamin Franklin's drawing of his chimney and fireplace. With its excellent labeling, the drawing shows the air flow.

The drawings of the tarantula and the digger wasp in Petrunkevitch's article. Seeing them at the size and proportion they would see each other dramatizes the world they inhabit and prepares the reader for the grizzly content of the article.

The cross-section of underground strata, showing various types of well construction. The cross-section enables you to visualize precisely how wells differ.

The photos of chimpanzees by Jane Goodall. All show the chimps in unusual, unexpected, or scientifically significant activities.

Crompton's representations of a spider web being spun. We can appreciate not only the method of creation, but also the irregularity of its regular beauty.

All of these pictorial graphics enhance the author's technical or scientific presentation. They all display something unusual or something that isn't usually seen or conceptualized. By doing this, they expand our imaginative horizons.

Schematics

Schematics are commonly used in technical writing, especially that of engineering. Figure 14-1 is a schematic representation of a cross-flow filter. The schematic illustrates the filtering mechanism and mounting arrangement. Figure 14–2 represents a wiring diagram for my home's modest entertainment center, which includes a television, a VCR, and a game computer.

It is common to think of a schematic in its simple sense, as a wiring diagram or as a chart of some process. But a schematic is a more sophisticated idea than this simple application would suggest. "Schematic" comes from the Greek *(schema),* meaning form, shape, or figure. A schema is a mental concept, or a cognitive perception.

FIGURE 14-1
Schematic Representation of Cross-flow Filter

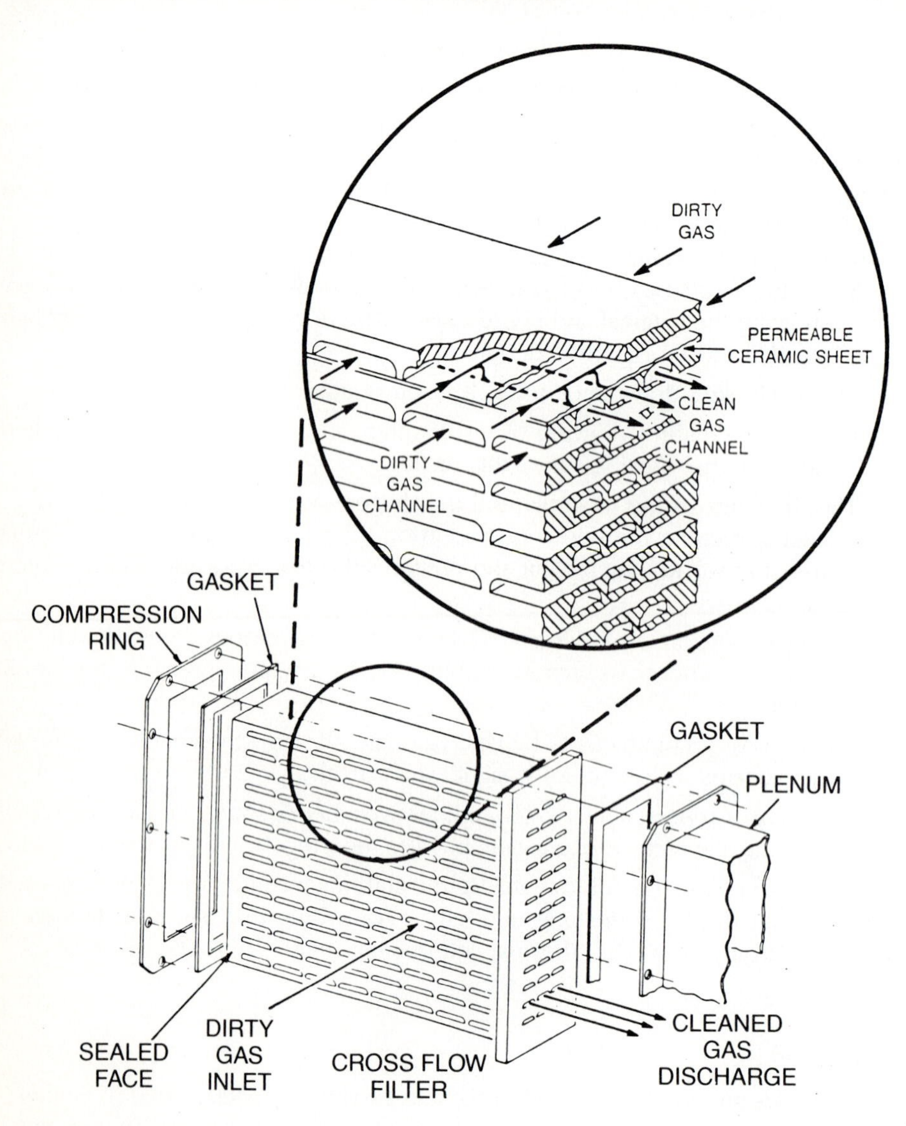

In an extended sense, a schematic is a diagrammatic representation of a typical situation, a representation that includes an underlying organizational plan or structure or a conceptual framework. Such a representation, usually abridged and generalized, corresponds to an established or formalized conception and shows the relationship between the individual parts of an object, mechanism, or concept.

That is, a schematic is a drawing that is *organized by an idea.* The schematic of the cross-flow filter shows the idea or concept of the flow of the gas and the collection of the particles.

FIGURE 14-2
Circuit Diagram

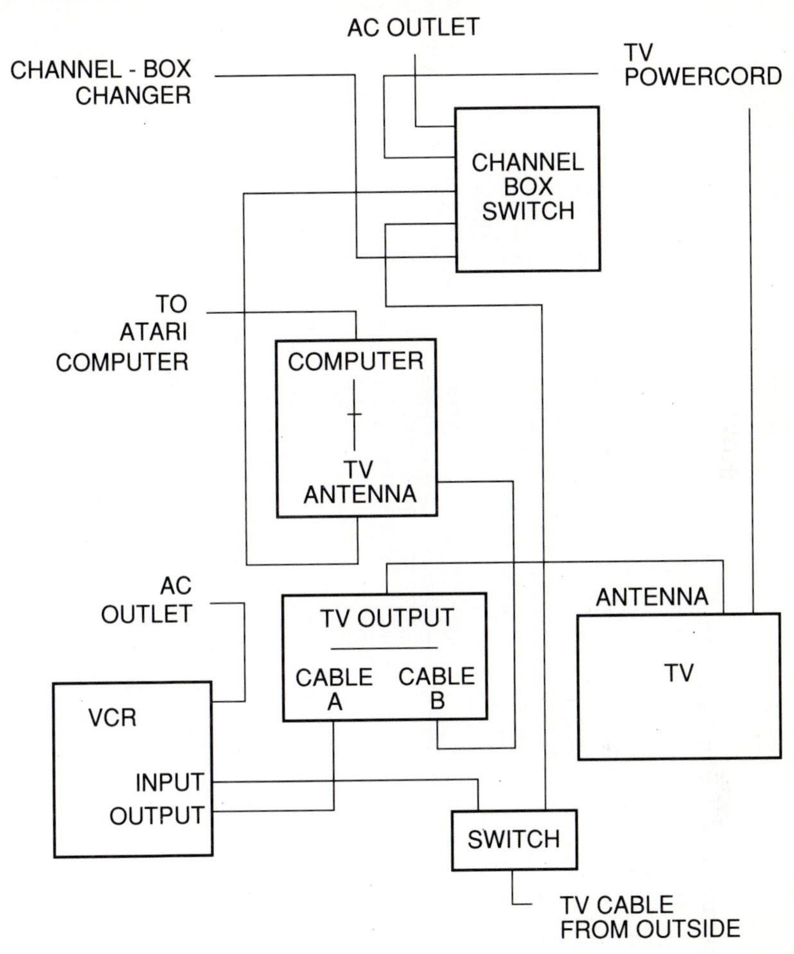

This schematic hooks together a VCR, a TV, an Atari game computer, and a 26-channel selector, so we can watch TV and tape another TV program simultaneously.

The intellectual power of a schematic lies in its ability to combine visual and conceptual material.

For instance, a drawing of a new bridge design is not, by itself, a schematic. But if you add engineering ideas to the drawing and if the point of the drawing is to show the idea, then it can be called a schematic. Such a drawing might include the lines of force acting on the bridge or a dramatic illustration of the amount of flexibility built into the bridge to allow it to respond to high water, strong winds, or other adverse conditions.

Figure 14-3 diagrams surface instrumentation used to monitor hydraulic fracturing experiments, which are performed to recover gas from very low permeability

FIGURE 14-3
Schematic of Surface Instrumentation for Stimulation Test

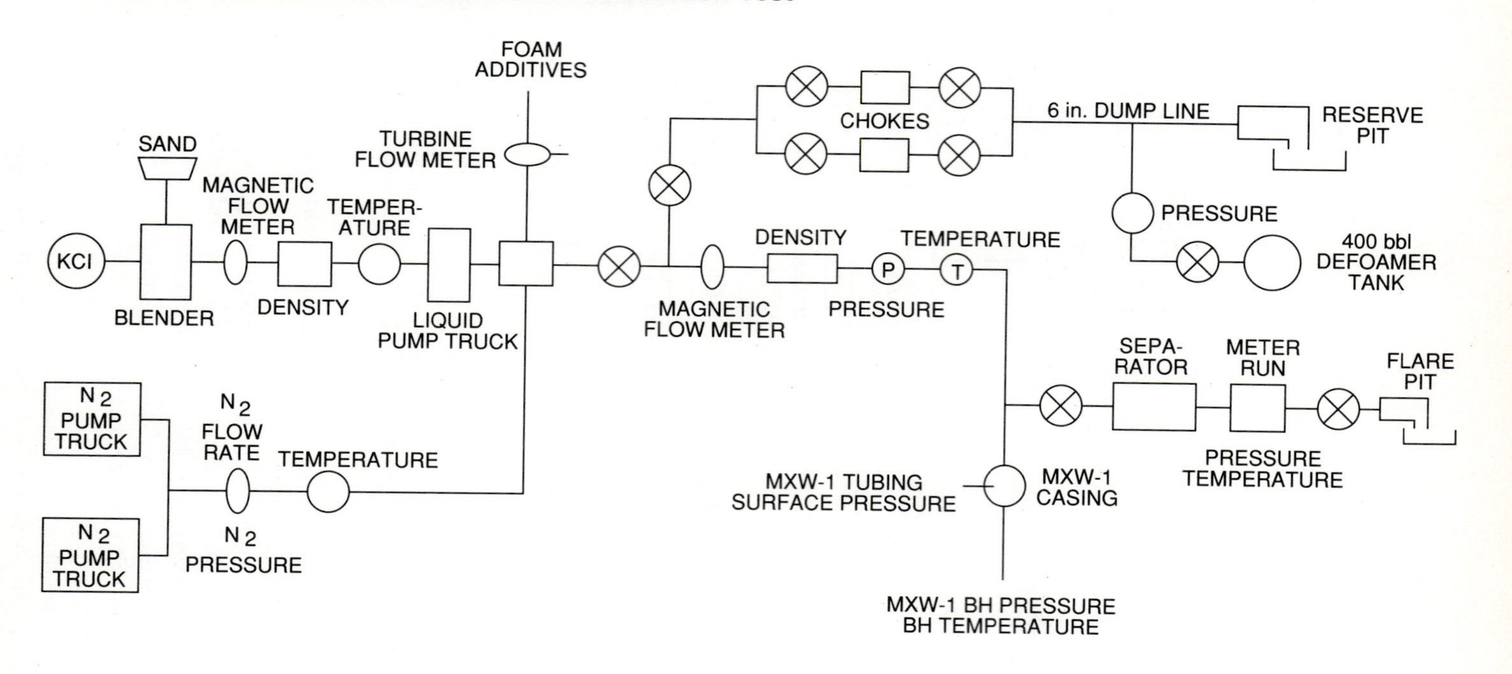

FIGURE 14-4
KILnGas Gasifier

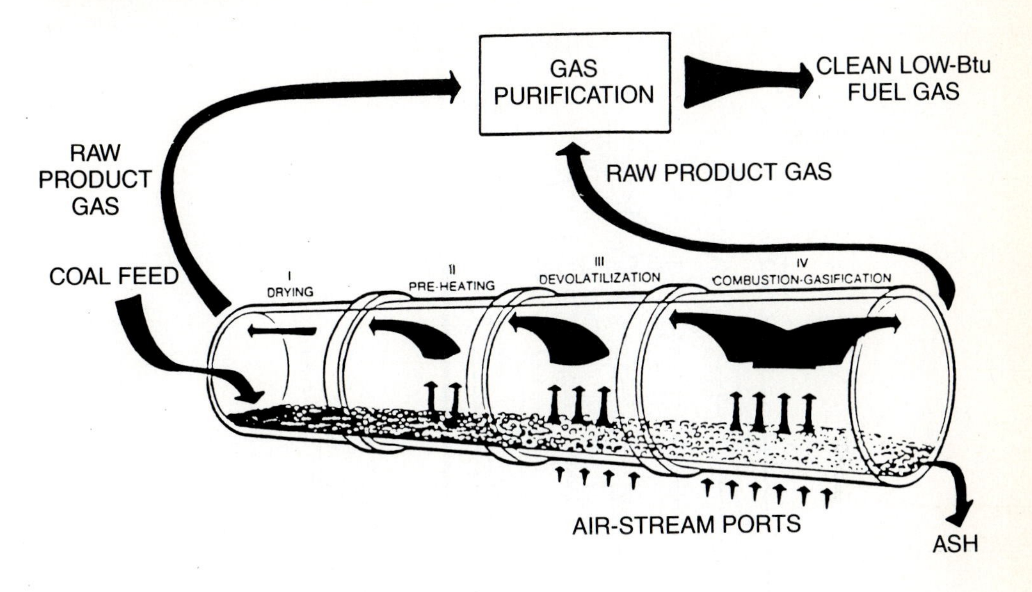

gas sands. This schematic shows the element-by-element connections in this instrumentation system.

Figure 14–4, a KILnGas Gasifier, is a conceptualized drawing of a gasification process in which gas is recovered from coal. This schematic is powerful because it shows, in simplified form, the process of obtaining gas from coal, as well as the steps of the process.

One of the most dramatic kinds of schematics is the cut-away, which displays the interior of a mechanism. This form of illustration shows the arrangement and organization of the inside of an important technical component. Figure 14–5 presents a cut-away of a single-pressure vessel (designed to clean gas obtained from a gasifier) that contains the kind of ceramic cross-filters seen before.

Quantitative Graphics: Graphs and Tables of Data

Engineers and scientists visualize numbers and data through graphs and tables of data. Tables show data collected in an experiment. Table 14-1 shows the gasification performance of an in-bed desulfurization test (a test designed to remove sulphur from coal by the use of sorbents [chemical sponges]. The information is organized in columns so the reader can easily see the improvements made.

In contrast to a table, which organizes information into stable groupings, a graph shows a *relationship* between two qualities as they change in respect to each other. In the graph in Figure 14–6, four novel desulfurization sorbents are shown plotted for temperature (the *x*-axis) and equilibrium of H_2S (the *y*-axis). The graph shows

FIGURE 14-5
Arrangement of Ceramic Cross-flow Filters in a Pressure Vessel

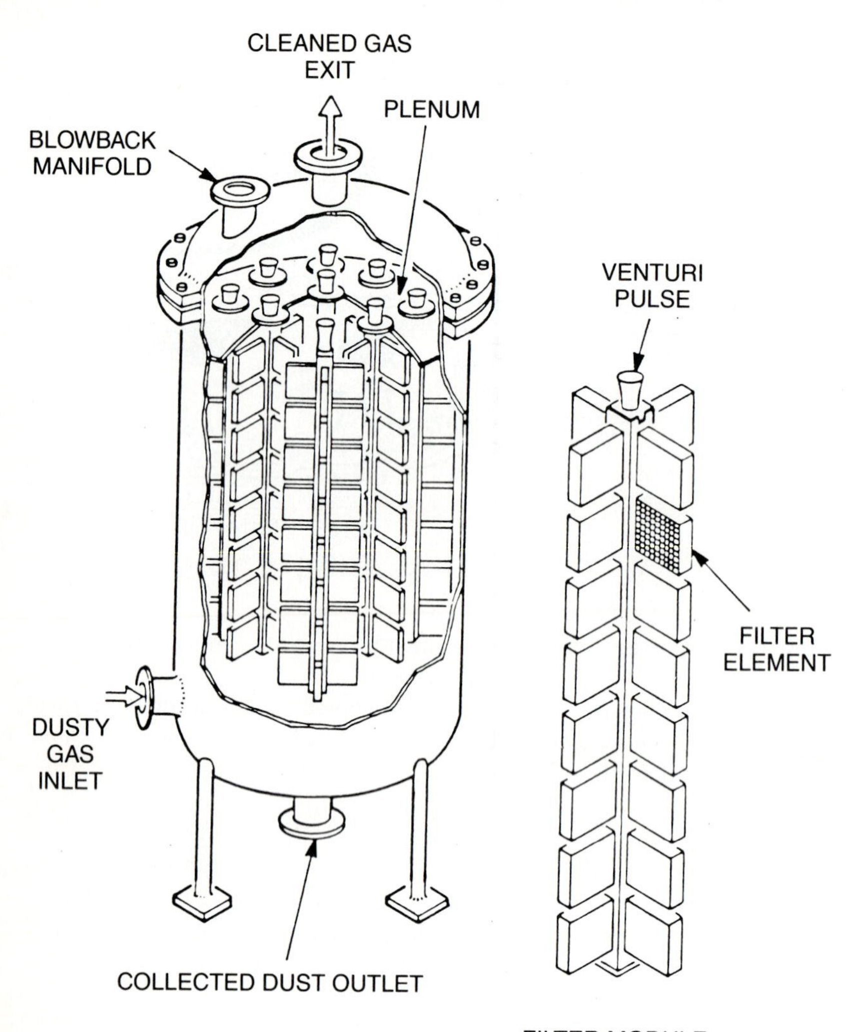

that the sorbent symbolized as CFA is superior in attaining low equilibrium levels of H_2S. Such graphs are used to display the results of tests and experiments.

TRUTH IN GRAPHICS

It is important that the graphics you use represent the truth, or a state of affairs as close to the truth as possible. Consider the following scenario: You are employed as the chief cost-control engineer for a municipality. The city council is considering next year's budget allocation for parks and recreation. The council wants you to review

Table 14-1 KRW Gasifier Performance Test Results

HOT GAS CLEANUP	PDU RESULT	REQUIREMENTS	BASIS
Sulfur Removal by Sorbent Addition	>90 Percent (Pittsburgh Coal)	>90 Percent	NSPS
	85 Percent (Wyoming Coal)	>85 Percent	NSPS
Cold Gas Efficiency	73 Percent (30 Percent Improvement)		
Heating Value	150 Btu/scf (40 Percent Improvement)		
Fines Elutriation	0.2 #/# Coal Feed (>50 Percent Reduction)		
Emissions			
Particulate	Virtually Zero	<0.03 $\#/10^6$ Btu	NSPS
NO_x	0.2 $\#/10^6$ Btu	0.60 $\#/10^6$ Btu	NSPS
Alkali	<20.0 ppb	1.00 ppm	Turbine
Lead	0.004 ppm	1.00 ppm	Turbine
Calcium	1.5 ppm	10.00 ppm	Turbine
Vanadium	0.007 ppm	0.50 ppm	Turbine

FIGURE 14-6
Equilibrium Hydrogen-sulfide Levels for Novel Desulfurization Sorbents

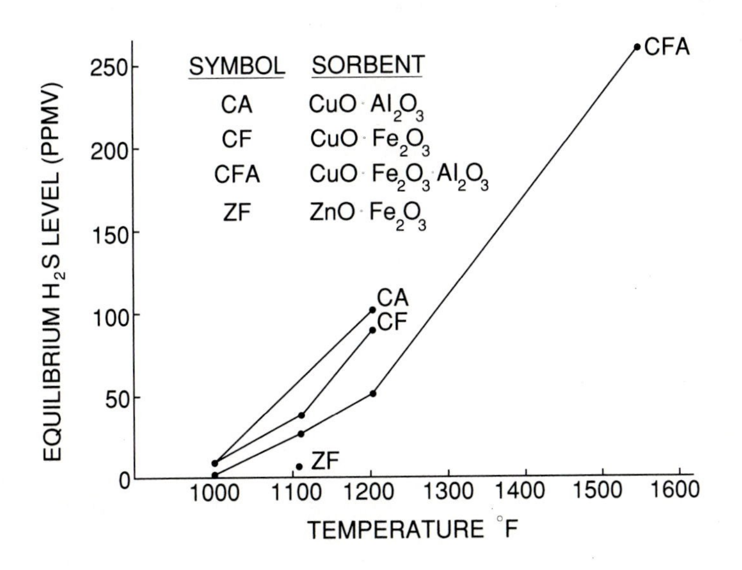

FIGURE 14-7
Graph of Dollar Allocation: Budget for Parks and Recreation, Heavensville, WV

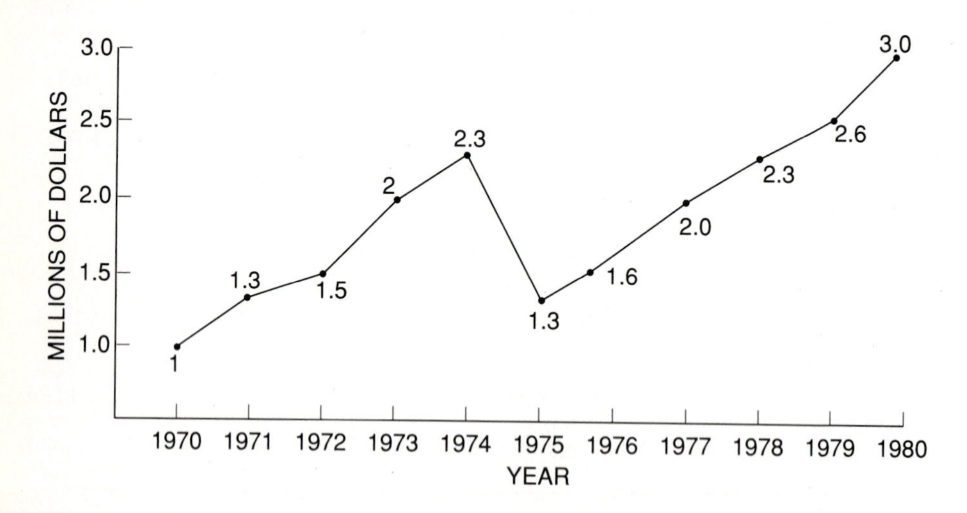

with them their funding for parks and recreation over the last 10 years so they can increase support for these services in the next fiscal year. To help your presentation, a budget officer in your department prepares the following graph (Figure 14–7): This graph indicates that the city increased its allocation 300 percent over 10 years.

Luckily for you, another financial expert in your office sees this graph while it's still in draft stage. She points out that the graph is meaningless because its dollars are not adjusted for inflation, for their real buying power. She prepares another

FIGURE 14–8
Graph of Dollars Adjusted for Inflation: Budget for Parks and Recreation, Heavensville, WV

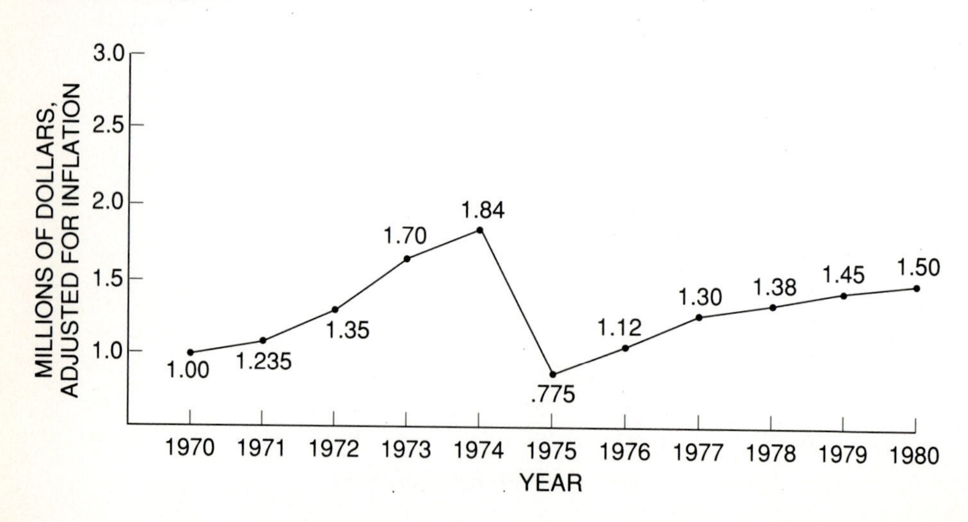

FIGURE 14-9
Graph of Inflation-adjusted Dollars Spent Per Resident: Budget for Parks and Recreation Dollars Spent Per Resident, Heavensville, WV

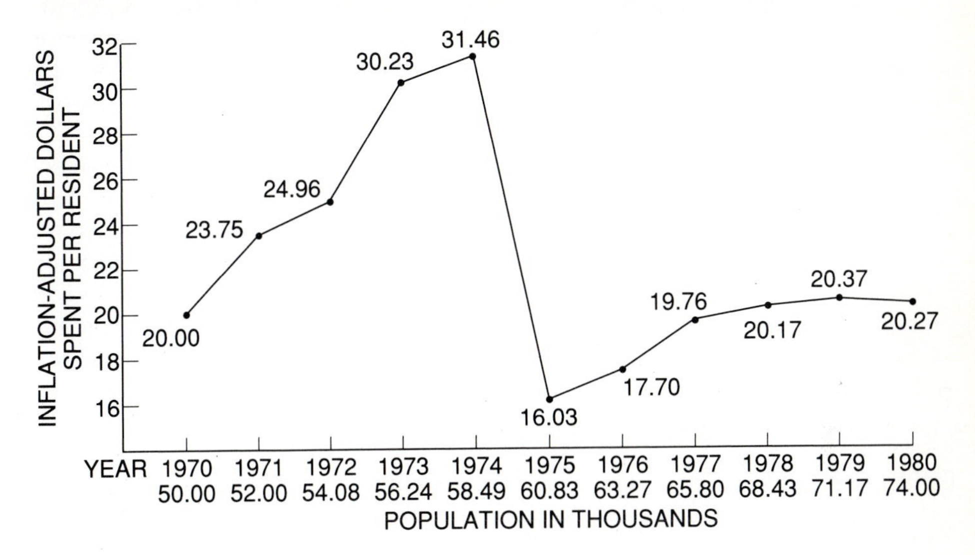

graph that, she claims, more nearly reflects the *truth:* the actual buying power of each allocated dollar (Figure 14–8): Instead of increasing 300 percent, the actual usable allocation has increased only 50 percent over the last decade.

Your financial group finally sits down to review your presentation. In the discussion, a third graph is suggested: one that uses inflation-adjusted dollars corrected to the dramatic increase in population that your area has enjoyed over the last 10 years. This third graph is then produced (Figure 14–9): You can now see that, when adjusted for inflation and the population increase, the actual budget has remained constant: the city council is spending just about the same number of dollars per resident in 1980 as it spent in 1970.

This is the *truth* that you can take to city council. Your presentation will reflect the actual, complex state of affairs and will not be marred by naive or misleading graphic representations. This is exactly the kind of presentation that you, as an engineer, want to make.

A FINAL WORD

It is difficult to exaggerate the importance of graphic elements in the world of scientific and technical writing. The fact is that, in the best documents, graphics and writing form an integrated whole, much as they do in the sophisticated software programs that allow an author to insert graphics in the text and to display the whole document on the screen. Graphics and text should form an organic whole, as they do in the examples displayed in this chapter and throughout this book.

For scientists and engineers, graphics can also carry an emotional charge and can initiate progress and innovation. An apparently modest table of test results, for instance, can represent the combined energies of several research scientists and engineers over several months or years. A seemingly simple mathematical description of an organic structure (as those descriptions by D'Arcy Thompson) can represent the most original idea of a scientist's entire career. And an accurate rendering of a failed mechanical part in an engineering design can lead to substantial innovation in design features and to the retooling of critical parts.

Pictorial graphics can even elevate the soul. Think of the sublime picture of our planet Earth, as seen from space: the mysterious beauty of our blue and white-streaked Earth seen against the cold black of the abyss of space has transformed the consciousness of all who have seen it. And this transformation is a result of extraordinary innovation in science and technology. Although the graphics you choose or produce might not move the soul or inspire innovation, they should always conform to the highest standards of accuracy and precision, and they should present a clear and important idea.

Reference List

Tufte, E. R. 1983. *The visual display of quantitative information.* Cheshire, Conn.: Graphic Press.

Turnbull, A. T., and R. N. Baird. 1980. *The graphics of communication.* 4th ed. New York: Holt, Rinehart & Winston.

APPENDIX A

The Recent Literature of Technology and Science

As future scientists and engineers, you need to be aware of the major developments, both theoretical and practical, in the history of science and technology. If you don't understand the past, you may be not only condemned to repeat its mistakes but also unable to take advantage of its extraordinary successes. You should not put yourself at the disadvantage of working in a historical vacuum.

You need to be acquainted, therefore, with the factual history of science and technology. And you also need to understand the philosophical issues being debated within science and technology:

How does science develop and progress?

How do innovations in technology come about?

What is the relationship between a scientist's or an engineer's world view and the work he or she produces?

Does knowledge in science and technology accumulate step-by-step or by a series of intuitive leaps?

Do scientists and engineers rely more on mathematical modeling, empirical research, or logical hunches to formulate new theories and approaches?

Intelligent scientists and engineers continually debate these questions.

Finally, you need to be aware of the extraordinary discoveries and inventions taking place right now. The following bibliography gives you ready access to many of these fascinating topics. Many of the readings included in this text come from this list.

RECENT BOOKS IN SCIENCE AND TECHNOLOGY: A BIBLIOGRAPHY

Archaeology

Brothwell, D. 1987. *The Bog Man and the archaeology of people.* Cambridge: Harvard University Press.

Page, R. I. 1987. *Runes.* Berkeley and Los Angeles: University of California Press/ British Museum.

Architecture

Fitchen, J. 1986. *Building construction before mechanization.* Cambridge: The MIT Press.

Oliver, P. 1987. *Dwellings: The house across the world.* Austin: University of Texas Press.

Rowe, P. G. 1986. *Design thinking.* Cambridge: The MIT Press.

Astronomy, Astrophysics, and Cosmology

Burke, J. 1985. *The day the universe changed.* Boston: Little, Brown.

Harrison, E. 1987. *Darkness at night: A riddle of the universe.* Cambridge: Harvard University Press.

Hawking, S. W. 1988. *A brief history of time: From the big bang to black holes.* New York: Bantam.

Koyré, A. 1957. *From the closed world to the infinite universe.* Baltimore: Johns Hopkins University Press.

Trefil, J. S. 1986. *The moment of creation: Big bang physics from before the first millisecond to the present universe.* New York: Scribner's.

Biology, Molecular Biology, DNA, and Evolution

Dawkins, R. 1986. *The blind watchmaker: Why the evidence of evolution reveals a universe without design.* New York: Norton.

Dawkins, R. 1976. *The selfish gene.* New York: Oxford University Press.

Gabriel, M. L., and S. Fogel. 1955. *Great experiments in biology.* Englewood Cliffs, N.J.: Prentice-Hall.

Gribbin, J. 1987. *In search of the double helix: Quantum physics and life.* New York: McGraw-Hill.

Judson, H. F. 1980. *The eighth day of creation: The makers of the revolution in biology.* New York: Touchstone, Simon & Schuster.

Wilson, E. O. 1984. *Biophilia: The human bond with other species.* Cambridge: Harvard University Press.

Cartography

Makower, J., ed. 1986. *The map catalog.* New York: Vintage Books-Tilden Press.

Chemistry

Atkins, P. W. 1987. *Molecules.* New York: Freeman, Scientific American.

Caglioti, L., and M. Giaconi. 1985. *The two faces of chemistry: The benefits and the risks of chemical technology.* Cambridge: The MIT Press.

Levi, P. 1984. *The periodic table.* Translated by R. Rosenthal. New York: Schocken Books.

Clocks and Time

Campbell, J. 1987. *Winston Churchill's afternoon nap: A wide-awake inquiry into the human nature of time.* New York: Simon & Schuster.

Landes, D. S. 1983. *Revolution in time.* Cambridge: Harvard University Press.

Fraser, J. T. 1987. *Time, The familiar stranger.* Amherst: University of Massachusetts Press.

Winfree, A. T. 1988. *The timing of biological clocks.* New York: Freeman, Scientific American.

Computers and Artificial Intelligence

Alexsander, I., and P. Burnett. 1987. *Thinking machines: The search for artificial intelligence.* New York: Knopf.

Brand, S. 1987. *The Media Lab: Inventing the future at MIT.* New York: Viking.

Cherniak, C. 1987. *Minimal rationality.* Cambridge: The MIT Press.

Forester, T. 1987. *High-tech society: The story of the information technology revolution.* Cambridge: The MIT Press.

Johnson, G. 1987. *Machinery of the mind: Inside the new science of artificial intelligence.* Redmond, Wash.: Microsoft Press.

Johnson-Laird, P. N. 1988. *The computer and the mind: An introduction to cognitive science.* Cambridge: Harvard University Press.

Minsky, M. 1985. *The society of mind.* New York: Simon & Schuster.

Roszak, T. 1987. *The cult of information: The folklore of computers and the true art of thinking.* New York: Pantheon.

Turkle, S. 1985. *The second self: Computers and the human spirit.* New York: Touchstone.

Discovery and Scientific Revolutions

Boorstin, D. J. 1983. *The discoverers.* New York: Random House.

Brown, K. A. 1988. *Inventors at work: Interviews with 16 notable American inventors.* Redmond, Wash.: Microsoft Press.

Bruner, J. S., and L. Postman. 1949. On the perception of incongruity: A paradigm. *Journal of Personality* XVIII: 206–23.

Cohen, I. B. 1980. *The Newtonian revolution.* Cambridge: Cambridge University Press.

———. 1986. *Revolution in science.* Cambridge: Harvard University Press.

Drake, S. 1980. *Galileo.* New York: Oxford University Press.

Eiseley, L. 1958. *Darwin's century: Evolution and the men who discovered it.* New York: Doubleday.

Goodfield, J. 1981. *An imagined world: A story of scientific discovery.* New York: Harper & Row.

Hanson, N. R. 1958. *Patterns of discovery.* Cambridge: Cambridge University Press.

Holton, G. 1975. *Thematic origins of scientific thought: Kepler to Einstein.* Cambridge: Harvard University Press.

Koestler, A. 1959. *The sleepwalkers: A history of man's changing vision of the universe.* New York: Macmillan.

———. 1964. *The act of creation: A study of the conscious and unconscious in science and art.* New York: Dell.

Kuhn, T. 1957. *The Copernican revolution: Planetary astronomy in the development of western thought.* Cambridge: Havard University Press.

———. 1970. *The structure of scientific revolutions.* 2d ed., enlarged. Chicago: University of Chicago Press.

———. 1977. *The essential tension: Selected studies in scientific tradition and change.* Chicago: University of Chicago Press.

———. 1978. *Black body theory and the quantum discontinuity, 1894–1912.* Oxford: Clarendon Press.

Langley, P., H. A. Simon, G. L. Bradshaw, and J. M. Zytkow. 1986. *Scientific discovery: Computational explorations of the creative process.* Cambridge: The MIT Press.

Shapiro, G. 1987. *A skeleton in the darkroom: Stories of serendipity in science.* New York: Harper & Row.

Scientific genius and creativity: Readings from Scientific American. 1987. Introduction by Owen Gingerich. New York: Freeman.

Temple, R. 1987. *The genius of China: 3,000 years of science, discovery, and invention.* New York: Simon & Schuster.

Ecology

Chase, A. 1987. *Playing God in Yellowstone: The destruction of America's first national park.* San Diego: Harcourt Brace Jovanovich.

Caufield, C. 1986. *In the rainforest: Report from a strange, beautiful, imperiled world.* Chicago: University of Chicago Press.

Horton, T. 1987. *Bay country.* Baltimore: Johns Hopkins University Press.

Kingsland, S. E. 1986. *Modeling nature: Episodes in the history of population biology.* Chicago: University of Chicago Press.

Lopez, B. 1988. *Crossing open ground.* New York: Scribner's.

Weir, D. 1987. *The Bhopal Syndrome: Pesticides, Environment, and Health.* San Francisco: Sierra Club Books.

Electricity

Meyer, H. W. 1971. *A history of electricity and magnetism.* Cambridge: The MIT Press.

Ethnology

Jenness, D. 1985. *Dawn in Arctic Alaska.* Chicago: University of Chicago Press.

Malaurie, J. 1985. *The last kings of Thule.* Translated by Adrienne Foulke. Chicago: University of Chicago Press.

Feminist Approaches

Alic, M. 1987. *Hypatia's heritage: A history of women in science from antiquity through the nineteenth century.* Boston: Beacon Press.

Bleier, R. 1984. *Science and gender: A critique of biology and its theories on women.* New York: Pergamon Press.

———, ed. 1986. *Feminist approaches to science.* New York: Pergamon Press.

Gornick, V. 1985. *Women in science: Recovering the life within.* New York: Simon & Schuster, Touchstone.

Haraway, D. 1976. *Crystals, fabrics, and fields: Metaphors of organicism in 20th-Century developmental biology.* New Haven: Yale University Press.

Hrdy, S. B. 1981. *The woman that never evolved.* Cambridge: Harvard University Press.

Herschberger, R. [1948] 1971. *Adam's rib.* Reprint. New York: Harper & Row.

Janeway, E. 1971. *Man's world, woman's place: A study in social mythology.* New York: Morrow.

Keller, E. F. 1983. *A feeling for the organism: The life and work of Barbara McClintock.* New York: Freeman.

———. 1985. *Reflections on gender and science.* New Haven: Yale University Press.

Kevles, B. 1986. *Females of the species: Sex and survival in the animal kingdom.* Cambridge: Harvard Univesity Press.

Game Theory and Creativity

Carse, J. P. 1986. *Finite and infinite games: A vision of life as play and possibility.* New York: Free Press.

von Oech, R. 1986. *A kick in the seat of the pants.* New York: Harper & Row.

General

Berry, W. 1981. *The gift of good land.* San Francisco: North Point Press.

Bodanis, D. 1986. *The secret house.* New York: Simon & Schuster.

Burton, R. 1987. *Eggs: Nature's perfect package.* New York: Facts on File.

Burtt, E. A. 1932. *The metaphysical foundations of modern physical science.* Rev. ed. Atlantic Highlands, N.J.: Humanities.

Butterfield, H. 1949. *The origins of modern science, 1300–1800.* London: Bell.

Calvin, W. H. 1986. *The river that flows uphill: A journey from the big bang to the big brain.* San Francisco: Sierra Club Books.

Crompton, J. 1951. *The life of the spider.* Boston: Houghton Mifflin.

Eiseley, L. 1957. *The immense journey.* New York: Random.

———. 1973. *The man who saw through time.* New York: Scribner's.

———. 1975. *All the strange hours: The excavation of a life.* New York: Scribner's.

Foucault, M. 1967. *Madness and civilization: A history of insanity in the Age of Reason.* Translated by R. Howard. London: Tavistock.

Gardner, M. 1981. *Science: Good, bad, and bogus.* Buffalo: Prometheus Books.

Gleick, J. 1987. *Chaos: Making a new science.* New York: Viking.

Gould, S. J. 1987. *An urchin in the storm: Essays about books and ideas.* New York: Norton.

Hill, C. 1965. *The intellectual origins of the industrial revolution.* New York: Oxford University Press.

Judson, H. F. 1987. *The search for solutions.* Rev. ed. Baltimore: Johns Hopkins University Press.

Kramer, M. 1987. *Three farms: Making milk, meat, and money from the American soil.* Cambridge: Harvard University Press.

McCormmach, R. 1983. *Night thoughts of a classical physicist.* New York: Avon.

McKenzie, A. E. E. 1973. *The major achievements of science: The development of science from ancient times to the present.* New York: Simon & Schuster.

Patent, D. H. 1987. *Wheat: The golden harvest.* New York: Dodd, Mead.

Peitgen, H. O., and P. H. Richter. 1986. *The beauty of fractals.* New York: Springer-Verlag.

Quine, W. V. 1987. *Quiddities: An intermittently philosophical dictionary.* Cambridge: Harvard University Press.

Raban, J. 1987. *Coasting.* New York: Simon & Schuster.

Sagan, C. 1976. *Broca's brain.* New York: Random House.

Thomas, L. 1984. *Late night thoughts on listening to Mahler's Ninth Symphony.* New York: Bantam.

Trefil, J. 1987. *Meditations at sunset: A scientist looks at the sky.* New York: Scribners. (See also his *A scientist at the seashore; The unexpected vista;* and *From atoms to quarks.*)

Weisskopf, V. F. 1979. *Knowledge and wonder: The natural world as man knows it.* Cambridge: The MIT Press.

Wertsch, J. V. 1986. *Vygotsky and the social formation of mind.* Cambridge: Harvard University Press.

Geology

Cloud, P. 1988. *Oasis in space: Earth history from the beginning.* New York: Norton.

Gould, S. J. 1987. *Time's arrow, time's cycle: Myth and metaphor in the discovery of geological time.* Cambridge: Harvard University Press.

Heather, D. C. 1979. *Plate tectonics.* Baltimore: Edward Arnold.

Laudan, R. 1988. *From mineralogy to geology: The foundations of a science.* Chicago: University of Chicago Press.

Menard, H. W. 1987. *Islands.* New York: Freeman, Scientific American.

Trefil, J. 1986. *Meditations at 10,000 feet: A scientist in the mountains.* New York: Macmillan.

Human Physiology

Forbes, G. B. 1987. *Human body composition: Growth, aging, nutrition, and activity.* New York: Springer-Verlag.

Mathematics

Davis, P. J., and R. Hersh. 1987. *Descartes' dream: The world according to mathematics.* San Diego: Harcourt Brace Jovanovich.

Hoffman, P. 1988. *Archimedes' revenge.* New York: Norton.

Ifrah, G. 1987. *From one to zero: A universal history of numbers.* New York: Penguin.

Kline, M. 1953. *Mathematics in western culture.* Harmondsworth, Eng.: Pelican.

Morrison, P. 1983. *Powers of ten.* New York: Freeman, Scientific American.

Rucker, R. 1987. *Mind tools: The five levels of mathematical reality.* Boston: Houghton Mifflin.

Medicine

Altman, L. K. 1987. *Who goes first?: The story of self-experimentation in medicine.* New York: Random House.

Colen, B. D. 1987. *Hard choices.* New York: Putnam's.

Gonzalez-Crussi, F. 1985. *Notes of an anatomist.* San Diego: Harcourt Brace Jovanovich.

Hall, S. S. 1987. *Invisible frontiers: The race to synthesize a human gene.* Boston: Morgan Entrekin, Atlantic Monthly.

Hunan hand and other ailments: Letters to the New England Journal of Medicine. 1987. Edited by Shirley Blotnick Moskow, foreword by Arnold S. Relman. Boston: Little, Brown.

Kramer, M. 1984. *Invasive procedures: A year in the world of two surgeons.* New York: Penguin.

Levi-Montalcini, R. 1988. *In praise of imperfection: My life and work.* New York: Basic Books.

Levine, H. 1987. *Life choices.* New York: Simon & Schuster.

Nilsson, L., et al. 1987. *The body victorious: The illustrated story of our immune system and other defenses of the human body.* New York: Delacorte Press.

Ober, W. B. 1987. *Bottoms up!: A pathologist's essays on medicine and the humanities.* Carbondale: Southern Illinois University Press.

Reiser, S. J. 1978. *Medicine and the reign of technology.* Cambridge: Cambridge University Press.

Rose, K. Jon. 1988. *The body in time.* New York: Wiley.

Selzer, R. 1978. *Mortal lessons: Notes on the art of surgery.* New York: Simon & Schuster.

———. 1986. *Taking the world in for repairs.* New York: Morrow.

———. 1987. *Confessions of a knife.* New York: Morrow.

———. 1987. *Rituals of Surgery.* New York: Morrow.

Thomas, L. 1975. *The lives of a cell: Notes of a biology watcher.* New York: Bantam.

Weissmann, G. 1987. *They all laughed at Christopher Columbus: Tales of medicine and the art of discovery.* New York: Time Books.

Microscopy

Burgess, J., M. Marten, and R. Taylor. 1987. *Microcosmos.* Cambridge: Cambridge University Press.

Natural History

Belt, T. 1987. *The naturalist in Nicaragua.* Chicago: University of Chicago Press. (A re-issue of a journal description of the tropics of more than 100 years ago.)

Darwin, C. [1881] 1945. *The formation of vegetable mould through the action of worms with observations on their habits.* London: Faber & Faber.

Evans, P. G. H. 1987. *The natural history of whales and dolphins.* New York: Facts on File.

Gould, S. J. 1984. *Hen's teeth and horse's toes: Further reflections in natural history.* New York: Norton.

———. 1985. *The flamingo's smile: Reflections in natural history.* New York: Norton.

Hsu, K. J. 1986. *The great dying.* San Diego: Harcourt Brace Jovanovich.

McPhee, J. 1979. *Coming into the country.* New York: Bantam.

———. 1987. *Rising from the plains.* New York: Farrar, Straus & Giroux.

Mowat, F. 1980. *And no birds sang.* Boston: Little, Brown.

———. 1984. *Never cry wolf.* New York: Bantam.

Raup, D. M. 1986. *The nemesis affair: A story of the death of dinosaurs and the ways of science.* New York: Norton.

Steinbeck, J., and E. F. Ricketts. 1941. *Sea of Cortez: A leisurely journal of travel and research, with a scientific appendix comprising materials for a source book on the marine animals of the Panamic faunal province.* New York: Viking.

Steinbeck, J. 1977. *The log from the Sea of Cortez.* New York: Penguin.

Neurobiology

Changeux, J.-P. 1986. *Neuronal man: The biology of mind.* New York: Oxford University Press.

Edelman, G. 1987. *Neural Darwinism: The theory of neuronal group selection.* New York: Basic Books.

Franklin, J. 1987. *Molecules of the mind: The brave new science of molecular psychology.* New York: Atheneum.

Snyder, S. H. 1987. *Drugs and the brain.* New York: Freeman, Scientific American.

Philosophy of Science

Arbib, M. A., and M. B. Hesse. 1986. *The construction of reality.* Cambridge: Cambridge University Press.

Churchland, P. M. 1986. *Scientific realism and the plasticity of mind.* Cambridge: Cambridge University Press.

Feyerabend, P. 1975. *Against method: Outline of an anarchistic theory of knowledge.* New York: Schocken.

Gutting, G., ed. 1980. *Paradigms and revolutions: Appraisals and applications of Thomas Kuhn's philosophy of science.* Notre Dame, Ind.: University of Notre Dame Press.

Hacking, I., ed. 1981. *Scientific revolutions.* New York: Oxford University Press.

Hanson, N. R. 1958. *Patterns of discovery.* Cambridge: Cambridge University Press.

Kuhn, T. 1957. *The Copernican revolution: Planetary astronomy in the development of western thought.* Cambridge: Harvard University Press.

———. 1970. *The structure of scientific revolutions.* 2d ed., enlarged. Chicago: University of Chicago Press.

———. 1977. *The essential tension: Selected studies in scientific tradition and change.* Chicago: University of Chicago Press.

———. 1978. *Black body theory and the quantum discontinuity, 1894–1912.* Oxford: Clarendon Press.

Merton, R. K. 1973. *The sociology of science: Theoretical and empirical investigations.* Edited by Norman W. Storer. Chicago: University of Chicago Press.

Munitz, M. K. 1987. *Cosmic understanding: Philosophy and science of the universe.* Princeton: Princeton University Press.

Nagel, E. 1961. *The structure of science: Problems in the logic of scientific expression.* New York: Harcourt, Brace & World.

Polanyi, Michael. 1958. *Personal knowledge: Towards a post-critical philosophy.* London: Routledge & Kegan Paul.

———. 1966. *The tacit dimension.* London: Routledge and Kegan Paul.

Popper, Sir K. R. 1959. *The logic of scientific discovery.* London: Hutchinson.

Toulmin, S. 1953. *The philosophy of science.* London: Hutchinson.

———. 1961. *Foresight and undersanding.* Bloomington: Indiana University Press.

———. 1972. *Human understanding.* Vol 1. Oxford: Clarendon Press.

Vygotsky, L. 1986. *Thought and language.* Rev. ed. Edited by Alex Kozulin. Cambridge: The MIT Press.

Physics

Bohren, C. F. 1987. *Clouds in a glass of beer: Simple experiments in atmospheric physics.* New York: Wiley.

Crease, R. P., and C. C. Mann. 1985. *The second creation: Makers of the revolution in twentieth-century physics.* New York: Macmillan.

Epstein, L. C. 1987. *Thinking physics is gedanken physics: Practical lessons in critical thinking.* San Francisco: Insight Press.

Feynman, R. P. 1986. *QED.* Princeton: Princeton University Press.

Glashow, S., with B. Bova. 1988. *Interactions: A journey through the mind of a particle physicist and the matter of this world.* New York: Warner Books.

Jungnickel, C., and R. McCormmach. 1986. *Intellectual mastery of nature: Theoretical physics from Ohm to Einstein.* Chicago: University of Chicago Press.

Kevles, D. J. 1987. *The physicists: The history of a scientific community in modern America.* Cambridge: Harvard University Press.

Mook, D. E., and T. Vargish. 1987. *Inside relativity.* Princeton: Princeton University Press.

Pais, Abraham. 1982. *Subtle is the Lord: Science and life of Albert Einstein.* New York: Oxford University Press.

———. 1987. *Inward bound: Of matter and forces in the physical world.* New York: Oxford University Press.

Peierls, R. 1985. *Bird of passage: Recollections of a physicist.* Princeton: Princeton University Press.

Preston, R. 1987. *First light: The search for the edge of the universe.* Boston: Atlantic Monthly.

Prigogine, I., and I. Stengers. 1984. *Order out of chaos: Man's new dialogue with nature.* Boston: Shambhala.

Regis, E. 1987. *Who got Einstein's office?: Eccentricity and genius at the Institute for Advanced Study.* Reading, Mass.: Addison-Wesley.

Riordan, M. 1987. *The hunting of the quark: A true story of modern physics.* New York: Simon & Schuster.

Schwinger, J. 1987. *Einstein's legacy: The unity of space and time.* New York: Freeman, Scientific American.

Sobel, M. 1987. *Light.* Chicago: University of Chicago Press.

Weaver, J. H. 1987. *The world of physics: A small library of the literature of physics from antiquity to the present.* 3 Vols. New York: Simon & Schuster.

Wilczek, F., and B. Devine. 1988. *Longing for the harmonies: Themes and variations from modern physics.* New York: Norton.

Wolf, F. Alan. 1981. *Taking the quantum leap: The new physics for nonscientists.* New York: Harper & Row.

Zee, A. 1986. *Fearful symmetry: The search for beauty in modern physics.* New York: Macmillan.

Primatology

Fossey, D. 1983. *Gorillas in the mist.* Boston: Houghton Mifflin.

Goodall, Jane. 1971. *In the shadow of man.* Boston: Houghton Mifflin.

———. 1986. *The chimpanzees of Gombe: Patterns of behavior.* Cambridge: Harvard University Press.

Jolly, A. 1985. *The evolution of primate behavior.* New York: Macmillan.

Morris, D. 1967. *The naked ape.* New York: McGraw-Hill.

Mowat, F. 1987. *Lost in the barrens.* New York: Bantam.

Strum, S. C. 1987. *Almost human: A journey into the world of baboons.* New York: Random House.

Psychiatry

Argyle, M. 1987. *The psychology of happiness.* New York: Methuen.

Goleman, D. 1985. *Vital lies, simple truths.* New York: Simon & Schuster.

Minuchin, S. 1986. *Family kaleidoscope: Images of violence and healing.* Cambridge: Harvard University Press.

Quinn, S. 1987. *A mind of her own: The Life of Karen Horney.* New York: Summit Books.

Sociobiology

Wilson, E. O. 1975. *Sociobiology: The new synthesis.* Cambridge: Harvard University Press.

———. 1978. *On human nature.* Cambridge: Harvard University Press.

———. 1986. *Biophilia.* Cambridge: Harvard University Press.

Technology and Engineering

Asimov, I. 1982. *Foundation's edge.* New York: Doubleday.

Beniger, J. R. 1986. *The control revolution: Technological and economic origins of the information society.* Cambridge: Harvard University Press.

Brandin, D. H., and M. A. Harrison. 1987. *The technology war: A case for competitiveness.* New York: Wiley.

Crowther, J. G. 1962. *Scientists of the industrial revolution.* Philadelphia: Cresset Press.

de Sola Pool, I. 1986. *Technologies of freedom.* Cambridge: Harvard University Press.

Drexler, K. E. 1987. *Engines of creation.* New York: Doubleday, Anchor Press.

Florman, S. C. 1977. *The existential pleasures of engineering.* New York: St. Martin's.

———. 1981. *Blaming technology: The irrational search for scapegoats.* New York: St. Martin's.

———. 1982. *Engineering and the liberal arts: A technologist's guide to history, literature, philosophy, art, and music.* Melbourne, Fla.: Krieger.

———. 1987. *The civilized engineer.* New York: St. Martin's.

Friedel, R., and P. Israel. 1985. *Edison's electric light: Biography of an invention.* New Brunswick, N.J.: Rutgers University Press.

Gordon, J. E. 1988. *The science of structures and materials.* New York: Freeman, Scientific American.

Hill, D. 1984. *A history of engineering in classical and medieval times.* LaSalle, Ill.: Open Court.

Hill, C. 1965. *The intellectual origins of the industrial revolution.* Oxford: Oxford University Press.

Kerridge, E. 1967. *The agricultural revolution.* Winchester, Mass.: Allen & Unwin.

Mumford, L. 1934. *Technics and civilization.* New York: Harcourt Brace.

Norman, D. A. 1988. *The psychology of everyday things.* New York: Basic Books.

Pacey, A. 1983. *The culture of technology.* Cambridge: The MIT Press.

Petroski, H. 1982. *To engineer is human.* New York: St. Martin's.

Rybczynksi, W. 1983. *Taming the tiger: The struggle to control technology.* New York: Viking.

Sporn, P. 1964. *Foundations of engineering.* New York: Macmillan.

Stopping time: The photographs of Harold Edgerton. 1987. Edited by Gus Kayafas. Text by Estelle Jussim. New York: Abrams.

Turner, R., and S. L. Goulden, eds. 1981. *Great engineers and pioneers in technology.* Vol. 1: *From antiquity through the Industrial Revolution.* New York: St. Martin's.

Usher, A. P. [1929] 1970. *A history of mechanical inventions.* Cambridge: Harvard University Press.

Wenk, E. 1987. *Tradeoffs: Imperatives of choice in a high-tech world.* Baltimore: Johns Hopkins University Press.

Williams, T. I. 1987. *The history of invention: From stone axes to silicon chips.* New York: Facts on File.

Winner, L. 1986. *The whale and the reactor: A search for limits in an age of high technology.* Chicago: University of Chicago Press.

Zubrowsky, B. 1987. *Wheels at work: Building and experimenting with models of machines.* New York: Morrow.

Visual Imaging and Illustration

Kim, S. 1985. *Inversions: A catalog of calligraphic cartwheels.* Cambridge: The MIT Press.

Marr, D. 1984. *Vision: A computational investigation into the human representation and processing of visual information.* New York: Freeman.

Richardson, G. T. 1987. *Illustrations.* Clifton, N.J.: Humana Press.

Tufte, E. R. 1983. *The visual display of quantitative information.* Cheshire, Conn.: Graphics Press.

APPENDIX B

Other Forms of Communication: Memos, Letters, and Oral Presentations

AUDIENCE AND LEVELS OF DICTION

Memos and letters are two of the workhorses of communication in industry and in the academic world. Both forms are designed for audiences outside of narrow specialties. A memo, for instance, might be read by members from all the divisions of a corporation. A letter might be written from an independent consulting engineer to the chairman of the board of a company. Consequently, writers of memos and letters must pay particular attention to the nature of the audience they are addressing and to writing within the sphere of mutual knowledge that the audience and the writer share.

How much context or background you supply, then, is the first thing to think about when you write a memo or a letter. After considering how much your audience knows and, therefore, how much background material you must put into your communication, you must also tailor your vocabulary to the audience. One of the most serious faults in writing for an uninitiated audience is using words that are not generally familiar. Malcolm W. Browne has written about this problem.

Wanted: Interpreters for the Frontiers of Science

Malcolm W. Browne

The Tower of Babel so annoyed the Almighty, the Bible tells us, that the Lord forced its builders to converse in a babble of mutually unintelligible languages.

A glance at any of the thousands of scientific journals published these days is enough to tell the story; the titles alone are enough to sow confusion among all but a few initiates. Typical of the genre was a recent paper entitled "Long-Term Potentiation in Dentate Gyrus:

Induction by Asynchronous Volleys in Separate Afferents." Another paper published in the past month bore this title: "Surface Extended X-Ray-Absorbtion Fine-Structure Study of the O(2×1)/Cu(110) System: Missing-Row Reconstruction, and Anisotropy in the Surface Mean Free Path and in the Surface Debye-Waller Factor."

With the help of appropriate reference books, a scientifically savvy outsider can figure out that the first paper is about some interesting experiments on rat nervous systems, and that the second has to do with the positions of atoms in a particular kind of crystal structure. But the fact that intimidating scientific titles can sometimes be partially translated into plain language does not imply that the paper on rats' nerves would be genuinely intelligible to, say, a condensed-matter physicist, or that the crystallography paper would make sense to a neurophysiologist.

One of the nice things about the frontiers of science a century ago was that they were more or less accessible even to ordinary people. For example, two papers published 100 years ago in the distinguished British journal Nature carried the straight-forward titles "Observations on Heredity in Cats With An Abnormal Number of Toes," and "The Law of Storms in the Eastern Seas." Eventoed cats and typhoons are subjects with which even a mediocre mind can come to grips.

Scientific writing remained relatively penetrable for a long time. In 1905 Einstein published his revolutionary Special Theory of Relativity, thereby shaking the very foundations of science. But despite the staggering intellectual power of Einstein's paper, it carried the unpretentious title "On the Electrodynamics of Moving Bodies," and its mathematics were within the easy grasp even of a high school algebra student.

But times have changed. For one thing, so many scientific papers are being published that every author must bear down on distinctive but often obscure minutiae if he hopes to impress a publisher. At the last national meeting of the American Chemical Society, some 3,600 papers were presented in the course of a few days, and the A.C.S. is but one of hundreds of scientific societies that publish papers. Official chemical records now describe more than 7.4 million compounds, each of which has generated its own bibliography of papers. Comparable information explosions have occurred in astronomy, microbiology, mathematics, climatology and countless other scientific disciplines.

As if this were not bad enough, there is a growing suspicion among many scientists that some of their colleagues deliberately try to impress their peers by confusing them. Dr. Robert Schoenfeld, the editor of an Australian chemistry journal who frequently snipes at scientific gobbledegook, sarcastically suggested that the simple word "table" should be renamed using standard chemical notation as "rectangular-r-1,c-2,c-3,c-4-tetrastick," which would probably be abbreviated as "RATS."

Dr. J. Scott Armstrong, an educator specializing in the psychology of marketing, described this kind of thing as "creative obfuscation," and conducted an experiment that showed even scientists to be susceptible to significant-sounding flimflam. In his experiment, Dr. Armstrong carefully coached an actor in the delivery of a purportedly scientific lecture entitled "Mathematical Game Theory as Applied to Physician Education." The investigator described the lecture as consisting entirely of "double talk, meaningless words, false logic, contradictory statement, irrelevant humor, and meaningless references to unrelated topics."

Dr. Armstrong's bogus scientist presented this lecture to three audiences of social workers, psychologists, psychiatrists, educators and administrators, none of whom detected the hoax. In a poll taken by Dr. Armstrong in which anonymous responses were sought from members of the audiences, most respondents described the lecture as "clear and stimulating." From this and other experiments, Dr. Armstrong concluded that "an unintelligible communication from a legitimate source in the recipient's area of expertise will increase the recipient's rating of the author's competence."

In other words, it can pay a scientific writer to be incomprehensible. Society and science, however, are losers.

Browne implies that it is the duty of scientists and engineers to make their knowledge and expertise available in a practical way to the general, educated public.

He points out that the most serious mistake is using vocabulary and systems of abbreviation or notation that no one but a specialist can understand. For instance, say that you were a petroleum engineer writing to the board of directors of a local church that had invested in an oil field in order to use the profits to fund a soup kitchen, and that you had to write to the board telling them that the method used for pumping out the oil must be changed. If you started your letter this way,

OIL RECOVERY, INC.
4 Devon Way
Dallas, TX 76543

April 3, 1989

Board of Directors
First Church of the Brethren
12 Cross Forks Road
Dallas, TX 76543

Dear Board of Directors:

Our computerized calculations of yield based on porosity, formation channeling, and hydrocarbon saturation indicate that your well (Well #28 in Oil Patch Field) will only produce the 26 million barrels of crude that remain locked in it if we begin to employ the tertiary recovery method of CO_2 injection.

[And so forth]

you would have lost your audience from the start and wasted an effort at communication. Although the technical language is second nature to you, the church's board of directors could understand only a letter that uses more common terminology and descriptions, especially if it stated the problem first, instead of the solution. Something like the following letter would be better:

OIL RECOVERY, INC.
4 Devon Way
Dallas, TX 76543

April 3, 1989

Board of Directors
First Church of the Brethren
12 Cross Forks Road
Dallas, TX 76543

Dear Board of Directors:

Well #28 in Oil Patch Field, the one you bought to raise money for the Soup Kitchen, isn't producing very much oil anymore,

```
though it still contains 26 million barrels. The oil is locked in
the rock formation and cannot be pumped out by regular drilling
methods.

In order to recover this oil, we need your go-ahead to cover the
expense of pumping carbon dioxide into the rock formation, in
order to force the oil to the surface.

[And so forth]
```

This kind of revised communication, whether in the form of a memo or a letter, can be understood by people who are not specialists in the field. Certainly, of course, there are times when you should be a specialist writing to other specialists, but in writing memos and letters—like reports and articles—you should always consider the level of knowledge of your audience, as well as the interests, involvement, and concerns of that audience.

THE MEMO

Memorandum, usually shortened to just *memo,* is a Latin word meaning "something that should be remembered." Most memos are short pieces of in-house communication, often used to accomplish a specific task. On some occasions, a memo may take a longer form, containing an interim report, a review of a problem, a request for suggestions, or other kinds of everyday business communication. The author of the memo writes as an authority on a problem or as an organizer of an effort.

By their very nature as an insider's document, memos assume that the audience has some knowledge about the background of the topic. Nevertheless, the memo should begin with a clear statement of the purpose of the communication, which may be to make an announcement, state a problem, request a decision, or give directions. Like an essay or a report, the memo must give its readers as much information as the they need to understand the announcement, solve the problem, make the decision, or complete the task. Because memos are brief, their tone should be straightforward and professional.

Format and Application

Most companies have specially printed stationery or a given format for memos. Whether you are using company forms or are creating your own memo, you should present your information in basically the same manner. Every memo begins with a list of the addressee, the author, the date, and a brief statement of the subject. Individuals' titles or affiliations can follow their names, when applicable. The body of the memo then follows, usually in block format. If typed by a secretary or an assistant, the memo will contain the author's initials and the typist's initials, separated by a slash or a colon, at the end of the body. Additional persons who should receive a copy of the memo are indicated on the line below the typist's initials and are listed following the abbreviation "cc:," which originally meant "carbon copy." Note the general format in the sample memo that follows.

COMPUTER CHIPS, INC.
2 Silicon Drive
Palo Alto, CA 98123

Manager
Research Division

TO: Marsha Benkowski

FROM: Nina Brent

DATE: April 10, 1989

SUBJECT: Request for Report

Please prepare a short report on two developments that I have recently learned about:

1. The U.S. Department of Commerce and the Defense Department are concerned that Japan is trying to take over the world market in the manufacture of computer chips.

2. International physicists report that new breakthroughs in the field of superconductivity (materials offering almost no electrical resistance) will transform the computer-chip industry.

Your report should analyze how these developments will affect our research and production of chips for our 64-bit computer market.

NB/nf
cc: Al Ludwig

THE LETTER

In contrast to memos, letters are written to an audience outside the organization. Because the writer and the reader share less of the mutual knowledge that insiders share, more context or explanation must be supplied. Keep your readers in mind at all stages of the writing process and think about what they need to know.

Format

While the memo is often limited to one page, the letter is generally one to two pages long. Like the memo, the letter has a standard format, which can be represented in boilerplate form. (Check with your organization for the format its administration prefers.)

Name of organization
address

date

name of audience (Mr. or Ms.)
title of division
name of company or affiliation
address

Dear Mr./Ms. audience:

Set context and review reason for the letter being written.

Give the actual information.

Be courteous and open in signing off.

Sincerely,

Name (signed)

Name (typed)

initials of writer and preparer
cc:

Here is a letter following this format.

ENVIRONMENT ENGINEERING, INC.
201 Owl Hollow Road
Morgantown, WV 26505

March 23, 1989

Ms. Helen Burkeheart
Director of Research
Wholesome Dairy Products, Inc.
3402 Pineview Drive
Morgantown, WV 26505

Dear Ms. Burkeheart:

For some time, your company has piped its waste effluent to the Morgantown Water Treatment Plant for clean-up and sludge disposal. Although the charges for this service and the yearly

Letter adapted with permission from the work of John Young, III, a student at West Virginia University.

pipeline maintenance costs have been high as compared to those throughout the industry, the cost of constructing and operating a private, on-site waste treatment or pretreatment plant has so far prevented our considering any alternative solutions.

However, we have just received from the Morgantown plant a new schedule of their service charges. Effective June 1 of next year, your basic cost per 1,000 gallons of effluent will increase by 12%. Furthermore, you are also faced with increasing the capacity (through replacement) of your two-mile pipeline, and the cost for labor and materials for this type of work have increased more than threefold during the past 10 years.

As your engineering consultant, we want to tell you about a new development in the field, a development that may solve your short- and long-term problems and reduce costs at the same time.

Recently, the successful field testing of a new kind of on-site treatment plant was announced by DuPont engineers. The plant combines the activated-sludge process with a carbon filtration system. Until now, we could not consider an activated-sludge plant because the rate for removing the waste was not high enough to permit us to dump the treated waste water directly into the Monongahela River. The new process, called PACT (Powdered Activated Carbon Treatment), is reported to be efficient enough to allow direct disposal in the public waterway.

The enclosed report contains engineering data on the new PACT process and compares the overall cost of this alternative to that of your present situation. Since our company concentrates on process analysis, the enclosed information about costs doesn't include tax considerations (which will have to be studied by your Finance Committee).

The initial figures indicate that this new process for waste treatment will save your company money while, at the same time, offering it more flexibility.

We will be glad to meet with you to discuss how best to tailor this plant to your needs.

Sincerely,

John Young

JPY/adp
Encl.: Background and Feasibility
Report
cc: John Ryan, President
Environment Engineering, Inc.

ORAL PRESENTATIONS

Knowing What The Occasion Calls For

When you're called upon to make an oral presentation, you must first find out what the occasion calls for. To do this, ask the person who has made the assignment exactly what's expected of you:

Is your presentation part of a series? If so, where do you fit in?

Are you supposed only to present information, or are you expected to make judgments and recommendations?

Who is coming: only your peers and immediate managers, or employees and managers from other sections?

Should you be prepared for hostile questioning, or is the meeting designed only for sharing information?

How much detail should you go into? How long do you have?

How much will your audience already know about your subject?

Before you plan your presentation, you should have clear answers to all of these questions. Because audience, environment, time allotment, purpose, and content all interrelate, your presentation must respond to each of them.

The key to a successful presentation is understanding what your audience needs or wants to know and then giving it to them. If your audience feels that you understand and respect their needs, they'll be more receptive. Imagine how it would go, for instance, if you were a sales representative for a computer company and tried to sell a new product by giving a presentation to middle-level managers, all of whom had years of experience with computers. If you spent the first 20 minute talking in general about computers and their capabilities, your audience would become bored, then angry, and you'd lose the contract before you even began to present your product's advantages. You would have given a canned sales pitch, instead of spending time tailoring your presentation to your audience.

How to Prepare

1. Write an outline of your presentation. Give a summary of it to the person who asked you to do the speech, for approval, if appropriate.

2. Fill in the outline. Rehearse it privately; then present it to a friend. Check how long it takes.

3. If it's possible, look at the room where you're going to make the presentation. Stand where you'll be standing and become familiar with the area. Try out the chair. Be sure the overhead projector works and that the room is reasonably sound proof. Being familiar with the room will make you more relaxed.

4. Check and proofread all transparencies, handouts, and visuals that you plan to use. Though other departments may have prepared these for you, you're responsible for their being effective and error free. Check them out far enough ahead so they can be revised if needed.

5. Plan your schedule so you have at least 30 minutes free before the presentation. You'll want to use the time to make sure that you have all of your material ·ether.

Being Realistic about Giving Oral Presentations

You'll probably be nervous thinking about being the center of attention. But your audience will not expect you to be as perfect or as brilliant as you expect yourself to be. After all, your presentation will be over in 10–15 minutes, and then your audience will go back to their normal routine. They probably won't remember your presentation in detail; they'll just remember the rough outline, your main points.

In most cases, your presentation is more important to you than it is to anyone else. Try to develop a realistic sense of how important your presentation is. Although you should perform well, you are not expected to introduce earth-shaking material.

Most audiences value five features in a presentation:

1. Brevity is not only the soul of wit; it also guarantees that you won't bore your audience. You are giving a presentation of factual information, not a Sunday sermon. So you can be brief and to the point. You should condense your content. Think about about your major points, and figure out how to present them with reasonable fullness.

2. Visual elements and handouts enhance any presentation. Transparencies (acetate sheets imprinted with your material) displayed through overhead projectors give your audience something to look at while you talk, and handouts give them something to hold. Displays featuring blow-up illustrations help your audience remember what you're saying. With their eyes and hands busy, your audience will pay more attention to what you say.

3. Announcing a clear organization for your presentation dispels your audience's fear that they'll be there all afternoon, or that they'll miss an important point. You should tell your audience what they can expect to hear, and then you should keep within your time limit. People respect you when you respect their schedules.

4. As far as possible, speak in a calm voice and at a regular pace. You've heard speakers that yell at you or that eat their words trying to talk too fast. Be relaxed and pretend you're trying to explain something to a friend.

5. Everyone likes a little humor in a presentation. In terms of general tone, it's wise not to be pompous and not to give the appearance that you're bringing the Tablets down from the mountain. Keeping your tone serious but also light tells your audience that you have a healthy perspective about your material, that you're not obsessively involved with it.

Although you shouldn't try to be a stand-up comic, a little humor never hurts. The best humor pokes gentle fun at yourself or at a harmless misuse that is latent within the material you're discussing (such comments fall under the category of "Let me tell you how this can go wrong."). But one word of caution: "off-color" jokes are always in bad taste.

How to Give the Presentation

1. Try to relax. Find a comfortable way of standing. You can walk around some, but don't tour the front of the room.

Above all, don't fidget or indulge in nervous habits (such as continually rearranging the collar on your jacket).

2. Have double-spaced notes in front of you. Know them well enough so you only have to look at them occasionally.

3. Be responsive during the presentation. Maintain a relaxed eye contact with your audience; look for people who might have questions.

Invite questions at regular intervals. Answer the questions briefly, and see if anyone else has other comments. Then begin your presentation again where you left off.

Leave time for a question-and-answer session at the end. Don't be defensive about seemingly hostile questions. Answer them matter-of-factly or ask the questioner to explain the question further (an effective disarming technique).

4. Remember Murphy's Law: if anything can go wrong, it will. Try and stay cool

When the fan of your overhead projector is so loud that you have to stand out in front of it instead of at your lectern and therefore have to hold your notes in your hand; or

When, in the middle of your presentation, two managers begin to argue among themselves about something you've presented, and you see yourself running out of time; or

When your presentation is canceled 20 minutes before you're supposed to start because of an emergency in the company.

Just realize that you don't have much control over these kinds of things and that you'll get another chance later.

5. As the meeting breaks up, talk informally with those who stay to talk, thus setting a good tone for future meetings.

The more you give oral presentations, the better you'll become at it. You will become more relaxed and, therefore, more effective at presenting complex material. As you become more professional and learn more about your specialty, you will be able to sift out the less important from the more important. Finally, you will learn other techniques by sitting through effective presentations at professional meetings, and you will eventually become a relaxed, proficient speaker.

APPENDIX C

Grammar, Punctuation, and Other Conventions

GRAMMAR

Subject and Verb Agreement

Use singular verbs with singular subjects and plural verbs with plural subjects.

Fluidized-bed *technology allows* coal to be burned with fewer noxious emissions.

Various *technologies* for enhanced oil recovery *are receiving* increased funding from the Department of Energy.

Errors in agreement can occur when the subject and verb are separated by one or more words.

Correct: The *solubility* of gases *increases* with pressure.

Incorrect: The *solubility* of gases *increase* with pressure.

Agreement of Nouns and Pronouns

Use a singular pronoun if its antecedent is singular, and a plural one if its antecedent is plural.

The *dog* begged for *its* food.

The coach thanked the *players* for *their* discipline and hard work.

Agreement between a Verbal Phrase and the Subject

Make the implied subject of a verbal phrase identical to the subject of the sentence.

Incorrect: *Having stored* the blood, *the lab* was closed.

The implied subject of "having stored" is some person, but the subject of the sentence is "the lab," an entity, not a person.

Correct: *Having stored* the blood, *the medical technician* closed the lab.

Parallelism

Use parallel grammatical structure for parallel elements.

Incorrect: She enjoys *swimming, playing* racquetball, and *to run.*

Correct: She enjoys *swimming, playing* racquetball, and *running.*

Restrictive and Nonrestrictive Clauses

Use *that* to introduce restrictive clauses and *which* to introduce nonrestrictive clauses. A restrictive clause specifically identifies its antecedent or referent. A nonrestrictive clause just adds interesting information.

Restrictive: The beams are sunk only six feet deep because of the rail lines *that* run under each end of the bridge.

This sentence means: The beams are sunk only six feet deep *because of those specific rail lines* running under each end of the bridge.

Nonrestrictive: The beams are sunk only six feet deep because of the rail lines, *which* were laid down before anyone considered building a bridge at that location.

Here, the "which" clause only adds interesting information; the rail lines are already assumed to have been adequately identified.

Restrictive clauses are not preceded by or enclosed with commas. Nonrestrictive clauses are preceded by a comma if they occur at the end of a sentence and are enclosed with commas if they appear within the sentence.

PUNCTUATION

Punctuating Sentences

The following guide shows how to punctuate sentences. Recall that coordinating conjunctions are *and, but, or, nor, for, so,* and *yet* and that adverbial conjunctions are *therefore, consequently, however, nevertheless, also, finally, thus,* and so on.

1. Two separate sentences: Subject + verb**.** Subject + verb.

   We went to the Air and Space Museum**.** I was surprised that Lindbergh's plane was so small.

2. One compound sentence: Subject + verb**, coor. conj.** subject + verb.

   We went to the Air and Space Museum**, and** we saw the movie on early, manned flight.

   Tyrell Biggs made $1.25 million for fighting Mike Tyson**, and**, bloodied and dazed, he earned it the hard way. —Nack, "Very Tough Night at the Office"

3. One sentence with a compound predicate: Subject + verb **coor. conj.** verb.

   We went to the Air and Space Museum **and** stayed all day.

   We went to the Air and Space Museum **but** left at 1 p.m.

4. Two separate sentences: Subject + verb**.** **adv. conj.,** subject + verb.

   We went to the Air and Space Museum**. However,** the lines were so long that we didn't get in.

Announcing Lists and Definitions

Use a colon (**:**) to announce lists and definitions.

A List

Psychiatric researchers have identified three of the bodily systems in which stress can be measured**:** blood pressure, heart rate, and skin temperature.

A Definition

Two days of the conference were devoted to presentations on entropy**:** a theory derived from the second law of thermodynamics, stating that the universe follows an irreversible path to disorder.

Using the Dash, instead of Parentheses or Commas, for Special Emphasis

System commands**—**those that activate the computer's operating system**—**are usually easy to learn.

Using Hyphens for Clarity

Use hyphens to join words that function as a compound adjective.

Most dealers in antique cars highly value the traditional, *12-cylinder* engine that Jaguar developed.

The manager of the *systems-engineering* division decided to initiate a *computer-based* outlining program for her engineers to use when writing reports.

Using Commas for Clarity

Enclose Interrupters

Meltdowns, *this author assures us*, are always possible.

The dentist, *working quickly and efficiently*, extracted the wisdom tooth before the patient's anesthesia wore off.

All the tests, *therefore*, were graded again.

Punctuate Introductory Elements

When your crew has finished drilling, the well should be 1285 ft deep.

Outside, the hotel was being surrounded by helmeted police.

When we drove by, the church door was open.

Signal Examples

The diet recommends a daily intake of 2 tbsp. of fiber, *such as wheat bran.*

In the last analysis, a sense of the right punctuation is more an art than a science, combining as it does a feel for the pulse of spoken English, an understanding of the grammatical principles of written English, and a broad exposure to the variations of punctuation found in excellent writers. Aside from the period, the comma is the most frequently used mark of punctuation. In the following article written for *Time,* Pico Iyer pays tribute to the nature and power of this ubiquitous little mark.

In Praise of the Humble Comma

Pico Iyer

The gods, they say, give breath, and they take it away. But the same could be said—could it not?—of the humble comma. Add it to the present clause, and, of a sudden, the mind is, quite literally, given pause to think; take it out if you wish or forget it and the mind is deprived of a resting place. Yet still the comma gets no respect. It seems just a slip of a thing, a pedant's tick, a blip on the edge of our consciousness, a kind of printer's smudge almost. Small, we claim, is beautiful (especially in the age of the microchip). Yet what is so often used, and so rarely recalled, as the comma—unless it be breath itself?

Punctuation, one is taught, has a point: to keep up law and order. Punctuation marks are the road signs placed along the highway of our communication—to control speeds, provide directions and prevent head-on collisions. A period has the unblinking finality of a red light; the comma is a flashing yellow light that asks us only to slow down; and the semicolon is a stop sign that tells us to ease gradually to a halt, before gradually starting up again. By establishing the relations between words, punctuation establishes the relations between the people using words. That may be one reason why schoolteachers exalt it and lovers defy it ("We love each other and belong to each other let's don't ever hurt each other Nicole let's don't ever hurt each other," wrote Gary Gilmore to his girlfriend). A comma, he must have known, "separates inseparables," in the clinching words of H.W. Fowler, King of English Usage.

Punctuation, then, is a civic prop, a pillar that holds society upright. (A run-on sentence, its phrases piling up without division, is as unslightly as a sink piled high with dirty dishes.) Small wonder, then, that punctuation was one of the first proprieties of the Victorian age, the age of the corset, that the modernists threw off: the sexual revolution might be said to have begun when Joyce's Molly Bloom spilled out all her private thoughts in 36 pages of unbridled, almost unperioded and officially censored prose; and another rebellion was surely marked when E. E. Cummings first felt free to commit "God" to the lower case.

Punctuation thus becomes the signature of cultures. The hot-blooded Spaniard seems to be revealed in the passion and urgency of his doubled exclamation points and question marks ("*¡Caramba! ¿Quien sabe?*"), while the impassive Chinese traditionally added to his so-called

---

From Iyer, Pico. 13 June, 1988. In praise of the humble comma. *Time,* 80.

inscrutability by omitting directions from his ideograms. The anarchy and commotion of the '60s were given voice in the exploding exclamation marks, riotous capital letters and Day-Glo italics of Tom Wolfe's spray-paint prose; and in Communist societies, where the State is absolute, the dignity—and divinity—of capital letters is reserved for Ministries, Sub-Committees and Secretariats.

Yet punctuation is something more than a culture's birthmark: it scores the music in our minds, gets our thoughts moving to the rhythm of our hearts. Punctuation is the notation in the sheet music of our words, telling us when to rest, or when to raise our voices; it acknowledges that the meaning of our discourse, as of any symphonic composition, lies not in the units but in the pauses, the pacing and the phrasing. Punctuation is the way one bats one's eyes, lowers one's voice or blushes demurely. Punctuation adjusts the tone and color and volume till the feeling comes into perfect focus: not disgust exactly, but distaste; not lust, or like, but love.

Punctuation, in short, gives us the human voice, and all the meanings that lie between the words. "You aren't young, are you?" loses its innocence when it loses the question mark. Every child knows the menace of a dropped apostrophe (the parent's "Don't do that" shifting into the more slowly enunciated "Do not do that"), and every believer, the ignominy of having his faith reduced to "faith." Add an exclamation point to "To be or not to be . . ." and the gloomy Dane has all the resolve he needs; add a comma, and the noble sobriety of "God save the Queen" becomes a cry of desperation bordering on double sacrilege.

Sometimes, of course, our markings may be simply a matter of aesthetics. Popping in a comma can be like slipping on the necklace that gives an outfit quiet elegance, or like catching the sound of running water that complements, as it completes, the silence of a Japanese landscape. When V.S. Naipaul, in his latest novel, writes, "He was a middle-aged man, with glasses," the first comma can seem a little precious. Yet it gives the description a spin, as well as a subtlety, that it otherwise lacks, and it shows that the glasses are not part of the middle-agedness, but something else.

Thus all these tiny scratches give us breadth and heft and depth. A world that has only periods is a world without inflections. It is a world without shade. It has a music without sharps and flats. It is a martial music. It has a jackboot rhythm. Words cannot bend and curve. A comma, by comparison, catches the gentle drift of the mind in thought, turning in on itself and back on itself, reversing, redoubling and returning along the course of its own sweet river music; while the semicolon brings clauses and thoughts together with all the silent discretion of a hostess arranging guests around her dinner table.

Punctuation, then, is a matter of care. Care for words, yes, but also, and more important, for what the words imply. Only a lover notices the small things: the way the afternoon light catches the nape of a neck, or how a strand of hair slips out from behind an ear, or the way a finger curls around a cup. And no one scans a letter so closely as a lover, searching for its small print, straining to hear its nuances, its gasps, its sighs and hesitations, poring over the secret messages that lie in every cadence. The difference between "Jane (whom I adore)" and "Jane, whom I adore," and the difference between them both and "Jane—whom I adore—" marks all the distance between ecstasy and heartache. "No iron can pierce the heart with such force as a period put at just the right place," in Isaac Babel's lovely words: a comma can let us hear a voice break, or a heart. Punctuation, in fact, is a labor of love. Which brings us back, in a way, to gods.

OTHER CONVENTIONS

Numbers

If a number is the first word in a sentence or if it is a single digit, write it out. Otherwise, use the standard Arabic system for representing numbers.

> *Three hundred wells* were drilled during *five* Arctic explorations, but only *18* of them finally produced oil.

Abbreviations

The use of correct abbreviations is essential in scientific and technical writing. Here are some common abbreviations.

TERM	ABBREVIATION
barrel	bbl
barrel per day	bbl/d
billion cubic feet	Bcf
British Thermal Unit	Btu
calorie	cal
centimeter	cm
decibel	dB
degree Celsius	°C
degree Fahrenheit	°F
degree Kelvin	K
electron volt	eV
foot	ft
feet per second	ft/s
gallon	gal
gram	g
horsepower	hp
kilogram	kg
liter	L
megawatt	MW
meter	m
miles per gallon	mpg
miles per hour	mph
millimeter	mm
minute	min
mole	mo
pound	lb
pound force per square inch	psi
psi, gauge	psig
revolutions per minute	rpm
specific gravity	sp gr
standard deviation	SD
ton	ton
volt	V
watt	W

These abbreviations were selected from a list compiled by Diana Murrell, Technical Writer, EG&G/WASC, Morgantown, WV.

Illustration Credits

Figures 2-1, 2-2 Illustrations from *Knowledge and Wonder* by Victor F. Weisskopf. Copyright © 1962, 1966 by Educational Services, Inc. Reprinted by permission of Doubleday, a division of Bantam, Doubleday, Dell Publishing Group, Inc. **3-2a–b** Reprinted with permission from *Technology Review,* copyright © 1982. **3-2c** From J. R. Birchfield, "Welding—In Collapse of Hyatt Walkways," *Welding Design and Fabrication,* 55 (June 1982). Penton Publishing, Cleveland, OH. **3-2d** Reprinted with permission. Engineering News Review, McGraw-Hill Information Systems Co. **5-1** A. E. E. McKenzie, "Observation of a Flea," *Major Achievements of Science,* 1960. Cambridge University Press. **5-3** Reprinted by permission from *Nature,* vol. 171, p. 737–38, copyright © 1953 Macmillan Magazine, Ltd. **5-4a–b** Courtesy of Volkswagen of North America, Inc. **5-5a–d** From "The Spider and the Wasp" by Alexander Petrunkevitch. Copyright © 1952 by Scientific American, Inc. All rights reserved. **5-6, 5-7** From *A Feeling for the Organism: The Life and Work of Barbara McClintock* by Evelyn Fox Keller. Copyright © 1983 W. H. Freeman and Company. Reprinted with permission. **5-8, 5-9** From *Water Systems Handbook: A Complete Text on Private Systems Design, Operation and Maintenance,* 9th ed. Published by the Water Systems Council, Chicago, IL, 1987. **5-10** University of Rochester, *Rochester Review.* With permission. **5-13–5-15** Reprinted with permission from *The New England Journal of Medicine,* 316, no. 22 (May 28, 1987): 1385. **6-1** Copyright © 1985 The New York Times Company. Reprinted by permission. **6-5–6-11** M. T. Benchaita, P. Griffith, and E. Rabinowicz from *The Journal of Engineering for Industry,* August 1983. © American Society for Mechanical Engineers. **7-1–7-4** Copyright © 1985 by The New York Times Company. Reprinted by permission. **7-6–7-8** D'Arcy Wentworth Thompson, *On Growth and Form,* 2nd ed. Cambridge University Press, 1963, New York. **8-1** From Edison's *Electric Light: Biography of an Invention* by Robert Friedel, Paul Israel with Bernard S. Finn. Copyright © 1986 by Rutgers, The State University. Photo courtesy of Edison National Historical Site, National Park Service, U.S. Department of Interior. **13-1a–b** © William Zinsser, 1980. **14-1–14-6** U.S. Government Printing Office.

Index

A 9
B 0
C 1
D 2
E 3
F 4
G 5
H 6
I 7
J 8